AF352971

Novel Approaches for Targeting Metalloproteinases

Novel Approaches for Targeting Metalloproteinases

Editor

Salvatore Santamaria

Basel • Beijing • Wuhan • Barcelona • Belgrade • Novi Sad • Cluj • Manchester

Editor
Salvatore Santamaria
Biochemical and
Physiological Sciences
University of Surrey
Guildford
United Kingdom

Editorial Office
MDPI
St. Alban-Anlage 66
4052 Basel, Switzerland

This is a reprint of articles from the Special Issue published online in the open access journal *Pharmaceuticals* (ISSN 1424-8247) (available at: www.mdpi.com/journal/pharmaceuticals/special_issues/Targeting_Metalloproteinases).

For citation purposes, cite each article independently as indicated on the article page online and as indicated below:

Lastname, A.A.; Lastname, B.B. Article Title. *Journal Name* **Year**, *Volume Number*, Page Range.

ISBN 978-3-0365-9709-6 (Hbk)
ISBN 978-3-0365-9708-9 (PDF)
doi.org/10.3390/books978-3-0365-9708-9

Contents

About the Editor

Salvatore Santamaria

Salvatore Santamaria, Ph.D., M.Sc. (Hons), B.Sc., is a British Heart Foundation Intermediate Basic Science Research Fellow and Lecturer in Cardiovascular Science at the University of Surrey, Guildford, United Kingdom. He obtained his M.Sc. in Biotechnology from the University of Pisa, Italy, in 2008. He later joined Prof. Hideaki Nagase's laboratory at Imperial College London, where he developed inhibitory antibodies against ADAMTS5, a key protease in osteoarthritis. He was awarded his Ph.D. in 2014. Following a postdoc at the University of Oxford, he rejoined Imperial College as a Postdoctoral Researcher in Dr. Josefin Ahnstrom's lab. In 2019, he was awarded the Young Investigator Award by the British Society for Matrix Biology. His current research interests focus on the regulation of ADAMTS proteases and proteoglycans.

Preface

Metalloproteinases play key roles in physiological processes such as embryonic development, wound healing, angiogenesis, and blood coagulation, as well as pathologies like arthritis, cardiovascular disease, and cancer. In the therapeutic context, either decreasing or increasing the activity of a particular metalloproteinase may be ideal. This can be achieved, for example, through the action of molecules able to bind to substrate-binding sites (exosites) present on the ancillary domains of these enzymes or interfere with their transcriptional or post-translational regulation. This Special Issue aims to provide a state-of-the-art perspective on in vitro, preclinical, and clinical approaches to modulating metalloproteinase activity for therapeutic purposes. Its eleven articles (five reviews and six studies) cover a wide range of topics, from the screening of small molecules to tissue-specific drug delivery, and are addressed to a broad audience of scientists in life and medical science.

Salvatore Santamaria
Editor

Editorial

Novel Approaches for Targeting Metalloproteinases

Salvatore Santamaria

Department of Biochemical Sciences, School of Biosciences, Faculty of Health and Medical Sciences, Edward Jenner Building, University of Surrey, Guildford GU2 7XH, UK; s.santamaria@surrey.ac.uk

Citation: Santamaria, S. Novel Approaches for Targeting Metalloproteinases. *Pharmaceuticals* **2023**, *16*, 1637. https://doi.org/10.3390/ph16121637

Received: 11 November 2023
Accepted: 20 November 2023
Published: 22 November 2023

With 187 genes, metalloproteinases represent the most abundant protease family in the human proteome [1]. These proteases are involved in a variety of biological processes such as embryonic development, tissue resorption and repair, cell differentiation, migration, and apoptosis. As a result of this broad range of activities, dysregulated metalloproteinase activity is one of the drivers and hallmarks of diseases such as cancer, cardiac failure, atherosclerosis, and arthritis. The geometry of the active site around the catalytic ion, which is generally zinc, is highly conserved within each metalloproteinase superfamily, thus making it extremely challenging to achieve selective modulation of metalloproteinase activity.

This Special Issue aims to provide a state-of-the-art perspective on in vitro, preclinical and clinical approaches to modulate metalloproteinase activity for therapeutic purposes. Its 11 articles (5 reviews and 6 studies) cover a wide range of topics, from the screening of small molecules to tissue-specific drug delivery.

Hou et al. (contribution 1) performed ultrahigh-throughput activity assays (>650,000 molecules) to identify inhibitors of meprin α and β, two zinc metalloproteinases involved in several diseases such as cancer, fibrosis, and Alzheimer's. The assay was based on the cleavage of a fluorescent peptide and optimized for a 1536-well plate format. Three different scaffolds (triazole-hydroxyacetamides, sulfonamide-hydroxypropanamides, and phenoxy-hydroxyacetamides) provided robust inhibition of meprin α, with good selectivity (>30-fold) over meprin β and other metalloproteinases. The most selective meprin α inhibitors contained hydroxamate as a zinc-binding group and therefore likely achieved their selectivity through optimized interactions with the protease subsites. Representative compounds were tested for their ability to affect the viability of skin fibroblasts and melanocytes. Little or no effect on cell viability was observed, suggesting a lack of cytotoxicity. Screening differently biased libraries or finely tuning assay conditions would aid the identification of non-chelating inhibitors. The lead compounds identified in this study were further optimized by Wang et al. using structure–activity relationship studies (contribution 2). The aryl triazole SR19855 exhibited a 10-fold selectivity for meprin α over meprin β and activity in the low-micromolar range. SR19855 was docked into a homology model of meprin α and β's active site, thus highlighting crucial interactions between the protease and inhibitor. This highlighted that both the phenyl and pyrimidine rings of SR19855 could not be removed without a significant decrease in meprin inhibitory activity. The best compound, SR24717, exhibited sub-micromolar inhibitory activity against meprin α, with a 100-fold selectivity over meprin β, thus outperforming most of the previously reported meprin inhibitors and confirming the success of this approach. It will be interesting to probe the biological role of meprins by testing the effect of compounds like SR24717 in multiple cell-based assays.

High conservation of the active site across members of the metzincin superfamily has so far hampered the development of selective matrix metalloproteinase (MMP) inhibitors. The only MMP currently approved for clinical use is Periostat® (doxycycline) for periodontitis. While monoclonal antibodies have achieved impressive results in terms of potency and selectivity [2], synthetic MMP inhibitors have failed in clinical trials due to lack of selectivity and poor pharmacokinetics [3]. To improve selectivity, hydrophilicity, and bioavailability, small molecule inhibitors have been modified through conjugation with carbohydrate moieties. These derivatives and their inhibitory profiles are widely discussed in the review by Cuffaro et al. (contribution 3). Carbohydrate-based compounds are a growing area in the metalloproteinase field that will likely generate new therapeutic opportunities in the near future.

Das et al. (contribution 4) comprehensively review alternative mechanisms for inhibiting MMP activity. These include targeting distantly located, poorly conserved substrate-binding sites (exosites), homodimer formation (in the case of MMP9 and MMP14), and zymogen activation. Selective inhibition can be achieved using peptides, small molecules, or monoclonal antibodies. The authors highlight that the MMP substrate repertoire is not limited to extracellular matrix (ECM) substrates, including, for example, chemokines and cytokines. To uncover the metalloproteinase degradome, proteomics techniques such as terminal amine isotopic labeling of substrates (TAILS) have been developed.

Gonçalves et al. (contribution 5) explore current approaches to target MMP2 in heart failure (HF). MMP2 is involved in the degradation of the cardiac ECM and components of the contractile sarcomeric apparatus such as troponin I, titin, and myosin light chain. Plasma MMP2 is considered a biomarker of HF, and rodent models have shown a link between dysregulated MMP2 activity and cardiac dysfunction. Starting from non-selective, first generation, zinc-chelating hydroxamate MMP inhibitors, the authors move to describe non-hydroxamate inhibitors, antibiotics of the tetracycline class, siRNA, statins, and antihypertensive drugs, as well as their applications in preclinical and clinical models of HF.

Skrzypiec-Spring et al. (contribution 6) tested the effect of β-blockers carvedilol, nebivolol, and metoprolol in a rat model of ischemia–reperfusion (IR) injury. These molecules were previously reported to inhibit expression of MMP2 [4,5]. IR induced activation of MMP2, which was reversed specifically by carvedilol but not by the other β-blockers, suggesting that the cardioprotective activity of carvedilol is partly mediated by inhibition of MMP2 activation. Supporting this, carvedilol caused a 73% decrease in the cardiac levels of the sarcomeric protein troponin I at the end of perfusion, a proxy for troponin I degradation. The exact mechanism of carvedilol action still warrants further investigation, but based on the results from the authors, no direct effect on protein expression seemed to be involved.

Palladini et al. (contribution 7) tested the effect of obeticholic acid (OCA) in a rat model of hepatic IR to assess modulation of MMP activity, since previous in vitro studies have shown that OCA can restore the balance between MMPs (in particular MMP2 and MMP9) and their endogenous inhibitors, the tissue inhibitors of metalloproteinases (TIMPs) and reversion-inducing cysteine-rich protein with kazal motifs (RECK) [6]. In the kidney cortex and medulla, OCA treatment resulted in reduced MMP9 dimer activity compared with vehicle-treated IR rats, but no significant changes in MMP2, TIMP1, TIMP2, or RECK were observed. Additionally, OCA reduced serum levels of creatinine, an index of renal function, and reduced levels of the proinflammatory cytokines tumor necrosis factor (TNF)-α and interleukin-6 (IL-6) in the kidney cortex. Overall, the results of this study provide evidence that OCA treatment may ameliorate hepatic renal syndrome by inhibiting fibrosis and restoring renal function.

Ciccone et al. (contribution 8) review the ability of natural compounds to modulate the activity of gelatinases MMP2 and MMP9 in the context of neurodegenerative and neuroinflammatory diseases, in particular Alzheimer's disease. After summarizing the pathological and physiological role of MMP2 and MMP9 in the nervous system, the authors describe bioactive drugs derived from marine organisms and their application as gelatinase inhibitors. They then move on to discuss molecules derived from terrestrial sources, highlighting crucial differences in the mode of action, i.e., direct versus indirect (for example, transcriptional) inhibition.

Laghezza et al. (contribution 9) identified bisphosphonic acid derivatives as MMP13 inhibitors. These molecules can be tested for the treatment of bone metastasis due to the ability of the bisphosphonic acid group to specifically target the bone. The most potent compounds exhibited activity in the low/sub-micromolar range against MMP13 and good selectivity over MMP8 and MMP9. However, selectivity over MMP2 still needs to be improved.

Compared to MMPs, not much is known on the biological functions of dipeptidyl peptidase III (DPP III), a zinc-dependent exopeptidase. Agić et al. (contribution 10) used a combination of in vitro and in silico approaches to identify DPP III inhibitors. They reported coumarin-based compounds with inhibitory activity in the micromolar range that can be used to probe the patho-physiological role of DPP III. Quantitative structure activity relationship analysis identified crucial substituents necessary for inhibitory activity. The most active compound was docked in the active site of DPP III to model crucial interactions with the enzyme.

Metalloproteinases are also involved in osteoarthritis, the most common degenerative joint disease, due to their ability to degrade important ECM components such as collagens and proteoglycans. The lack of disease-modifying drugs prompted the development of innovative approaches to delay/arrest disease progression, as discussed in the review by McClurg et al. (contribution 11). Such strategies either aim to inhibit cartilage degradation (via their action on metalloproteinases such as A Disintegrin-like and metalloproteinase with thrombospondin motif 5 or MMP13) or promote cartilage anabolism (through administration of growth factors). The authors further described and critically discussed a number of methods to specifically target drug delivery to the cartilage, a poorly vascularized tissue, for example, through conjugation with peptides directed against chondrocytes or specific ECM components such as type II collagen or aggrecan.

The studies described in this Special Issue will undoubtedly stimulate further research in this area, thus increasing the druggability of metalloproteinases.

Funding: The work in the author's laboratory is supported by the British Heart Foundation (FS/IBSRF/20/25032) and the University of Surrey Faculty Research Support Fund.

Acknowledgments: I wish to thank all the authors for their high-quality contributions and the reviewers for providing critical feedback.

Conflicts of Interest: The author declares no conflict of interest.

List of Contributions

1. Hou, S.; Diez, J.; Wang, C.; Becker-Pauly, C.; Fields, G.B.; Bannister, T.; Spicer, T.P.; Scampavia, L.D.; Minond, D. Discovery and Optimization of Selective Inhibitors of Meprin α (Part I). *Pharmaceuticals* **2021**, *14*, 203. https://doi.org/10.3390/ph14030203.
2. Wang, C.; Diez, J.; Park, H.; Spicer, T.P.; Scampavia, L.D.; Becker-Pauly, C.; Fields, G.B.; Minond, D.; Bannister, T.D. Discovery and Optimization of Selective Inhibitors of Meprin α (Part II). *Pharmaceuticals* **2021**, *14*, 197. https://doi.org/10.3390/ph14030197.
3. Cuffaro, D.; Nuti, E.; D'Andrea, F.; Rossello, A. Developments in Carbohydrate-Based Metzincin Inhibitors. *Pharmaceuticals* **2020**, *13*, 376. https://doi.org/10.3390/ph13110376.
4. Das, N.; Benko, C.; Gill, S.E.; Dufour, A. The Pharmacological TAILS of Matrix Metalloproteinases and Their Inhibitors. *Pharmaceuticals* **2020**, *14*, 31. https://doi.org/10.3390/ph14010031.
5. Gonçalves, P.R.; Nascimento, L.D.; Gerlach, R.F.; Rodrigues, K.E.; Prado, A.F. Matrix Metalloproteinase 2 as a Pharmacological Target in Heart Failure. *Pharmaceuticals* **2022**, *15*, 920. https://doi.org/10.3390/ph15080920.
6. Skrzypiec-Spring, M.; Urbaniak, J.; Sapa-Wojciechowska, A.; Pietkiewicz, J.; Orda, A.; Karolko, B.; Danielewicz, R.; Bil-Lula, I.; Woźniak, M.; Schulz, R.; et al. Matrix Metalloproteinase-2 Inhibition in Acute Ischemia-Reperfusion Heart Injury-Cardioprotective Properties of Carvedilol. *Pharmaceuticals* **2021**, *14*, 1276. https://doi.org/10.3390/ph14121276.

7. Palladini, G.; Cagna, M.; Di Pasqua, L.G.; Adorini, L.; Croce, A.C.; Perlini, S.; Ferrigno, A.; Berardo, C.; Vairetti, M. Obeticholic Acid Reduces Kidney Matrix Metalloproteinase Activation Following Partial Hepatic Ischemia/Reperfusion Injury in Rats. *Pharmaceuticals* **2022**, *15*, 524. https://doi.org/10.3390/ph15050524.
8. Ciccone, L.; Vandooren, J.; Nencetti, S.; Orlandini, E. Natural Marine and Terrestrial Compounds as Modulators of Matrix Metalloproteinases-2 (MMP-2) and MMP-9 in Alzheimer's Disease. *Pharmaceuticals* **2021**, *14*, 86. https://doi.org/10.3390/ph14020086.
9. Laghezza, A.; Piemontese, L.; Brunetti, L.; Caradonna, A.; Agamennone, M.; Loiodice, F.; Tortorella, P. (2-Aminobenzothiazole)-Methyl-1,1-Bisphosphonic Acids: Targeting Matrix Metalloproteinase 13 Inhibition to the Bone. *Pharmaceuticals* **2021**, *14*, 85. https://doi.org/10.3390/ph14020085.
10. Agić, D.; Karnaš, M.; Šubarić, D.; Lončarić, M.; Tomić, S.; Karačić, Z.; Bešlo, D.; Rastija, V.; Molnar, M.; Popović, B.M.; et al. Coumarin Derivatives Act as Novel Inhibitors of Human Dipeptidyl Peptidase III: Combined In Vitro and In Silico Study. *Pharmaceuticals* **2021**, *14*, 540. https://doi.org/10.3390/ph14060540.
11. McClurg, O.; Tinson, R.; Troeberg, L. Targeting Cartilage Degradation in Osteoarthritis. *Pharmaceuticals* **2021**, *14*, 126. https://doi.org/10.3390/ph14020126.

References

1. Puente, X.S.; Sánchez, L.M.; Gutiérrez-Fernández, A.; Velasco, G.; López-Otín, C. A genomic view of the complexity of mammalian proteolytic systems. *Biochem. Soc. Trans.* **2005**, *33 Pt 2*, 331–334. [CrossRef] [PubMed]
2. Santamaria, S.; de Groot, R. Monoclonal antibodies against metzincin targets. *Br. J. Pharmacol.* **2019**, *176*, 52–66. [CrossRef] [PubMed]
3. Fields, G.B. Mechanisms of Action of Novel Drugs Targeting Angiogenesis-Promoting Matrix Metalloproteinases. *Front. Immunol.* **2019**, *10*, 1278. [CrossRef] [PubMed]
4. Skrzypiec-Spring, M.; Haczkiewicz, K.; Sapa, A.; Piasecki, T.; Kwiatkowska, J.; Ceremuga, I.; Wozniak, M.; Biczysko, W.; Kobierzycki, C.; Dziegiel, P.; et al. Carvedilol Inhibits Matrix Metalloproteinase-2 Activation in Experimental Autoimmune Myocarditis: Possibilities of Cardioprotective Application. *J. Cardiovasc. Pharmacol. Ther.* **2018**, *23*, 89–97. [CrossRef] [PubMed]
5. Ceron, C.S.; Rizzi, E.; Guimarães, D.A.; Martins-Oliveira, A.; Gerlach, R.F.; Tanus-Santos, J.E. Nebivolol attenuates prooxidant and profibrotic mechanisms involving TGF-β and MMPs, and decreases vascular remodeling in renovascular hypertension. *Free Radic Biol. Med.* **2013**, *65*, 47–56. [CrossRef] [PubMed]
6. Ferrigno, A.; Palladini, G.; Di Pasqua, L.G.; Berardo, C.; Richelmi, P.; Cadamuro, M.; Fabris, L.; Perlini, S.; Adorini, L.; Vairetti, M. Obeticholic acid reduces biliary and hepatic matrix metalloproteinases activity in rat hepatic ischemia/reperfusion injury. *PLoS ONE* **2020**, *15*, e0238543. [CrossRef] [PubMed]

 *pharmaceuticals*

Article

Discovery and Optimization of Selective Inhibitors of Meprin α (Part I)

Shurong Hou [1], Juan Diez [2], Chao Wang [1], Christoph Becker-Pauly [3], Gregg B. Fields [4], Thomas Bannister [1], Timothy P. Spicer [1], Louis D. Scampavia [1] and Dmitriy Minond [2,5,*]

[1] Department of Molecular Medicine, The Scripps Research Molecular Screening Center, Scripps Research, Jupiter, FL 33458, USA; SHou@scripps.edu (S.H.); CWang@scripps.edu (C.W.); tbannist@scripps.edu (T.B.); spicert@scripps.edu (T.P.S.); scampl@scripps.edu (L.D.S.)

[2] Rumbaugh-Goodwin Institute for Cancer Research, Nova Southeastern University, 3321 College Avenue, CCR r.605, Fort Lauderdale, FL 33314, USA; mdiezi@yahoo.com

[3] Unit for Degradomics of the Protease Web, Institute of Biochemistry, University of Kiel, Rudolf-Höber-Str.1, 24118 Kiel, Germany; cbeckerpauly@biochem.uni-kiel.de

[4] Department of Chemistry & Biochemistry and I-HEALTH, Florida Atlantic University, 5353 Parkside Drive, Jupiter, FL 33458, USA; fieldsg@fau.edu

[5] Dr. Kiran C. Patel College of Allopathic Medicine, Nova Southeastern University, 3301 College Avenue, Fort Lauderdale, FL 33314, USA

* Correspondence: dminond@nova.edu

Citation: Hou, S.; Diez, J.; Wang, C.; Becker-Pauly, C.; Fields, G.B.; Bannister, T.; Spicer, T.P.; Scampavia, L.D.; Minond, D. Discovery and Optimization of Selective Inhibitors of Meprin α (Part I). *Pharmaceuticals* 2021, 14, 203. https://doi.org/10.3390/ph14030203

Academic Editor: Salvatore Santamaria

Received: 14 December 2020
Accepted: 20 January 2021
Published: 28 February 2021

Publisher's Note: MDPI stays neutral with regard to jurisdictional claims in published maps and institutional affiliations.

Abstract: Meprin α and β are zinc-dependent proteinases implicated in multiple diseases including cancers, fibrosis, and Alzheimer's. However, until recently, only a few inhibitors of either meprin were reported and no inhibitors are in preclinical development. Moreover, inhibitors of other metzincins developed in previous years are not effective in inhibiting meprins suggesting the need for de novo discovery effort. To address the paucity of tractable meprin inhibitors we developed ultrahigh-throughput assays and conducted parallel screening of >650,000 compounds against each meprin. As a result of this effort, we identified five selective meprin α hits belonging to three different chemotypes (triazole-hydroxyacetamides, sulfonamide-hydroxypropanamides, and phenoxy-hydroxyacetamides). These hits demonstrated a nanomolar to micromolar inhibitory activity against meprin α with low cytotoxicity and >30-fold selectivity against meprin β and other related metzincincs. These selective inhibitors of meprin α provide a good starting point for further optimization.

Keywords: meprin α; meprin β; zinc metalloproteinase; uHTS

1. Introduction

Meprin α and meprin β are zinc-dependent proteinases implicated in multiple diseases including cancers [1], fibrosis [2,3], and Alzheimer's [4,5]. Meprins cleave multiple cytokines and adhesion molecules thus contributing to inflammation and migration of inflammatory cells [6]. Chronic inflammation can lead to the excess deposition of collagen I resulting in fibrosis [7,8]. Meprins have been shown to cleave procollagen I leading to its maturation and deposition in skin and lung [3,9]. The roles of meprins in various processes are mediated via the cleavage of biological molecules. There are examples of common substrates that meprin α and meprin β share amongst themselves [10] and with other proteases [4]. This complicates the understanding of their respective roles in the specific disease scenarios and, as a consequence, their value as targets for drug discovery. To validate either meprin as a target in any particular disease, target modulation by combination of genomic (e.g., knockdown, overexpression) [11] and pharmacologic means (e.g., small molecules) [12] could be useful. However, due to the relatively recent discovery of meprins' involvement in pathologic conditions there are very few reports of small molecule inhibitor discovery efforts for these enzymes. Kruse et al. [13] reported several known metzincin inhibitors that are capable of inhibiting meprins with some degree of selectivity. However, these inhibitors were not selective for other metzincins, which made their utilization for studying the roles of meprins in various diseases difficult. Our group had reported the first low nanomolar meprin β inhibitors, NFF449 and PPNDS (Figure 1, Ki = 22 nM and 8 nM, respectively), with ~100-fold selectivity against meprin α and good selectivity against adamalysins and matrixins [14]. Ramsbeck et al. (2017) reported the low nanomolar selective meprin β inhibitor, 11 g, with 46-fold selectivity against meprin α (Figure 1, IC50 = 2735 nM and 60 nM for meprin α and β, respectively) with good selectivity against adamalysins and matrixins [15]. They also reported improved compounds based on the same scaffold [16] (Figure 1). The best compounds from this series, 8 h and 8i, are 27-fold and 15-fold selective against meprin α (IC50 = 23 nM and 626 nM for 8 h and 24 nM and 368 nM for 8i, for meprin β and α, respectively). A measure of 200 µM of either inhibitor had only limited effect on MMP and ADAM activity, but IC50 values were not reported. Tan et al. (2018) reported the first selective inhibitors of meprin α, 10d and 10e, with 18- and 19-fold selectivity against meprin β [17] (Figure 1). Herein we report the results of a large-scale parallel high-screening throughput effort to discover novel inhibitors of meprin α and meprin β.

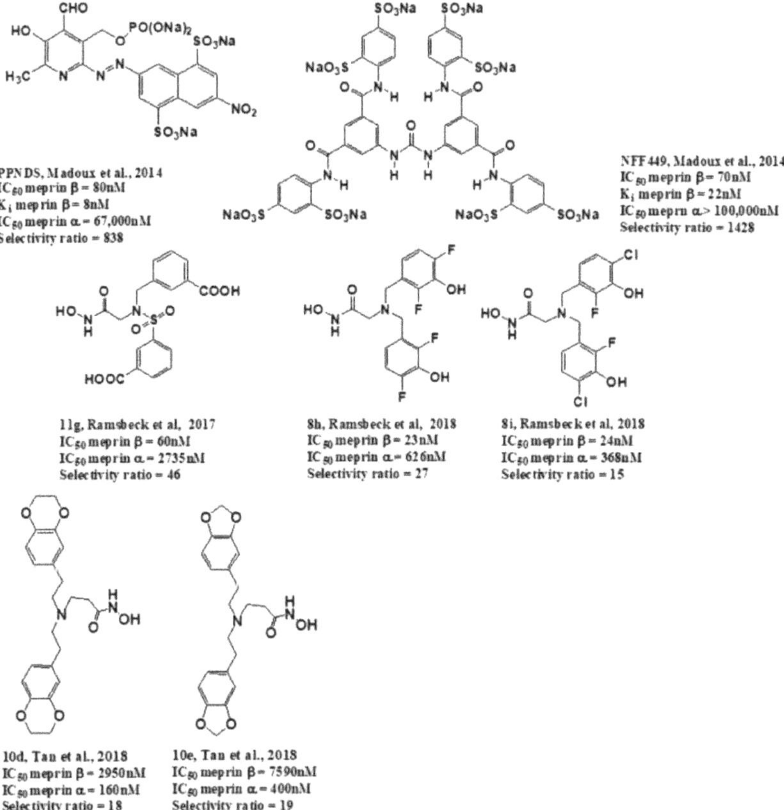

Figure 1. Synthetic selective meprin inhibitors described to date.

2. Results

2.1. Assay Miniaturization and Optimization in 1536 Well Plate Format

The meprin α and meprin β assays, which utilize the substrates (Mca)-YVADAPK-(K-ε-Dnp) and (Mca)-EDEDED-(K-ε-Dnp), respectively, have been described previously [14]. To enable an ultra-high-throughput screening (uHTS) campaign, we proceeded to miniaturize both assays to 1536 well plate format (wpf). First, we recapitulated the assays in 1536 well plate using reagents at the same concentrations as in 384 well plate format assays by scaling the volume down by the factor of 2.5. This resulted in the final volume of the assays of 4 μL. The meprin α assay in 1536 well plates demonstrated a lower signal-to-basal (S/B) ratio than in 384 well plates (1.85 vs. 2.3, respectively), but a better Z' value (0.76 vs. 0.6, respectively), suggesting that the assay is very suitable for large-scale HTS [18]. Actinonin's IC50 values were within 2-fold of each other (5.7 nM and 11 nM for 1536 and 384 well plate format, respectively) (Figure 2A and Table 1).

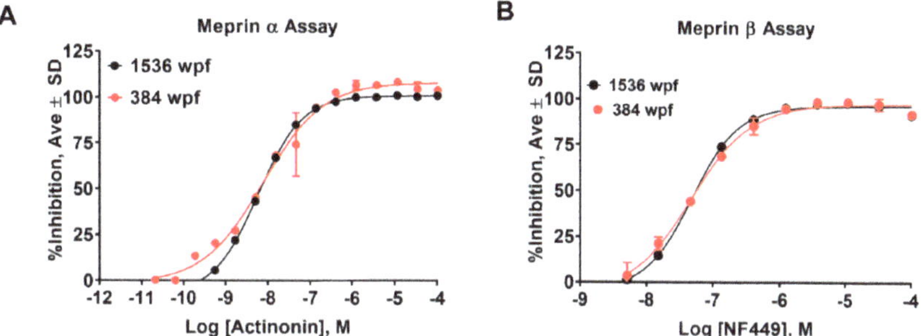

Figure 2. Assay recapitulation in 1536 well plate format. Concentration response studies in 384 and 1536 well plate formats showed similar potency of pharmacological controls for (**A**) meprin α (actinonin) and (**B**) meprin β (NF449) assays. Both assays were performed in triplicate.

Table 1. Comparison of meprin α and meprin β assay parameters in 384 and 1536 well plate formats.

Assay	S/B	Z'	Actinonin IC$_{50}$, nM	NFF449 IC$_{50}$, nM
Meprin α 384 wpf	2.3	0.6	11	>100,000
Meprin α 1536 wpf	1.85	0.76	5.7	>100,000
Meprin β 384 wpf	4.4	0.9	22,000	53
Meprin β 1536 wpf	6.9	0.91	9750	48

Meprin β assay exhibited greater S/B in 1536 wpf than in 384 wpf (6.9 vs. 4.4, respectively), while Z' factor values were identical at 0.9. NFF449 IC50 values were 48 nM and 53 nM for 1536 and 384 wpf, respectively (Figure 2B and Table 1). Despite excellent Z' values in the 1536 wpf in both assays, we wanted to ensure an optimal balance between robustness and sensitivity; in particular with meprin α.

First, both assays were run for 180 min at three different enzyme concentrations including the concentrations at which the assays were recapitulated in 1536 wpf (1.3 nM and 0.05 nM for meprin α and meprin β, respectively). QC parameters (Z' and S/B) and IC50 values of pharmacological controls (actinonin and NFF449) were calculated at 30, 60, and 90 min of the reaction time. The meprin α assay displayed the best S/B values after 90 min of reaction time using 1.3 nM enzyme; however, the reaction progress curve was not linear at the 90 min time point (Figure 3A). This suggested that while longer reaction times and higher than 1.3 nM enzyme concentration may lead to somewhat better S/B values, the assay sensitivity may suffer due to a nonlinear relationship between signal and proteolysis inhibition. Therefore, to ensure optimal assay sensitivity, we chose 60 min reaction end point and 1.3 nM meprin α as final assay conditions for the primary HTS campaign.

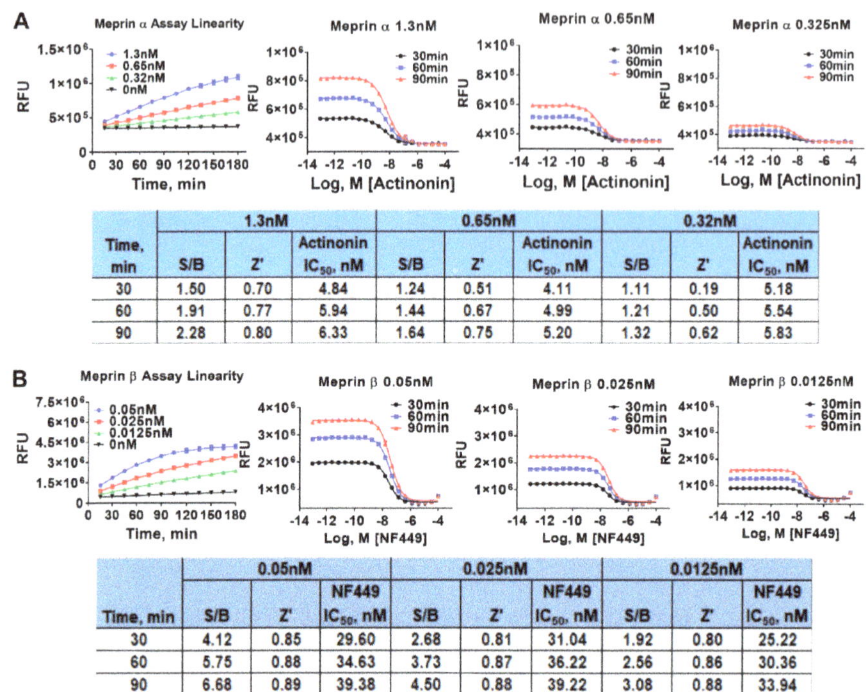

Time, min	1.3nM			0.65nM			0.32nM		
	S/B	Z'	Actinonin IC$_{50}$, nM	S/B	Z'	Actinonin IC$_{50}$, nM	S/B	Z'	Actinonin IC$_{50}$, nM
30	1.50	0.70	4.84	1.24	0.51	4.11	1.11	0.19	5.18
60	1.91	0.77	5.94	1.44	0.67	4.99	1.21	0.50	5.54
90	2.28	0.80	6.33	1.64	0.75	5.20	1.32	0.62	5.83

Time, min	0.05nM			0.025nM			0.0125nM		
	S/B	Z'	NF449 IC$_{50}$, nM	S/B	Z'	NF449 IC$_{50}$, nM	S/B	Z'	NF449 IC$_{50}$, nM
30	4.12	0.85	29.60	2.68	0.81	31.04	1.92	0.80	25.22
60	5.75	0.88	34.63	3.73	0.87	36.22	2.56	0.86	30.36
90	6.68	0.89	39.38	4.50	0.88	39.22	3.08	0.88	33.94

Figure 3. Enzyme concentration and time of reaction optimization experiments in 1536 wpf. (**A**) Meprin α 1536 wpf assay optimization study. (**B**) Meprin β 1536 wpf assay optimization study. Experiments repeated twice, n=4. S/B-signal-to-basal ratio.

The meprin β assay progress curve was hyperbolic rather than linear at 0.05 nM and 0.025 nM enzyme; therefore, we chose 0.0125 nM enzyme concentration where assay linearity was demonstrated (Figure 3B). Z' and S/B values were acceptable at 60 min reaction end point (0.86 and 2.6, respectively). IC50 values of NFF449 were not significantly affected by the variations of reaction length and meprin β concentrations.

Next, we performed substrate optimization to achieve balanced assay conditions defined as [S]/KM = 1 [19]. In order to do that, we first determined kinetic parameters of proteolysis of meprin α and meprin β substrates by the respective enzymes (Figure 4A,B). Meprin α and meprin β proteolysis exhibited similar KM values (2.4 $\pm$ 0.3 μM and 2.7 $\pm$ 0.7 μM, respectively) suggesting the need for optimization of both assays' substrate concentration. Meprin β exhibited >20-fold faster turnover of its substrate than meprin α (6.4 $\pm$ 0.06 s^{-1} versus 0.29 $\pm$ 0.06 s-1, respectively) which is consistent with >100-fold difference in enzyme concentrations for meprin α and meprin β assays (1.3 nM versus 0.0125 nM, respectively). To optimize substrate concentrations, both assays were run for 90 min at three different substrate concentrations (10, 5, and 2.5 μM) which included the concentration at which the assays were recapitulated in 1536 wpf (10 μM for both meprin α and meprin β) and the concentration approximating [S]/KM = 1 condition (2.5 μM). Enzyme concentrations were fixed at 1.3 nM for meprin α and 0.0125 nM for meprin β. QC parameters (Z' and S/B) and IC50 values of pharmacological controls (actinonin and NFF449) were calculated at 40, 60, and 90 min of the reaction time (Figure 4C,D). The 2.5 μM substrate condition resulted in increased apparent potency of pharmacological controls for both assays (2-fold for actinonin in the meprin α assay and 3-fold for NFF449 in the meprin β assay). This suggested that 2.5 μM substrate concentrations result in greater assay sensitivity. Assay QC parameters (S/B and Z') at 2.5 μM substrate concentrations did not differ significantly from assays run at 10 µM substrate concentrations; therefore, we chose 2.5 µM substrate concentrations as a final assay condition.

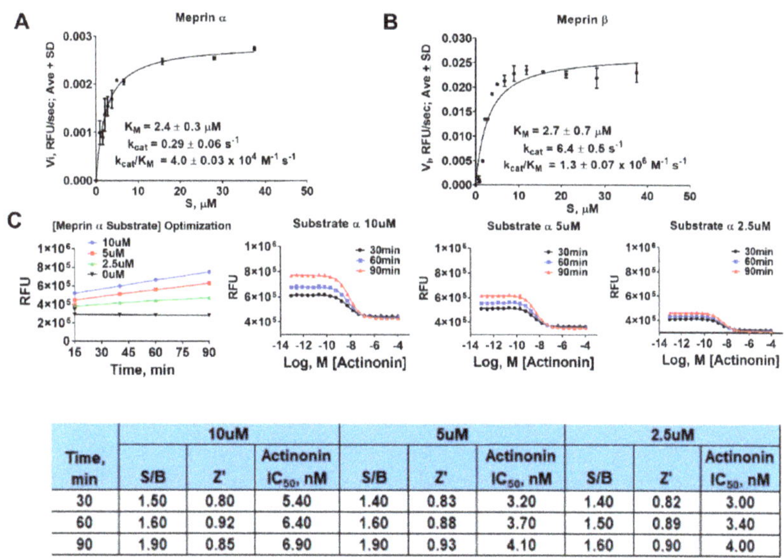

Time,	10uM			5uM			2.5uM		
min	S/B	Z'	Actinonin IC$_{50}$, nM	S/B	Z'	Actinonin IC$_{50}$, nM	S/B	Z'	Actinonin IC$_{50}$, nM
30	1.50	0.80	5.40	1.40	0.83	3.20	1.40	0.82	3.00
60	1.60	0.92	6.40	1.60	0.88	3.70	1.50	0.89	3.40
90	1.90	0.85	6.90	1.90	0.93	4.10	1.60	0.90	4.00

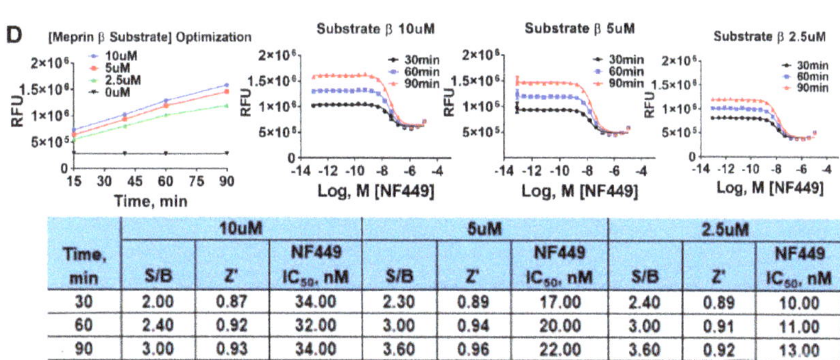

Time,	10uM			5uM			2.5uM		
min	S/B	Z'	NF449 IC$_{50}$, nM	S/B	Z'	NF449 IC$_{50}$, nM	S/B	Z'	NF449 IC$_{50}$, nM
30	2.00	0.87	34.00	2.30	0.89	17.00	2.40	0.89	10.00
60	2.40	0.92	32.00	3.00	0.94	20.00	3.00	0.91	11.00
90	3.00	0.93	34.00	3.60	0.96	22.00	3.60	0.92	13.00

Figure 4. Substrate concentration optimization experiments in 1536 wpf. Results of kinetic studies of (**A**) meprin α and (**B**) meprin β hydrolysis of respective substrates. (**C**) Meprin α 1536 wpf assay optimization study. (**D**) Meprin β 1536 wpf assay optimization study. Experiments repeated twice, *n* = 4.

2.2. Online Robotic Pilot Study

To ascertain the readiness of the assays for a large-scale screening effort, a small pilot screen was conducted using Kalypsys GNF integrated online robotic platform (San Diego, CA, USA) [20]. Overall, ~39,000 compounds were tested using 31 assay plates in both meprin α and meprin β assays. Both assays performed well on the Kalypsys robotic system, as the meprin α assay average Z' and S/B were 0.88 ± 0.03 and 2.9 ± 0.07, respectively, while the meprin β assay average Z' and S/B were 0.91 ± 0.03 and 4.5 ± 0.17, respectively. The number of hits identified in the meprin α and meprin β assays were 169 and 260, respectively, which constituted 0.43% and 0.67% hit rates, respectively. After removal of duplicates, Venn analysis showed that 37 compounds inhibited both meprins, while there were 129 compounds selectively inhibiting meprin α and 220 compounds selectively inhibiting meprin β, suggesting that selective probes for both enzymes could be discovered. This also suggested that both assays were ready for large scale effort.

2.3. Primary HTS Campaign

Primary HTS campaigns were conducted using The Scripps Research Institute proprietary library of 649,570 compounds using both meprin α and meprin β assays [21]. Overall, 522 plates were used for each assay with excellent QC parameters (average Z' = 0.86 ± 0.04 and average S/B = 2.8 ± 0.09 for meprin α assay and average Z' = 0.88 ± 0.03 and average S/B = 4.4 ± 0.27 for meprin β assay). IC50 values of control compounds were reproducible with literature and our preliminary experiments (meprin α actinonin IC50 = 2.9 ± 0.12 nM, n = 11 plates; meprin β NF449 IC50 = 10.4 ± 0.85 nM, n = 11 plates). Using hit cutoffs derived from the average and 3 standard deviations of the activity of all samples tested which were 10.76% and 14.33% for the meprin α and meprin β assays, 5064 and 4929 hits were identified which constituted hit rates of 0.78% and 0.76%, respectively. It was noted that the majority of meprin α hits exhibited a percentage inhibition close to the hit cutoff, whereas meprin β hits were distributed evenly in the range of 20−100% inhibition (Figure 5A,B).

After removal of a handful of duplicates, Venn analysis showed that 1416 compounds inhibited both meprins, while there were 3632 compounds selectively inhibiting meprin α and 3470 compounds selectively inhibiting meprin β (Figure 5C). Correlational analysis showed 48 and 39 compounds selectively inhibiting meprin α and meprin β, respectively, with a percentage inhibition ≥ 50 (Figure 5D).

2.4. Hit Confirmation and Prioritization

For the confirmation assays all compounds that inhibited either of the meprins with >20% inhibition were selected. Confirmation assays were done at a single concentration point in triplicate. Out of 2378 total compounds tested in confirmation assays, only 206 confirmed activity against meprin α and 1097 confirmed activity against meprin β constituting 8.7% and 46.1% confirmation rate for meprin α and meprin β, respectively. The low confirmation rate for meprin α was not unexpected due to the majority of meprin α hits from the primary campaign being close to the hit cutoff (Figure 5A).

Venn analysis showed that 81 compounds inhibited both meprins, while there were 125 compounds selectively inhibiting meprin α and 1016 compounds selectively inhibiting meprin β (Figure 5E). Correlational analysis showed 19 and 12 compounds selectively inhibiting meprin α and β, respectively, with ≥50% inhibition (Figure 5F). Overall, 827 compounds exhibited >20% inhibition.

It was also noted that the majority of the most active hits for each enzyme were potential Zn-binders due to the presence of hydroxamate and reverse hydroxamate moieties. Compounds acting via Zn binding may be undesirable due to clinical trial failures observed previously based on a lack of selectivity, toxicity, and metabolic instability. To prioritize selectivity, we introduced additional assays to help with triaging the compounds to ascertain that we are not biasing for nonselective compounds. We utilized ADAM10, MMP-8, and MMP-14 as the most relevant counter targets. The counter screens were conducted in triplicate using the same 2378 compounds that were tested in confirmation assays.

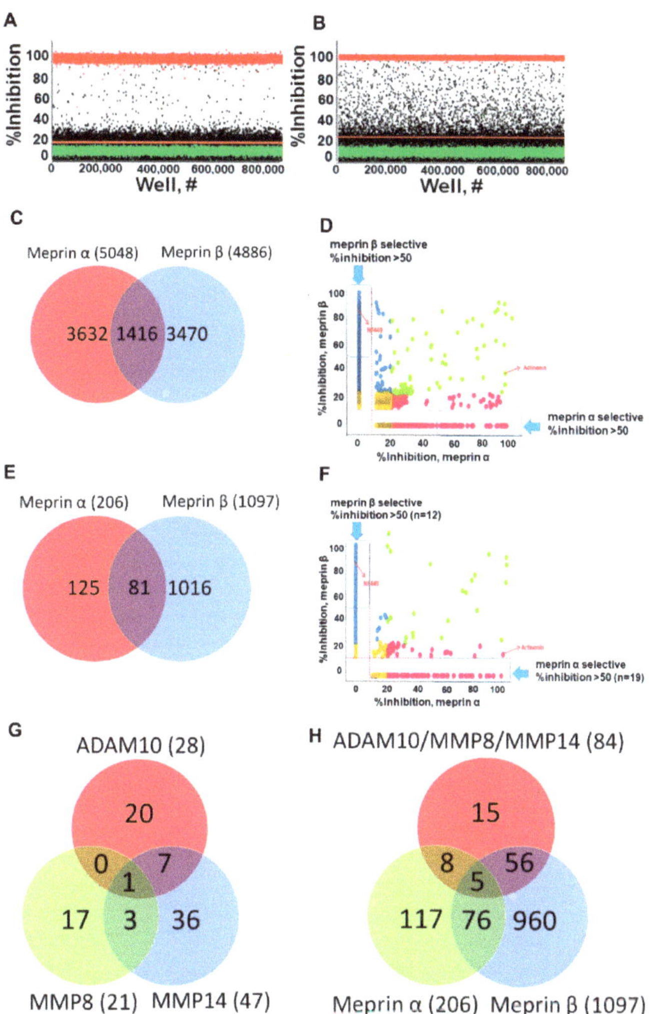

Figure 5. Primary uHTS campaigns. Scatter plots of (**A**) meprin α and (**B**) meprin β primary campaigns. Overall, >650,000 compounds were screened in singlicate against each target. (**C**) Venn diagram of meprin α and meprin β uHTS hits shows 1416 nonselective hits, 3632 meprin α and 3470 meprin β nominally selective hits. (**D**) Correlation plot of meprin α and meprin β actives demonstrates distribution of hits. (**E**) Venn diagram of meprin α and meprin β confirmation assays shows confirmed 81 confirmed nonselective hits, 125 meprin α and 1016 meprin β confirmed selective hits. Each confirmation assay was performed in triplicate. (**F**) Correlation plot of meprin α and meprin β actives demonstrates distribution of hits. (**G**) Venn diagram of meprin α and β hits screen against the counter targets MMP-8, MMP-14, and ADAM10. The 84 meprin actives inhibited one of three counter targets. (**H**) Venn diagram of meprin α and β hits versus three combined counter targets. A total of 117 compounds selectively inhibited meprin α while 960 compounds selectively inhibited meprin β.

Venn analysis showed that 84 meprin actives inhibited at least one counter target (Figure 5G), while there were 117 compounds selectively inhibiting meprin α and 960 compounds selectively inhibiting meprin β (Figure 5H) and 14 and 75 compounds selectively inhibiting meprin α and meprin β, respectively, with ≥50% inhibition. Cheminformatics analysis of the Scripps HTS assay database containing hundreds of biological assay results showed that 660 out of 1237 confirmed hits were not promiscuous; meaning they hit in less than 5 other assays. Out of these 660 compounds 536 were meprin α active and 195 were meprin β active. Medicinal chemistry triage suggested that 289 compounds out of 536 meprin α actives were tractable, while out of 195 meprin β actives 180 were tractable, which constitutes 469 total tractable compounds. Removal of 62 duplicates left us with 407 unique compounds of which 404 were available for concentration response studies. Despite the majority of top actives from the 2378 primary HTS hits being potential Zn-binders, the hit rate in counter screens was < 2.0% (Figure 5G,H) suggesting low metzincin promiscuity of meprin hits.

We conducted concentration response studies of 404 compounds in meprin α and β assays using 10-point 3:1 serial dilutions starting at the highest concentration of 17.4 μM in triplicate. Out of 404 tested compounds, 13 exhibited IC50 values < 1 μM and 47 < 5 μM in in both meprin α and meprin β assays.

To pick compounds for further characterization and probe development we used a cutoff of IC50 values < 10 μM against either meprin and 10-fold selectivity window for meprin α or meprin β. Additionally, we picked the top selective compounds with IC50 values < 10 μM that had no apparent Zn-binding moieties. More specifically, we prioritized selective compounds without apparent Zn-binding groups (hydroxamates, carboxylates, etc.). Using these criteria, we selected 46 compounds. Interestingly, the majority (42) were selective for meprin β and only 4 were selective for meprin α. These 46 potentially non-Zn-binding compounds were clustered in [21] distinct scaffolds. The most populated scaffold had 9 members suggesting its amenability to medicinal chemistry.

The second group of compounds was chosen based on selectivity between main target (either meprin α or β) and four other tested metzincins (either meprin α or meprin β, ADAM10, MMP-8, and MMP-14) and potency towards the main target (either meprin α or meprin β) regardless of the presence of Zn binders. These criteria yielded 41 compounds belonging to 17 distinct clusters. Interestingly, the majority (32) were selective for meprin α and only 9 were selective for meprin β, which is the opposite trend from non-Zn-binders.

2.5. Hit Potency, Selectivity, and Cytotoxicity

We were able to procure 64 out of 87 selected compounds from commercial sources, which we tested in triplicate, 10-point, 3:1 serial dilution concentration response format starting at the highest concentration of 17.4 μM against both meprin α and meprin β. In addition to meprins, we also tested 64 hits against related metzincins (MMP-2, MMP-3, MMP-8, MMP-9, MMP-10, MMP-14, ADAM10, and ADAM17) to ascertain general nonpromiscuity against zinc-dependent proteases.

The top nine compounds exhibited IC50 values ≤ 1 μM against meprin α (Table 2). Examination of the structures of meprin α top hits revealed that they fall into four groups (Figure 6), thiadiazole-hydroxyacetamides (SR19849, SR19848, SR19847), triazole-hydroxyacetamides (SR19850, SR19855), sulfonamide-hydroxypropanamides (SR162808, SR162799), and phenoxy-hydroxyacetamides (SR1220670, SR1596857). SR162808 was the most potent and selective inhibitor of meprin α with an IC50 value of 0.446 μM and >30-fold selectivity against meprin β and other tested metzincins (Table 2). Both sulfonamide-hydroxypropanamides (SR162808 and SR162799) exhibited sub-micromolar IC50 values for meprin α inhibition and >30-fold selectivity against meprin β and other tested metzincins.

Table 2. Selectivity testing of meprin α top HTS hits. All units are IC$_{50}$, μM.

Compound ID	Structure	Meprin α	Meprin β	MMP2	MMP3	MMP8	MMP9	MMP10	MMP14	ADAM17
19847	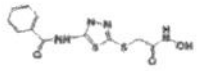	0.892	1.43	2.87	>17	>17	>17	>17	>17	>17
19848	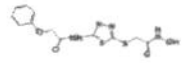	0.335	0.385	>17	>17	>17	>17	>17	>17	>17
19849	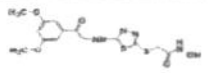	0.218	0.287	>17	>17	>17	>17	>17	>17	8.01
19850	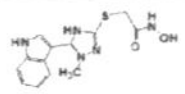	0.564	17	17	>17	>17	>17	>17	>17	5.03
19855	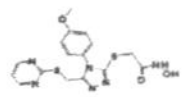	1.3	>17	>17	>17	>17	>17	>17	>17	>17
1596857	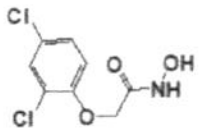	1.18	>17	>17	>17	>17	>17	>17	>17	>17
220670	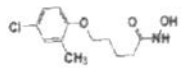	1.12	>17	>17	>17	4.20	17	>17	>17	>17
162799	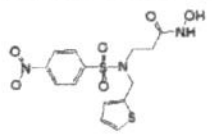	0.564	>17	>17	>17	>17	>17	>17	>17	>17
162808	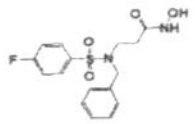	0.446	>17	17	>17	>17	>17	>17	>17	>17

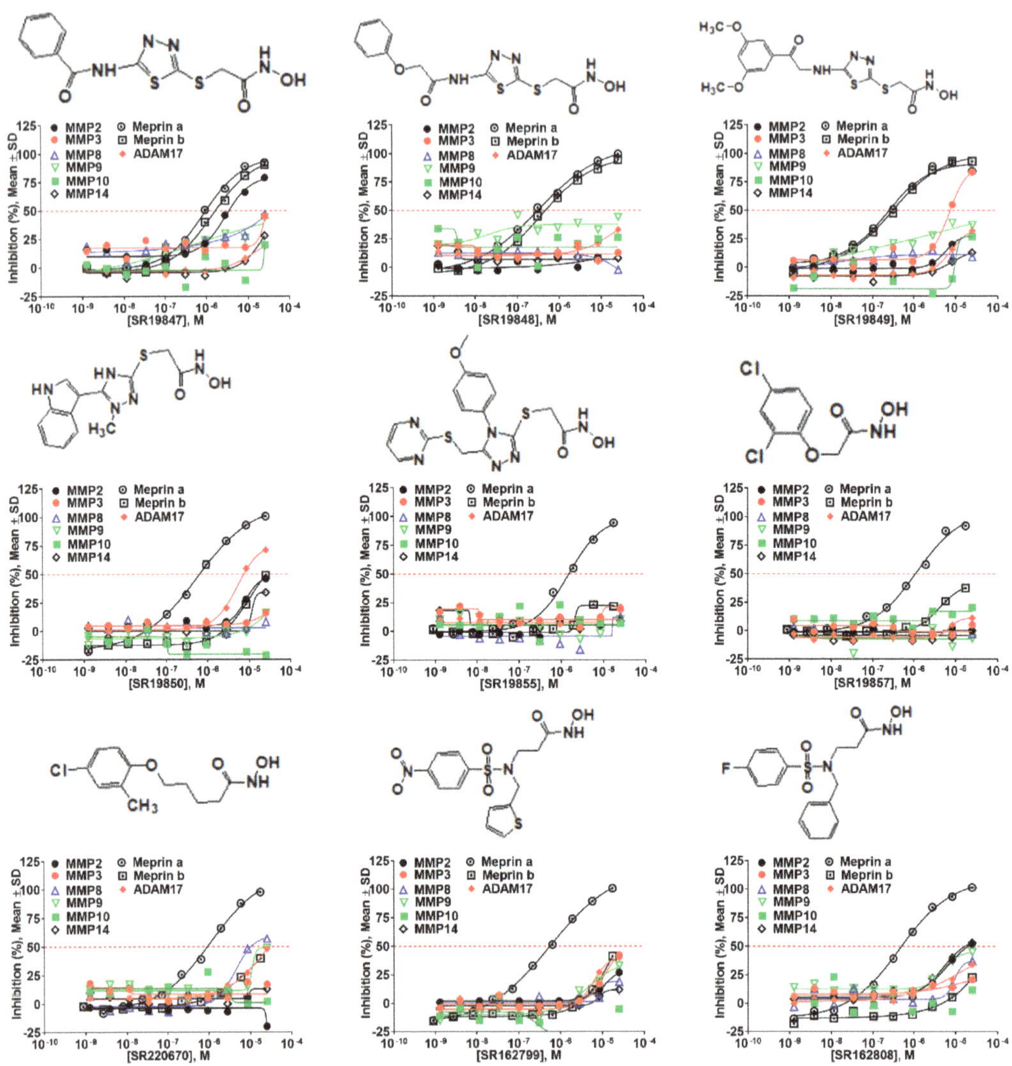

Figure 6. Results of concentration response studies of top potent and selective meprin α inhibitors.

The top meprin β inhibitors belonged to two structural families (Figure 7 and Table 3), isobutyryl-tetrahydronaphthalen-amides (SR910128, SR910130, and SR910140) and nitrofuran-containing compounds (SR207820 and SR412882). Compound SR355996 was the only representative of the bis-nitrobenzoic acid scaffold.

We also tested representative compounds from each scaffold for effects on skin fibroblast and melanocyte viability to ascertain cytotoxicity towards various skin cell types. Overall, hits showed either no or very little effect on cell viability (Figure 8) suggesting a lack of general cytotoxicity and amenability of hit chemotypes for the development into in vitro probe for biological studies.

Table 3. Selectivity testing of meprin β top HTS hits. All units are IC$_{50}$, μM.

Compound ID	Structure	Meprin α	Meprin β	MMP-2	MMP-3	MMP-8	MMP-9	MMP-10	MMP-14	ADAM17
SR207820		>17	1.5	>17	17	>17	>17	>17	>17	>17
SR412882		>17	3.5	>17	15	>17	>17	9.7	10.5	17
SR910128		>17	1.0	>17	>17	3.1	4.5	>17	>17	>17
SR910130		>17	2.0	>17	>17	3.0	9.9	>17	>17	>17
SR910140		>17	1.6	>17	>17	4.0	10	>17	>17	>17
SR355996		>17	0.97	>17	>17	>17	17	>17	>17	>17

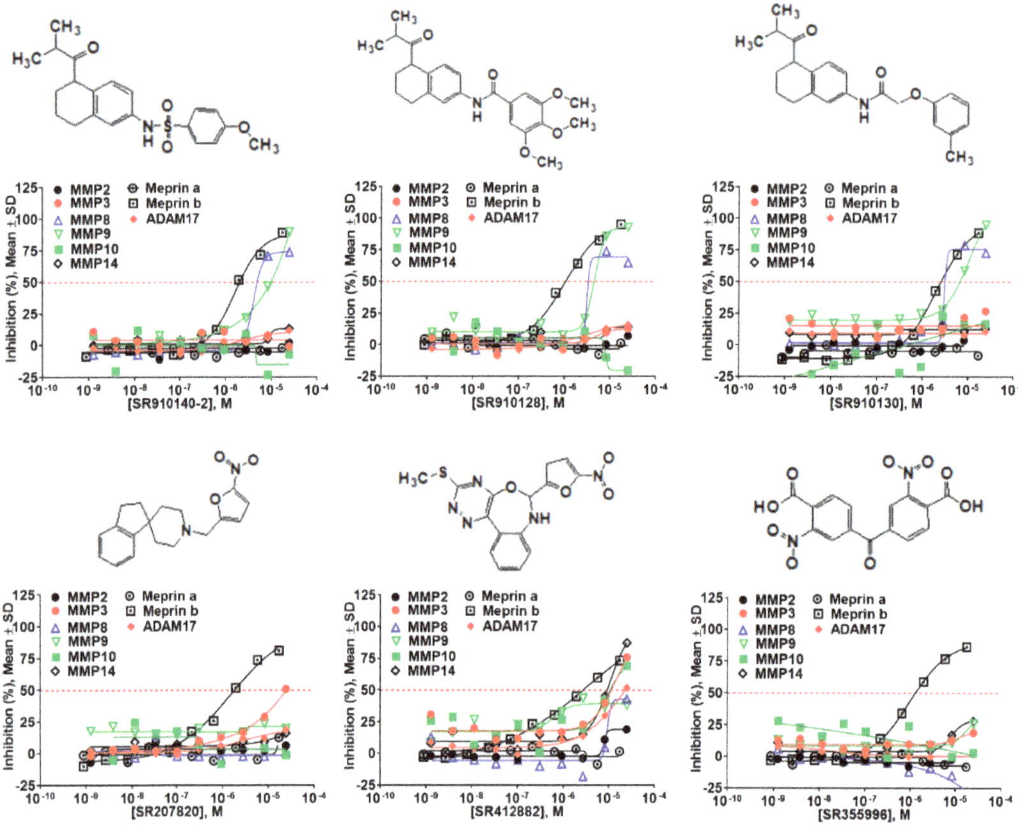

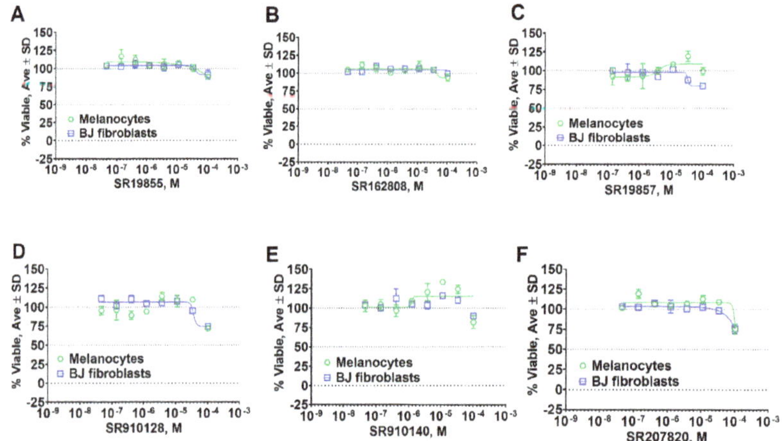

Figure 7. Results of concentration response studies of top potent and selective meprin β inhibitors.

Figure 8. Results of cytotoxicity studies of representative meprin α and meprin β inhibitors. (**A–C**) Meprin α inhibitors. (**D–F**) Meprin β inhibitors.

3. Discussion

As the result of the uHTS effort we discovered and characterized several novel scaffolds with activity against meprin α and meprin β. All top selective meprin α HTS hits contain a hydroxamate moiety, whereas meprin β hits lack one. Based on the presence of the hydroxamate moiety in meprin α inhibitors it is likely that they act via binding of the active site zinc atom as was demonstrated for numerous other metzincins. Tan et al. [17] proposed the interaction model whereby the hydroxamic moiety of an analog of compounds 10d and 10e (Figure 1) binds zinc and carboxylate moieties interact with residues of the S1 and S1$'$ subsites. Based on this model, the selectivity of 10d and 10e is derived from differences between meprin α and meprin β S1 and S1$'$ subsites. Both 10d and 10e structures are symmetric with a central hydroxamate moiety connected via propyl linkers to either terminal benzodioxanes or benzodioxols. Our HTS hits are unlikely to interact with both subsites as the hydroxamate is terminal in all cases. Similar to 10d and 10e, most of the hits (Table 2) have at least one other electronegative moiety in addition to the hydroxamate that could be interacting with positively charged residues in either subsite of meprin α. However, only five (SR19855, SR1596857, SR220670, SR162799, and SR162808) out of nine hits show selectivity for meprin α suggesting that additional interactions may be responsible for selectivity against meprin β.

The most selective and potent meprin α HTS hit, SR162808, exhibited more than 30-fold selectivity against meprin β and other metzincins (Table 2) and no cytotoxicity (Figure 8B). For comparison, 10d and 10e exhibit 18-fold and 19-fold selectivity, respectively (Table 4). Unfortunately, nothing has been reported about their effects on cell viability. For in vivo probe or drug lead development significant selectivity and toxicity windows are extremely important; therefore, SR162808 represents a good starting point for a medicinal chemistry optimization effort.

Table 4. Comparison of SR162808 and compounds 10d and 10e from [17]. All units are IC$_{50}$, μM.

ID	Meprin α	Meprin β	Selectivity Fold
SR162808	0.30	>17	38
10d	0.16	2.95	18
10e	0.40	7.59	19

The HTS-based approach to metalloproteinase or any other inhibitor discovery has inherent limitations and strengths. The strength of HTS is in its ability to assess multiple chemical scaffolds for activity against the target of choice and the ability to select the most promising scaffolds for further optimization. In the event that the chosen scaffold is not amenable to optimization, the researchers can go back to the HTS campaign results and choose additional scaffolds. The main limitation of the HTS approach is that it relies heavily on the composition of the HTS library that is being screened. In our case, we found several novel scaffolds active and selective for both meprins, however, most of these scaffolds contain zinc-binding moieties such as hydroxamates. While hydroxamates have good binding affinity to zinc of the active site, they have their limitations (reviewed in [22]). More specifically, the zinc-binding property of hydroxamates can lead to off-target interactions with other members of metzincin superfamily (e.g., adamalysins, matrixins), unfavorable pharmacokinetic properties, dose-limiting toxicity and metabolic instability, to name a few. However, despite these limitations, there are multiple examples of hydroxamates being used in the clinic (reviewed in [23]). Examples include, but are not limited to histone deacetylase inhibitors (HDACI) panobinostat and bellinostat, deferoxamine, a chelating agent used to treat iron or aluminum toxicity [24].

In conclusion, an HTS campaign led to the discovery of [5] selective meprin α hits belonging to three different chemotypes: triazole-hydroxyacetamides (SR19855), sulfonamide-hydroxypropanamides (SR162808 and SR162799), and phenoxy-hydroxyacetamides (SR1220670 and SR1596857). The chemical diversity of the HTS hits, a good metzincin selectivity profile, and low cytotoxicity suggest that these hits can be developed into more potent compounds for in vivo studies. Further medicinal chemistry optimization and in vivo studies will help determine the value of these scaffolds as probes or leads for drug development. The main value of current HTS hits is in their good selectivity for main target meprin α against a counter target meprin β and related metzincins. This property will help differentiate the roles of these two enzymes in biological and pathobiological processes.

4. Materials and Methods

4.1. Reagents

MMP-1, MMP-2, MMP-8, MMP-9, MMP-10, MMP-13, MMP-14, ADAM10, ADAM17 and Mca-KPLGL-Dpa-AR-NH2 fluorogenic peptide substrates were purchased from R&D Systems (cat # 901-MP, 902-MP, 908-MP, 911-MP, 910-MP, 511-MM, 918-MP, 936-AD, 930-ADB, and ES010, respectively). All common chemicals were purchased from Sigma. NFF449 was purchased from Tocris (cat# 1391) and actinonin was from Sigma-Aldrich (cat# 01809).

4.2. HTS Substrate Synthesis

Meprin α and meprin β substrates ((Mca)-YVADAPK-(K-ε-Dnp) and (Mca)-EDEDED-(K-ε-Dnp), respectively) [25] were synthesized utilizing Fmoc solid-phase methodology on a peptide synthesizer. All peptides were synthesized as C-terminal amides to prevent diketopiperazine formation [26]. Cleavage and side-chain deprotection of peptide-resins was for at least 2 h using thioanisole-water-TFA (5:5:90). The substrates were purified and characterized by preparative RP HPLC and characterized by MALDI-TOF MS and analytical RP HPLC.

4.3. Meprins Expression Protocol

Recombinant human meprin α and meprin β were expressed using the Bac-to-Bac expression system (Gibco Life Technologies, Paisley, UK) as described [27–29]. Media and supplements were obtained from Gibco Life Technologies. Recombinant Baculoviruses were amplified in adherently growing *Spodoptera frugiperda* (Sf)9 insect cells at 27 °C in Grace's insect medium supplemented with 10% fetal bovine serum, 50 units/mL penicillin, and 50 μg/mL streptomycin. Protein expression was performed in 500 mL suspension cultures of BTI-TN-5B1-4 insect cells growing in Express Five SFM supplemented with 4 mM glutamine, 50 units/mL penicillin, and 50 μg/mL streptomycin in Fernbach flasks using a Multitron orbital shaker (INFORS AG, Bottmingen, Switzerland). Cells were infected at a density of 2×10^{6} cells/mL with an amplified viral stock at a MOI of ~10. Protein expression was stopped after 72 h, and recombinant meprins were further purified from the media by ammonium sulfate precipitation (60% saturation) and affinity chromatography (Streptactin for Strep-tagged meprin α and Ni-NTA for His-tagged meprin β). Meprins were activated by trypsin, which was removed afterwards by affinity chromatography using a column containing immobilized chicken ovomucoid, a trypsin inhibitor.

4.4. Meprin α and Meprin β Assays in 384 Well Plate

Both assays followed the same general protocol [14]. A 5 μL measure of $2 \times$ enzyme solution (2.6 and 0.1 nM for meprin α and meprin β, respectively) in assay buffer (50 mM Hepes, 0.01% Brij-35, pH 7.5) was added to solid bottom black 384 low volume plates (Nunc, cat# 264705). Next, 75 nL of test compounds or pharmacological control (actinonin or NFF449) were added to corresponding wells using a 384-pin tool device (V&P Scientific, San Diego, CA, USA). After 30 min incubation at RT, the reactions were started by addition of 5 μL of $2 \times$ solutions of substrates (20 μM, mepin α substrate Mca-YVADAPK-K(Dnp) or meprin β substrate Mca-EDEDED-K(Dnp)). Reactions were incubated at RT for 1 h, after which the fluorescence was measured using the Synergy H4 multimode microplate reader (Biotek Instruments) (λexcitation = 324 nm, λemission = 390 nm).

Three parameters were calculated on a per-plate basis: (a) the signal-to-background ratio (S/B); (b) the coefficient for variation (CV; CV = (standard deviation/mean) $\times$ 100)) for all compound test wells; and (c) the Z- or Z'-factor [18]. Z takes into account the effect of test compounds on the assay window, while Z' is based on controls.

4.5. Determination of Kinetic Parameters of Meprin α and Meprin β Mediated Proteolysis of Their Respective Substrates

Substrate stock solutions were prepared at various concentrations in HTS assay buffer (50 mM Hepes, 0.01% Brij-35, pH 7.5). Assays were conducted by incubating a range of substrate concentrations (2–50 μM) with various meprin concentrations at 25 °C. Fluorescence was measured on a multimode microplate reader Synergy H1 (Biotek Instruments, Winooski, VT, USA) using λexcitation = 324 nm and λemission = 393 nm. Rates of hydrolysis were obtained from plots of fluorescence versus time, using data points from only the linear portion of the hydrolysis curve. The slope from these plots was divided by the fluorescence change corresponding to complete hydrolysis and then multiplied by the substrate concentration to obtain rates of hydrolysis in units of μM/s. Kinetic parameters were calculated by nonlinear regression analysis using the GraphPad Prism 8.0 suite of programs.

4.6. Meprin α and Meprin β Assays in 1536 Well Plate Format

Both assays followed the same general protocol. A 2 μL portion of 2 $\times$ enzyme solution (1.3 and 0.0125 nM for meprin α and meprin β, respectively) in assay buffer (50 mM Hepes, 0.01% Brij-35, pH 7.5) was added to solid bottom black 1536 low volume plates (Corning cat# 7261). Next, 30 nL of test compounds or pharmacological control (actinonin or NFF449) was added to corresponding wells using a 1536 pin tool device (V&P Scientific, San Diego). After 30 min incubation at RT, the reactions were started by addition of 2 μL of 2x solutions of substrates (20 μM, meprin α substrate Mca-YVADAPK-K(Dnp) or meprin β substrate Mca-EDEDED-K(Dnp)). Reactions were incubated at RT for 1 h, after which the fluorescence was measured using the Viewlux multimode microplate reader (Perkin Elmer) (λexcitation = 324 nm, λemission = 390 nm).

Three parameters were calculated on a per-plate basis: (a) the signal-to-background ratio (S/B); (b) the coefficient for variation (CV; CV = (standard deviation/mean) $\times$ 100)) for all compound test wells; and (c) the Z- or Z'-factor [18]. Z takes into account the effect of test compounds on the assay window, while Z' is based on controls.

4.7. uHTS Campaign

The miniaturized 1536-well plate format meprin α and meprin β assays were used to screen a collection of approximately 650,000 compounds (The Scripps Research library, La Jolla, CA, USA) on the automated Kalypsys/GNF platform at The Scripps Research Molecular Screening Center (SRMSC, Jupiter, FL, USA: http://hts.florida.scripps.edu/). Both uHTS campaigns were run separately but in a similar manner. Briefly, the first step was the primary screen of all test compounds as singlicates against the meprin α and meprin β target at a final concentration of 7.0 μM. Next, compounds selected as primary hits were cherry-picked and retested in triplicate against the primary screen target and its antitarget (meprin α for the meprin β screening effort, and vice versa) at a same final concentration of 7.0 μM. The additional counter screen assays against related metzincins (MMP-8, MMP-14, and ADAM10) were performed in triplicate at a final concentration of 7.0 μM. The final step was the titration of selected hits as 10-point, 1:3 serial dilutions in both the target and antitarget assay, starting at a final nominal concentration of 17 μM. For all the aforementioned assays, actinonin and NFF449, for meprin α and meprin β, respectively, at a final concentration of 1 μM, were used as a positive control and reference for 100% inhibition. Wells treated with DMSO only were used as negative controls and 0% inhibition reference. The percentage inhibition of each well was then normalized as follows:

$$\%\_Inhibition = (RFU\_Test\_Compound - MedianRFU\_Low\_Control) / (MedianRFU\_High\_Control - MedianRFU\_Low\_Control) * 100 \qquad (1)$$

where "Test_Compound" refers to wells containing test compound, "High_Control" is defined as wells treated with either actinonin or NFF449 ($n = 24$) and "Low_Control" as wells containing DMSO only ($n = 24$). All data generated during this effort were uploaded to the SRMSC's institutional screening database (Assay Explorer, Symyx). Sample to background (S/B) ratios, as well as Z and Z' values, were calculated on a per plate basis as described 14. Curve fitting and resulting IC50 determinations were performed as previously reported [30].

4.8. ADAM10 and ADAM17 Assays

Both assays followed the same general protocol. A 2.5 μL portion of 2 × enzyme solution (20 nM) in assay buffer (10 mM HEPES, 0.001% Brij-35, pH 7.5) was added to solid bottom black 1536 plates (Greiner, cat# 789075). Next, test compounds and pharmacological controls were added to corresponding wells using a 1536 pin tool device (V&P Scientific, San Diego, CA, USA). After 30 min incubation at RT, the reactions were started by addition of 2.5 μL of 2x solutions of substrate (R&D Systems cat#: ES010, Mca-KPLGL-Dpa-AR-NH2, 20 μM). Reactions were incubated at RT for 2 h, after which the fluorescence was measured using a PerkinElmer Viewlux multimode microplate imager (λexcitation = 324 nm, λemission = 390 nm). Final concentration of test compounds in assays was 7.0 μM.

4.9. MMP Assays

All assays followed the same general protocol. A 5 μL portion of 2 × enzyme solution (5 nM) in assay buffer (50 mM Tricine, 50 mM NaCl, 10 mM CaCl2, 0.05% Brij-35, pH 7.5) was added to solid bottom black 384 plates (Nunc, cat# 264705). Next, test compounds and pharmacological controls were added to corresponding wells using a 384-pin tool device (V&P Scientific, San Diego, CA, USA). After 30 min incubation at RT, the reactions were started by addition of 5 μL of 2x solutions of substrate (R&D Systems cat#: ES010, Mca-KPLGL-Dpa-AR-NH2, 20 μM). Reactions were incubated at RT for 1 h, after which the fluorescence was measured using the Synergy H4 multimode microplate reader (Biotek Instruments) (λexcitation = 324 nm, λemission = 390 nm).

4.10. Cell Toxicity Studies

Test compounds were solubilized in 100% DMSO and added to polypropylene 384 well plates (Greiner cat# 781280). The 1250 of BJ skin fibroblasts (ATCC CRL-2522) and primary melanocytes (ATCC PCS-200-013) were plated in 384 well plates in 8 μL of serum free media (HybriCare for BT474, EMEM for HEK293). Test compounds and pharmacological assay control (lapatinib) were prepared as 10-point, 1:3 serial dilutions starting at 10 mM, then added to the cells using the pin tool mounted on the Integra 384. Plates were incubated for 72 h at 37 °C, 5% CO_2 and 95% relative humidity. After incubation, 8 μL of CellTiter-Glo (Promega cat# G7570) was added to each well and incubated for 15 min at room temperature. Luminescence was recorded using a Biotek Synergy H1 multimode microplate reader. Viability was expressed as a percentage relative to wells containing media only (0%) and wells containing cells treated with DMSO only (100%). Three parameters were calculated on a per plate basis: (a) the signal-to-background ratio (S/B); (b) the coefficient for variation (CV; CV = (standard deviation/mean) × 100)) for all compound test wells; and (c) the Z'-factor. IC50 values were calculated by fitting normalized data to sigmoidal log versus response equation utilizing nonlinear regression analysis from GraphPad Prizm 8.

Author Contributions: D.M. designed and oversaw the study, developed HTS assays, performed kinetic studies, performed post-HTS biochemical in vitro characterization of hits, wrote the first manuscript draft and prepared figures. J.D. performed cytotoxicity studies. G.B.F. synthesized and purified meprin α and meprin β substrates and edited the manuscript. T.B. and C.W. triaged HTS hits via cheminformatics. C.B.-P. expressed meprin α and meprin β. S.H., L.D.S. and T.P.S. performed uHTS campaign. All authors reviewed the manuscript. All authors have read and agreed to the published version of the manuscript.

Funding: This work was supported by the National Institutes of Health (AR066676 and CA249788 to DM). This work was also supported by the Deutsche Forschungsgemeinschaft (DFG) grant SFB877 "Proteolysis as a Regulatory Event in Pathophysiology" (project A9 and A15) (both to C.B.-P.). We thank Pierre Baillargeon and Lina DeLuca (Lead Identification, Scripps Florida) for compound management.

Institutional Review Board Statement: Not applicable.

Informed Consent Statement: Not applicable.

Data Availability Statement: All data are available upon request.

Conflicts of Interest: The authors declare no conflict of interest.

Abbreviations

uHTS	ultrahigh-throughput screening
MMP	matrix metalloprotease
ADAM	a disintegrin and metalloprotease

References

1. Peters, F.; Becker-Pauly, C. Role of meprin metalloproteases in metastasis and tumor microenvironment. *Cancer Metastasis Rev.* **2019**, *38*, 347–356. [CrossRef]
2. Prox, J.; Arnold, P.; Becker-Pauly, C. Meprin α and meprin β: Procollagen proteinases in health and disease. *Matrix Biol.* **2015**, *44*, 7–13. [CrossRef]
3. Broder, C.; Arnold, P.; Vadon-Le Goff, S.; Konerding, M.A.; Bahr, K.; Muller, S.; Overall, C.M.; Bond, J.S.; Koudelka, T.; Tholey, A.; et al. Metalloproteases meprin alpha and meprin β are C- and N-procollagen proteinases important for collagen assembly and tensile strength. *Proc. Natl. Acad. Sci. USA* **2013**, *110*, 14219–14224. [CrossRef]
4. Becker-Pauly, C.; Pietrzik, C.U. The Metalloprotease Meprin β Is an Alternative β-Secretase of APP. *Front. Mol. Neurosci.* **2016**, *9*, 159. [CrossRef] [PubMed]
5. Scharfenberg, F.; Armbrust, F.; Marengo, L.; Pietrzik, C.; Becker-Pauly, C. Regulation of the alternative β-secretase meprin β by ADAM-mediated shedding. *Cell. Mol. Life Sci.* **2019**, *76*, 3193–3206. [CrossRef] [PubMed]
6. Ohler, A.; Debela, M.; Wagner, S.; Magdolen, V.; Becker-Pauly, C. Analyzing the protease web in skin: Meprin metalloproteases are activated specifically by KLK4, 5 and 8 vice versa leading to processing of proKLK7 thereby triggering its activation. *Biol. Chem.* **2010**, *391*, 455–460. [CrossRef]
7. Bhogal, R.K.; Stoica, C.M.; McGaha, T.L.; Bona, C.A. Molecular aspects of regulation of collagen gene expression in fibrosis. *J. Clin. Immunol.* **2005**, *25*, 592–603. [CrossRef] [PubMed]
8. Trojanowska, M.; LeRoy, E.C.; Eckes, B.; Krieg, T. Pathogenesis of fibrosis: Type 1 collagen and the skin. *J. Mol. Med.* **1998**, *76*, 266–274. [CrossRef] [PubMed]
9. Biasin, V.; Wygrecka, M.; Marsh, L.M.; Becker-Pauly, C.; Brcic, L.; Ghanim, B.; Klepetko, W.; Olschewski, A.; Kwapiszewska, G. Meprin β contributes to collagen deposition in lung fibrosis. *Sci. Rep.* **2017**, *7*, 1–12. [CrossRef] [PubMed]
10. Arnold, P.; Otte, A.; Becker-Pauly, C. Meprin metalloproteases: Molecular regulation and function in inflammation and fibrosis. *Biochim. Biophys. Acta Mol. Cell Res.* **2017**, *1864*, 2096–2104. [CrossRef]
11. Guidi, A.; Mansour, N.R.; Paveley, R.A.; Carruthers, I.M.; Besnard, J.; Hopkins, A.L.; Gilbert, I.H.; Bickle, Q.D. Application of RNAi to Genomic Drug Target. Validation in Schistosomes. *PLoS Negl. Trop. Dis.* **2015**, *9*, e0003801. [CrossRef] [PubMed]
12. Farand, J.; Kropf, J.E.; Blomgren, P.; Xu, J.; Schmitt, A.C.; Newby, Z.E.; Wang, T.; Murakami, E.; Barauskas, O.; Sudhamsu, J.; et al. Discovery of Potent and Selective MTH1 Inhibitors for Oncology: Enabling Rapid Target (In)Validation. *ACS Med. Chem. Lett.* **2019**, *11*, 358–364. [CrossRef] [PubMed]
13. Kruse, M.N.; Becker, C.; Lottaz, D.; Kohler, D.; Yiallouros, I.; Krell, H.W.; Sterchi, E.E.; Stocker, W. Human meprin α and β homo-oligomers: Cleavage of basement membrane proteins and sensitivity to metalloprotease inhibitors. *Biochem. J.* **2004**, *378*, 383–389. [CrossRef] [PubMed]
14. Madoux, F.; Tredup, C.; Spicer, T.P.; Scampavia, L.; Chase, P.S.; Hodder, P.S.; Fields, G.B.; Becker-Pauly, C.; Minond, D. Development of high throughput screening assays and pilot screen for inhibitors of metalloproteases meprin α and β. *Biopolymers* **2014**, *102*, 396–406. [CrossRef]

15. Ramsbeck, D.; Hamann, A.; Schlenzig, D.; Schilling, S.; Buchholz, M. First insight into structure-activity relationships of selective meprin β inhibitors. *Bioorg. Med. Chem. Lett.* **2017**, *27*, 2428–2431. [CrossRef] [PubMed]

16. Ramsbeck, D.; Hamann, A.; Richter, G.; Schlenzig, D.; Geissler, S.; Nykiel, V.; Cynis, H.; Schilling, S.; Buchholz, M. Structure-Guided Design, Synthesis, and Characterization of Next-Generation Meprin β Inhibitors. *J. Med. Chem.* **2018**, *61*, 4578–4592. [CrossRef] [PubMed]

17. Tan, K.; Jager, C.; Schlenzig, D.; Schilling, S.; Buchholz, M.; Ramsbeck, D. Tertiary-Amine-Based Inhibitors of the Astacin Protease Meprin alpha. *ChemMedChem* **2018**, *13*, 1619–1624. [CrossRef] [PubMed]

18. Zhang, J.H.; Chung, T.D.; Oldenburg, K.R. A Simple Statistical Parameter for Use in Evaluation and Validation of High. Throughput Screening Assays. *J. Biomol. Screen.* **1999**, *4*, 67–73. [CrossRef] [PubMed]

19. Copeland, R. *Evaluation of Enzyme Inhibitors in Drug Discovery*, 1st ed.; John Wiley and Sons: Hoboken, NJ, USA, 2005; pp. 113–117.

20. Madoux, F.; Dreymuller, D.; Pettiloud, J.P.; Santos, R.; Becker-Pauly, C.; Ludwig, A.; Fields, G.B.; Bannister, T.; Spicer, T.P.; Cudic, M.; et al. Discovery of an enzyme and substrate selective inhibitor of ADAM10 using an exosite-binding glycosylated substrate. *Sci. Rep.* **2016**, *6*, 1–17. [CrossRef]

21. Baillargeon, P.; Fernandez-Vega, V.; Sridharan, B.P.; Brown, S.; Griffin, P.R.; Rosen, H.; Cravatt, B.; Scampavia, L.; Spicer, T.P. The Scripps Molecular Screening Center and Translational Research Institute. *SLAS Discov. Adv. Life Sci. R D* **2019**, *24*, 386–397. [CrossRef] [PubMed]

22. Fields, G.B. The Rebirth of Matrix Metalloproteinase Inhibitors: Moving Beyond the Dogma. *Cells* **2019**, *8*, 984. [CrossRef] [PubMed]

23. Liu, W.; Liang, Y.; Si, X. Hydroxamic acid hybrids as the potential anticancer agents: An Overview. *Eur. J. Med. Chem.* **2020**, *205*, 112679. [CrossRef] [PubMed]

24. Wu, M.; Gao, R.; Dang, B.; Chen, G. The Blood Component Iron Causes Neuronal Apoptosis Following Intracerebral Hemorrhage via the PERK Pathway. *Front. Neurol.* **2020**, *11*. [CrossRef] [PubMed]

25. Broder, C.; Becker-Pauly, C. The metalloproteases meprin α and meprin β: Unique enzymes in inflammation, neurodegeneration, cancer and fibrosis. *Biochem. J.* **2013**, *450*, 253–264. [CrossRef] [PubMed]

26. Fields, G.B.; Lauer-Fields, J.L.; Liu, R.-q.; Barany, G. Principles and Practice of Solid-Phase Peptide Synthesis. In *Synthetic Peptides: A User's Guide*, 2nd ed.; Grant, G.A., Ed.; W.H. Freeman & Co.: New York, NY, USA, 2001; pp. 93–219.

27. de Jong, G.I.; Buwalda, B.; Schuurman, T.; Luiten, P.G. Synaptic plasticity in the dentate gyrus of aged rats is altered after chronic nimodipine application. *Brain Res.* **1992**, *596*, 345–348. [CrossRef]

28. Becker-Pauly, C.; Howel, M.; Walker, T.; Vlad, A.; Aufenvenne, K.; Oji, V.; Lottaz, D.; Sterchi, E.E.; Debela, M.; Magdolen, V.; et al. The alpha and beta subunits of the metalloprotease meprin are expressed in separate layers of human epidermis, revealing different functions in keratinocyte proliferation and differentiation. *J. Investig. Dermatol.* **2007**, *127*, 1115–1125. [CrossRef]

29. Becker, C.; Kruse, M.N.; Slotty, K.A.; Kohler, D.; Harris, J.R.; Rosmann, S.; Sterchi, E.E.; Stocker, W. Differences in the activation mechanism between the alpha and beta subunits of human meprin. *Biol. Chem.* **2003**, *384*, 825–831. [CrossRef] [PubMed]

30. Smith, E.; Chase, P.; Niswender, C.M.; Utley, T.J.; Sheffler, D.J.; Noetzel, M.J.; Lamsal, A.; Wood, M.R.; Conn, P.J.; Lindsley, C.W.; et al. Application of Parallel Multiparametric Cell-Based FLIPR Detection Assays for the Identification of Modulators of the Muscarinic Acetylcholine Receptor 4 (M$_4$). *J. Biomol. Screen.* **2015**, *20*, 858–868. [CrossRef] [PubMed]

pharmaceuticals

Article

Discovery and Optimization of Selective Inhibitors of Meprin α (Part II)

Chao Wang [1,2], Juan Diez [3], Hajeung Park [1], Christoph Becker-Pauly [4], Gregg B. Fields [5], Timothy P. Spicer [1,6], Louis D. Scampavia [1,6], Dmitriy Minond [2,7] and Thomas D. Bannister [1,2,*]

[1] Department of Molecular Medicine, Scripps Research, Jupiter, FL 33458, USA; CWang@scripps.edu (C.W.); hajpark@scripps.edu (H.P.); spicert@scripps.edu (T.P.S.); scampl@scripps.edu (L.D.S.)

[2] Department of Chemistry, Scripps Research, Jupiter, FL 33458, USA; dminond@nova.edu

[3] Rumbaugh-Goodwin Institute for Cancer Research, Nova Southeastern University, 3321 College Avenue, CCR r.605, Fort Lauderdale, FL 33314, USA; mdiezi@yahoo.com

[4] The Scripps Research Molecular Screening Center, Scripps Research, Jupiter, FL 33458, USA; cbeckerpauly@biochem.uni-kiel.de

[5] Unit for Degradomics of the Protease Web, Institute of Biochemistry, University of Kiel, Rudolf-Höber-Str.1, 24118 Kiel, Germany; fieldsg@fau.edu

[6] Department of Chemistry & Biochemistry and I-HEALTH, Florida Atlantic University, 5353 Parkside Drive, Jupiter, FL 33458, USA

[7] Dr. Kiran C. Patel College of Allopathic Medicine, Nova Southeastern University, 3301 College Avenue, Fort Lauderdale, FL 33314, USA

* Correspondence: tbannist@scripps.edu

Citation: Wang, C.; Diez, J.; Park, H.; Becker-Pauly, C.; Fields, G.B.; Spicer, T.P.; Scampavia, L.D.; Minond, D.; Bannister, T.D. Discovery and Optimization of Selective Inhibitors of Meprin α (Part II). *Pharmaceuticals* **2021**, *14*, 197. https://doi.org/10.3390/ph14030197

Academic Editor: Salvatore Santamaria

Received: 14 December 2020
Accepted: 24 February 2021
Published: 27 February 2021

Publisher's Note: MDPI stays neutral with regard to jurisdictional claims in published maps and institutional affiliations.

Abstract: Meprin α is a zinc metalloproteinase (metzincin) that has been implicated in multiple diseases, including fibrosis and cancers. It has proven difficult to find small molecules that are capable of selectively inhibiting meprin α, or its close relative meprin β, over numerous other metzincins which, if inhibited, would elicit unwanted effects. We recently identified possible molecular starting points for meprin α-specific inhibition through an HTS effort (see part I, preceding paper). Here, in part II, we report further efforts to optimize potency and selectivity. We hope that a hydroxamic acid meprin α inhibitor probe will help define the therapeutic potential for small molecule meprin α inhibition and spur further drug discovery efforts in the area of zinc metalloproteinase inhibition.

Keywords: meprin α; meprin β; zinc metalloproteinase; medicinal chemistry; probe development

1. Introduction

Meprin α is a zinc metalloproteinase (metzincin) that has been implicated in multiple diseases, including fibrosis [1,2], and cancers [3]. Understanding meprin α's precise role, alone or in combination with other metzincins, including its close relative meprin β, has been difficult to establish due to the lack of known selective inhibitors. While many preliminary reports for meprin inhibitors have emerged [4–8], no compounds with suitable potency, selectivity, and drug-like attributes for in vivo use are known. We recently reported an HTS campaign to identify lead meprin α and meprin β inhibitors [9]. We then began SAR studies to understand the pharmacophore for meprin α inhibition, to enhance potency of the best leads, and to widen or maintain their selectivity over other metzincins. These efforts led to the identification of a potential probe molecule that may allow us in the future to ascertain the therapeutic potential for small molecule meprin α inhibition in the treatment of fibrosis or cancers.

Zinc metalloproteinases are ubiquitous in human biology and their dysregulation in certain disease states has spurred many researchers to seek modulators, usually inhibitors, of their function. The primary difficulty in most drug discovery efforts is to gain selectivity for the zinc metalloproteinase of interest. Most inhibitors are characterized by having groups that tightly bind zinc ions, often through bidentate coordination. Compounds that bind strongly to zinc metalloproteinases in this way can be found, though a frequently encountered liability is poor target selectivity, since coordination to zinc, and perhaps to other metal ions of similar size, is often the main driver of potency. As a consequence, many such inhibitors block off-target zinc-and even non-zinc metalloproteinase, including enzymes that are required for a myriad of essential biological functions of healthy cells, and they thus display intolerably high toxicity.

A common strategy is to avoid metal coordinating groups altogether. A second strategy is to employ a metal coordinating group that is: (1) consistent with drug-like properties, and (2) is merely one part of an organic molecule that also has several other functional groups that contribute substantially to its binding energy for the target, and not for other targets, thus imparting an acceptable level of target selectivity. Ideally this will translate to low toxicity in in vivo use. In our efforts here, our starting point was a screening hit containing a hydroxamic acid, R–CO–NH–OH, and we sought to alter other regions of the screening hit to widen its selectivity.

In our HTS effort (see part I), we sought any type of inhibitor, with or without metal coordinating groups. Only leads with known Zn coordinating functional groups were found to have suitable potency for follow-up, however, and any optimization effort required that we address selectivity concerns at an early stage. The hydroxamic acid compound SR19855 (Figure 1) was a preferred meprin α inhibitor hit, having both significant potency (low-micromolar inhibition of meprin α) and hints of target selectivity. In screens for selectivity, we saw 13-fold selectivity for meprin α inhibition over inhibition of the most closely-related enzyme, meprin β, and also at least 10-fold selectivity over a larger panel of metzincins, including MMP-2, MMP-3, MMP-8, MMP-9, MMP-10, MMP-14, ADAM10, and ADAM17. Given that a hydroxamic acid group can tightly bind zinc ions in all of these targets, we surmised that other regions of SR19855 were contributing to meprin α binding and to target selectivity. Further, we wished to further widen this selectivity window, with increases in potency as well, so that a probe molecule might be identified for use in order to isolate the effect of meprin α-specific inhibition in relevant biological and pharmacological contexts.

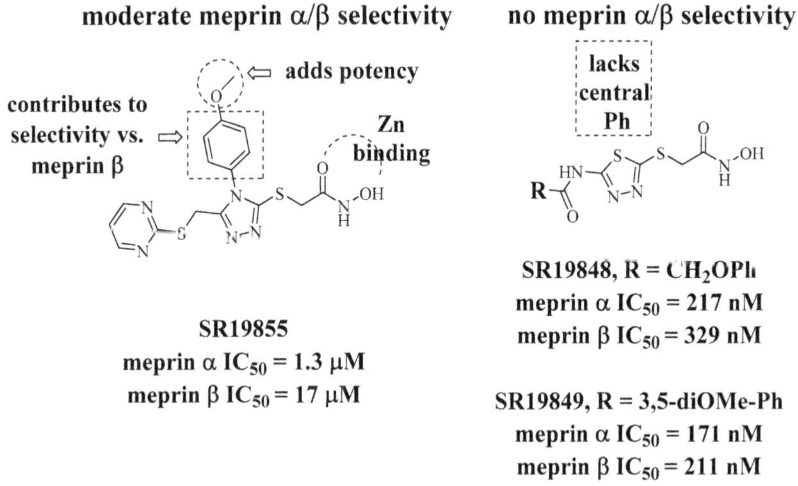

Figure 1. HTS hit, the aryl triazole SR19855 (left), and related, less selective thiadiazoles.

2. Results

2.1. Modeling of Lead Series

Our HTS effort and follow-up SAR-by-purchase efforts identified a number of moderately potent meprin α inhibitors bearing hydroxamic acid groups. Among these were related aryl triazoles and thiadiazoles (Figure 1). Notably, the thiadiazoles were unlike the aryl triazoles in terms of isoform selectivity: only the aryl triazoles were >10-fold selective for meprin α over meprin β. We considered isoform selectivity to be critically important, since low ligand selectivity limits the usefulness of metalloprotease inhibitors as isoform-specific probes. We surmised that the additional aromatic ring in the aryl triazoles (left) may be an important selectivity element favoring binding to meprin α over meprin β, and perhaps over other metzincins as well. In the thiadiazole series, similar compounds purchased with a thiazole core (one N replaced with CH) were much less potent, suggesting that the nitrogen atoms in the heterocyclic core are important.

We wished to understand the possible basis for meprin α binding and SR19855 selectivity. Using coordinates for a meprin β X-ray crystal structure, we constructed a homology model for meprin α and docked several HTS hits, including the aryl triazoles. The hydroxamate of SR19885, not surprisingly, is modeled to form a strong bidentate interaction with the Zn ion in both docking models (Figure 2). This binding anchors the ligand into each active site. Overall, geometry of binding for the backbone of the ligand is similar for the meprin enzymes. Important differences are apparent for the side chains, however, especially the interactions of the central *para*-methoxy phenyl group in meprin α. As shown in the left panel of Figure 2, Y149 of meprin α makes a face-on π-π interaction with the phenyl ring. R177 residue makes cation-π interactions with both phenyl and pyrimidine rings of the ligand. Although the primary sequences of the meprin isoforms are similar in the binding pocket, these interactions are absent or are much weaker in meprin β with the exception of the hydroxamate/Zn interaction and an H-bond interaction to Y211 (Y149 in meprin α) that are preserved. These findings suggest that the aryl triazole core of the ligand should be preserved, for selectivity purposes, during our SAR studies aimed to increase potency and selectivity for meprin α.

Meprin α   Meprin β

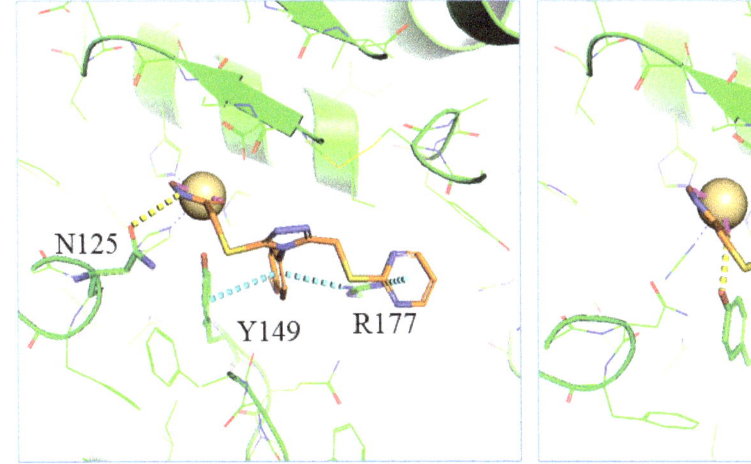

π–π interaction with Y149
cation-π interactions (two) with R177
H-bond to N125
bidentate salt bridge to Zn ion

No π-π interaction,
no cation-π interactions
H-bond to N125
bidentate salt bridge to Zn ion

Figure 2. SR19855 docked to meprins α and β.

In analyzing data from the HTS effort and for a few purchased analogs, we noted that both the phenyl and pyrimidine rings of SR19855 could not be deleted without a substantial drop in meprin α inhibitory activity. This finding agreed with our binding model (Figure 2). Together, these observations led us to explore the effect of a pyrimidine substituents and the replacement or alternative substitution of the central aryl group, with the aim of finding electronic and/or steric factors that would maximize meprin α inhibitory activity and maintain or (preferably) increase target selectivity.

2.2. Synthetic Strategy

To further study compounds in the series, we initiated an internal synthesis effort. A versatile 6-step synthesis used is shown in Figure 3, which is adaptable for different pyrimidine X groups and for substituted phenyl rings, or a replacement heterocycle. This strategy is based upon the initial 3 steps of a literature method [10]. Briefly, the alkylation of a thiopyrimidine **1** (step 1) to give ester **2** is followed by treatment with hydrazine (step 2) to form a hydrazine amide **3**, which is acylated with an isothiocyanate **4** (step 3), which following an immediate acidic workup (step 4), gives the mercaptotriazole **6**. This potentially oxidatively sensitive intermediate is then alkylated (step 5) to give ester **7**, which can be safely stored under inert conditions and purified by chromatography, if needed. Many of the earlier intermediates are obtained in high purity through precipitation or crystallization, with minimal chromatography, as outlined in the Experimental Section. Finally, the hydroxamic acid group is installed by treatment with hydroxylamine (step 6) to give the final test hydroxamic acid product **8**.

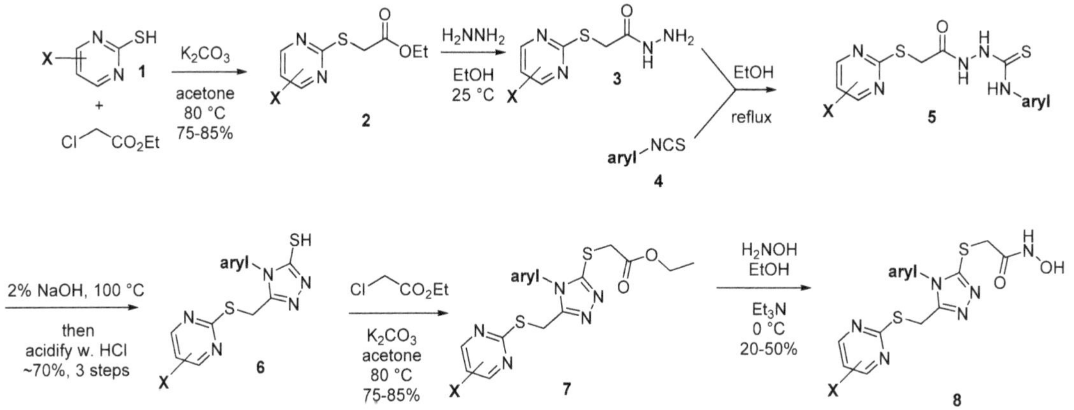

Figure 3. General synthesis scheme.

2.2.1. A. Pyrimidine Substitution

Holding the other groups constant, we varied the X substituents on the pyrimidine (Table 1), from the relatively electron-deficient unsubstituted pyrimidine, to a bulkier but neutral pyrimidine (3,5-dimethyl), to a more electron-rich 3,5-dimethoxy pyrimidine. We used symmetric disubstituted compounds due to their ready availability. Potency varied only in a small range, as all three compounds are single digit micromolar meprin α binders. Selectivity assessment using the meprin β inhibition assay, however, led us to prioritize 3,5-dimethoxy pyrimidines for further follow-up, as SR24144 showed complete lack of inhibition of meprin β at 100 μM. This contrasts with significant meprin β inhibition by SR19855 (IC$_{50}$ = 17 μM) and modest but complete meprin β inhibition by SR24319 (IC$_{50}$ = 60 μM). We felt that the maximization of selectivity, which was our primary criteria for probe development, offset a decrease in meprin α inhibition (meprin α IC$_{50}$ = 8.7 μM), *provided* that meprin α inhibition could be increased by making other modifications, such as replacements for the central methoxyphenyl ring.

Table 1. Pyrimidine substitution: dimethoxylation disfavors meprin β inhibition.

	X	ID	Meprin α IC$_{50}$, μM	Meprin β IC$_{50}$, μM
	H	SR19855	1.3 ± 0.1	17 ± 3
	Me	SR24139	4.1 ± 0.8	60 ± 7
	OMe	SR24144	8.7 ± 1.1	>100

2.2.2. Phenyl Substitution

Replacement of the central methoxyphenyl ring (Table 2) indeed allowed us to regain meprin α binding affinity without sacrificing selectivity vs. meprin β. Moving the *para* methoxy group in SR19855 to the meta or ortho position, however, reduced meprin α affinity and, more concerning, gave unwanted binding to meprin β (IC$_{50}$ = 59 and 52 μM for SR24460 and SR24459, respectively). Reducing electron density with a pyridyl ring present or with CF$_3$, Br, F, or Me groups replacing OMe maintained meprin β selectivity but had only minor effects on meprin α binding, with SR24403, with a CF$_3$ group, being the best alternative (meprin α IC$_{50}$ = 5.4 μM). A larger group, a 2-naphthyl ring system (SR24467), gave a modest increase in meprin α binding (IC$_{50}$ = 2.0 μM) at the expense of higher molecular weight and elevated lipophilicity. We saw enhanced potency when a *para* amino substituted phenyl was used, with the 4-morpholine analog SR26466 having an IC$_{50}$ = 1.1 μM, the 4-diethylamino analog 24,465 having an IC$_{50}$ = 600 nM, and the 4-dimethylamino analog SR24717 being nearly equipotent, having an IC$_{50}$ = 660 nM. IC$_{50}$ values for each compound were determined in triplicate and the average value is given in Table 2, with SD factor as an error. Selectivity over meprin β was maintained in all analogs other than the *meta*-methoxy and *ortho*-methoxy compounds (IC$_{50}$ vs. meprin β in all other cases was >70 μM). To test our presumption that the hydroxamic acid is essential, we also tested certain ester synthetic precursors to the hydroxamates (not shown) and saw no activity (meprin α IC$_{50}$ > 100 μM). As shown in Table 2, SR26465, with a diethylamino group present, and SR24717, with a with a diethylamino group present, are essentially equivalent with respect to meprin α/β affinity and a selectivity ratio >100. We chose SR24417 as the lead for further evaluation.

Table 2. SAR studies in the central aromatic ring.

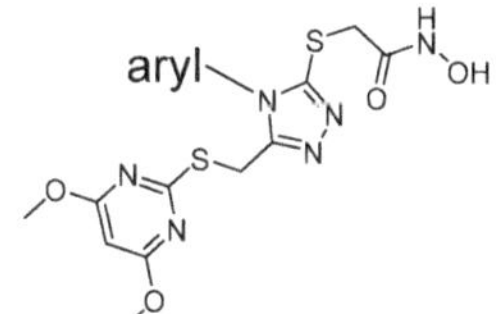

aryl	ID	Meprin α IC$_{50}$, μM	Meprin β IC$_{50}$, μM
MeO (para)	SR19855	1.3 ± 0.1	17 ± 3
MeO (meta)	SR24460	10.3 ± 1.1	59 ± 10
OMe (ortho)	SR24459	12.1 ± 3.9	52 ± 1
pyridyl	SR24718	10.1 ± 2.3	>100

Table 2. *Cont.*

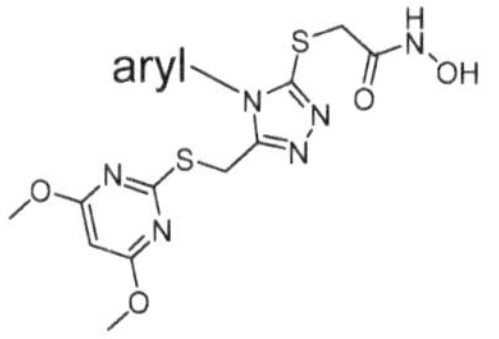

aryl	ID	Meprin α IC$_{50}$, μM	Meprin β IC$_{50}$, μM
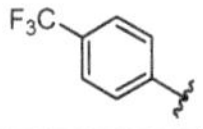 F$_3$C-phenyl	SR24003	5.4 ± 1.2	>100
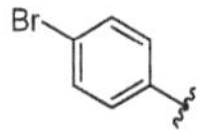 Br-phenyl	SR24462	11.2 ± 2.6	>100
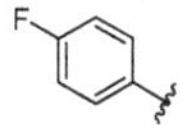 F-phenyl	SR24463	10.4 ± 2.8	>100
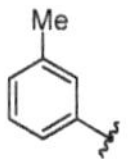 Me-phenyl	SR24467	10.3 ± 2.2	>100
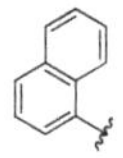 naphthyl	SR26467	2.0 ± 0.3	>100
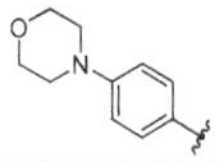 morpholino-phenyl	SR26466	1.1 ± 0.1	>100
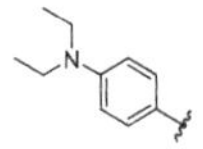 diethylamino-phenyl	SR26465	0.60 ± 0.06	75 ± 15
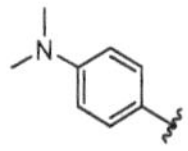 dimethylamino-phenyl	SR24717	0.66 ± 0.03	70 ± 5

2.3. SR24717 Characterization in In Vitro Assays

We tested a freshly re-synthesized, analytically pure sample of SR24717 against meprin α, meprin β, and related metzincins. IC$_{50}$ values for SR24717 were batch independent: IC$_{50}$ values for meprin α inhibition for multiple batches ranged from 490–670 nM. IC$_{50}$ values for meprin β inhibition were consistently at ~70 μM or above as well. Results shown in Figure 4 are for one scale-up batch, with results in triplicate. There is full meprin α inhibition and only partial meprin β inhibition, at the higher doses tested (Figure 4A). Assay results with related metzincins demonstrated excellent broad selectivity of SR24717 (IC$_{50}$ metzincins > 100 μM for all tested enzymes, Figure 4B).

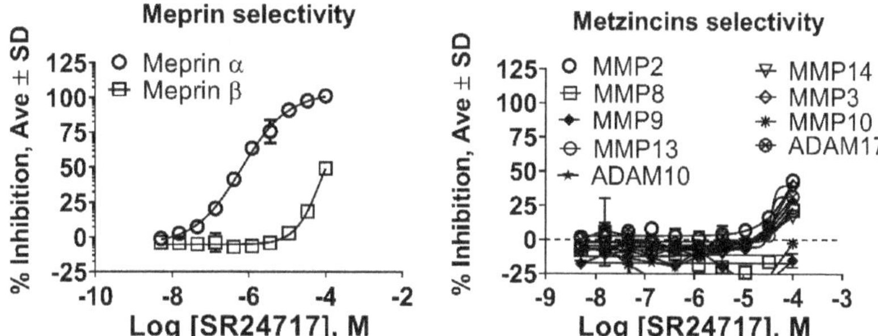

Figure 4. Characterization of a potential meprin α probe SR24717 against meprin β and related metzincins. (**A**) Dose response study of re-synthesis batch of SR24717 with meprin α and β; (**B**) Dose response study of re-synthesis batch of SR24717 with metzincins shows good selectivity profile. All experiments performed as 10 point 3-fold concentration-response curves in triplicate. All units are IC_{50}, μM.

We further characterized SR24717 for the mode of inhibition of meprin α. Pre-incubation of SR24717 with meprin α for 0–3 h showed no change in IC_{50} values, suggesting that SR24717 is not a time-dependent inhibitor (Figure 5A). This allowed us to use the steady-state assumption in our follow-up experiments. Varying substrate concentration in meprin α decreased the apparent potency of SR24717 (Figure 5B,C), suggesting a competitive mode of inhibition. Indeed, both linear (Figure 5D) and non-linear (Figure 5E) models showed good fit to the competitive inhibition. Global fit to the competitive model of inhibition using GraphPad Prism showed $K_i = 300.8$ nM.

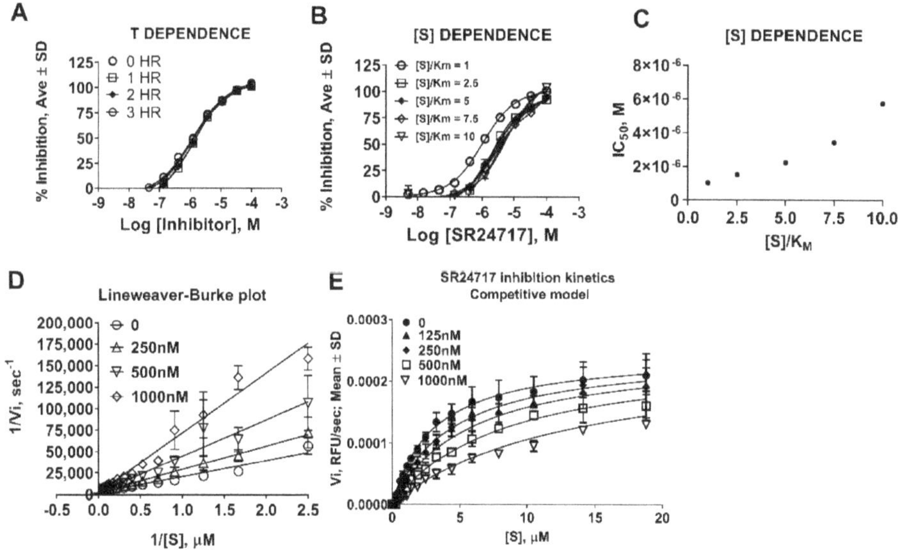

Figure 5. Characterization of mode of meprin α inhibition by potential probe. SR24717. (**A**) Time-dependence study of inhibition of meprin α-mediated proteolysis by SR24717. (**B**) Concentration-dependence study of meprin α inhibition-mediated proteolysis by SR24717. (**C**) Re-plot of (**B**) shows increase of IC_{50} values correlating with increase of substrate concentration suggestive of competitive inhibition mechanism by SR24717. (**D**) Lineweaver-Burke plot shows lines of best fit crossing at Y-axis, suggesting a competitive inhibition mechanism by SR24717. (**E**) Global fit of meprin α-mediated proteolysis in the presence of SR24717 to a competitive model using non-linear regression shows good fit.

We tested SR24717 for effects on viability and cytotoxicity of several cell types. Only at a concentration of 100 μM of SR24717 were negative effects on viability and cytotoxicity observed (Figure 6), suggesting that SR24717 may be a useful probe for studying the biological role of meprin α in in vitro and in vivo systems.

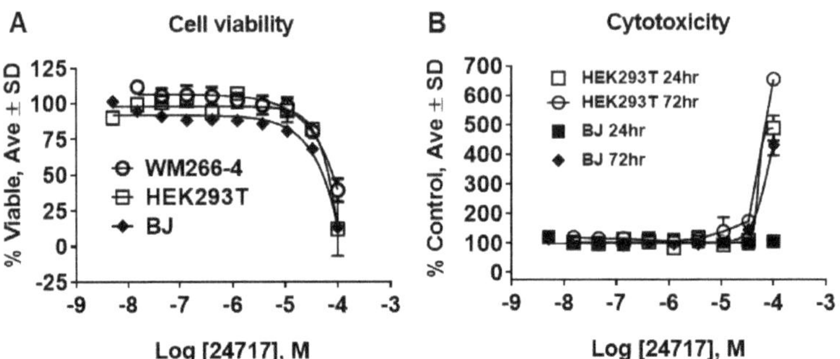

Figure 6. Characterization of SR24717 for effects on cell viability and cytotoxicity. (**A**) Effect of SR24717 on viability of WM266-4 (skin melanoma), HEK293 (kidney), and BJ (skin fibroblasts) cells. CellTiter Glo™ assay (Promega) was used after 72 h treatment. (**B**) Effect of SR24717 on viability of HEK293 (kidney) and BJ (skin fibroblasts) cells. The CellTox Green™ assay (Promega) was used. All experiments performed as 10-point 3-fold concentration-response curves in triplicate.

3. Discussion

While we had initially hoped that HTS efforts would identify non-chelating, non-hydroxamic acid meprin α or meprin β inhibitor leads, we have demonstrated that the two aromatic groups in the lead series of meprin α inhibitors are also very important, and perhaps uniquely positioned, for meprin α binding. Thus, Zn chelation is not the overwhelming driver of ligand affinity. SR24717 has unexpectedly high selectivity vs. 10 other zinc metalloproteases (Figure 4), indicative of potential use as a probe molecule in an in vitro setting. The selectivity is noteworthy in comparison to known inhibitors (Table 3), with structures depicted in Figure 7.

The meprin α/β fold selectivity (ratio of IC_{50} values) had been ~13 for the HTS hit SR19855 and was improved to 106 for SR24717. Actinonin is a meprin inhibitor that has long been commonly used in the literature, but its target selectivity is quite poor. It is a repurposed antibiotic agent that potently inhibits many metzincins and in cell-based environments, though it is significantly cytotoxic. Its wide use is purely historical: it was one of the first inhibitors of meprin α with some selectivity for meprin β, but off-target activity confounds the interpretations for any of actinonin's effects, as they cannot be cleanly ascribed to meprin α inhibition. This is illustrated in Table 3, where we see sub-micromolar activity for actinonin vs. MMP-2, MMP-8, MMP-9, MMP-13, ADAM17 (and also MMP1, see footer), with low micromolar activity vs. meprin β and ADAM10. This contrasts greatly with similar data for SR24717.

At the time of initial submission of this manuscript, the most potent meprin α in-hibitors known were bis-benzyl glycine hydroxamates **10d** and **10e**, both from the Rams-beck labs [8], which displayed meprin α/β fold selectivity ratios of 18 and 19, respectively. The same group very recently gave an update on this series and related analogs [11], re-porting more balanced pan tertiary amine inhibitors such as compound **1c** and also much more potent, conformationally constrained meprin α inhibitors **2c** and **2d**. Interestingly, compounds **2c** and **2d**, like SR24717, append two aromatic groups to a 5-membered ring core (Figure 7). Though selectivity IC_{50} values were not reported, most (as shown in Table 3) were weak inhibitors at 10 μM and showed more substantial inhibition at 200 μM. The most potent off-target for inhibitors **2c** and **2e** was ovastacin, with K_i values of 66 nM and 196 nM, respectively [11].

Table 3. Selectivity testing of SR24717 and literature meprin α inhibitors.

ID	Meprin α	Meprin β	MMP2	MMP3	MMP8	MMP9	MMP10	MMP13	MMP14	ADAM10	ADAM17
SR24717 [a]	0.66 ± 0.03	70 ± 5	>100	>100	>100	>100	>100	>100	>100	>100	>100
Actinonin [b]	0.004	4.8 ± 0.5	0.09	NR	0.19	0.1 ± 0.01	NR	0.1	>1000	2.2 ± 0.2	0.2 ± 0.02
10d [8] [c]	0.16 ± 0.001	2.95 ± 0.35	95% 26%			91% 61%		86% 55%		86% 61%	86% 57%
10e [8]	0.40 ± 0.03	7.59 ± 0.01	94% 65%			91% 21%		98% 47%		80% 56%	80% 41%
1c [11] [d]	0.19 ± 0.001	0.03 ± 0.35	95%			91%		86%		86%	86%
2c [11] [e]	0.004 ± 0.001	0.813 ± 0.001	87%			86%		80%		80%	62%
2e [11] [f]	0.003 ± 0.001	0.199 ± 0.001	104%			81%		75%		73%	61%

All units are IC_{50}, µM, % values indicate % activity remaining at 10 µM (top) and at 200 µM (bottom). All assays performed in triplicate, with error bars $\pm$ SD. a. Multiple synthesis of SR24717 batches were tested, shown are results of one representative batch. b. Actinonin also has $IC_{50} = 0.1$ µM vs. MMP1. c. Compound **10d** has $IC_{50} = 1.15$ µM vs. ovastacin [11]. d. Compound **1c** has $IC_{50} = 0.49$ µM vs. ovastacin [11]. e. Compound **2c** has $K_i = 66$ nM vs. ovastacin [11]. f. Compound **2e** has $K_i = 196$ nM vs. ovastacin [11].

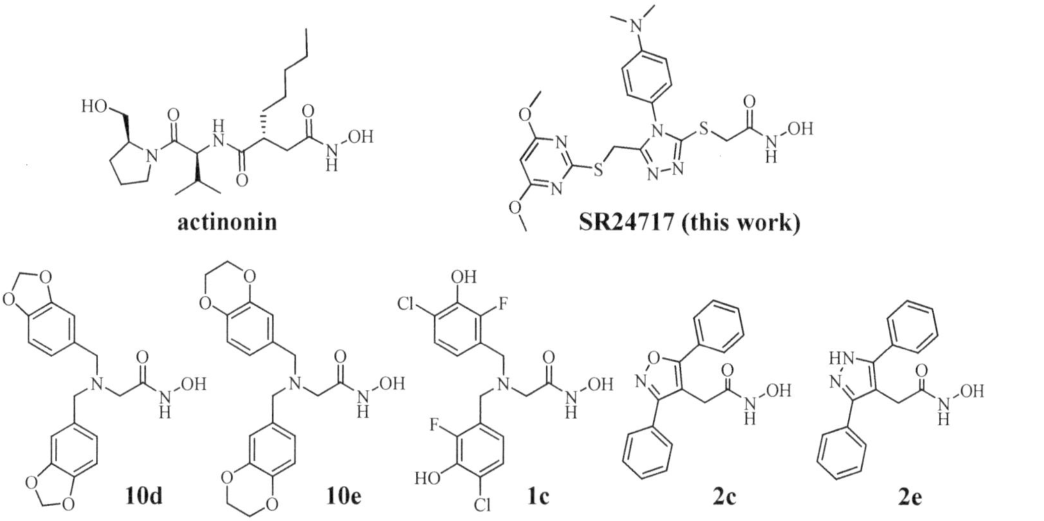

actinonin

SR24717 (this work)

10d

10e

1c

2c

2e

Figure 7. Structures of literature meprin inhibitors: actinonin, SR24717, **10d** [8]; **10e** [8], **1c** [11], **2c** [11], **2e** [11].

Our modeling study, depicted in Figure 2, suggests that two cation-π interactions are responsible for meprin α potency and selectivity. Our structure–activity relationship study supports the model, in that the 3,5-dimethoxypyrimidine and 4-dimethylaminophenyl groups of SR24717 work in concert to augment potency and selectivity. Though SR24717 is merely ~2-fold more potent than the HTS hit SR19855 in binding meprin α, it has selectivity advantages (~106-fold vs. meprin β), with consistently lower IC_{50} and/or lower maximum inhibition of nine other metzincins, when tested even at a high concentration (100 µM, see Figure 4 and Table 3).

We have developed a versatile, scalable, and operationally straightforward route to producing the SR24717 and its analogs. Future work includes characterization of all properties that are relevant for in vivo use, including the evaluation and optimization of DMPK properties, while maintaining or improving both target potency and meprin α selectivity.

4. Materials and Methods

4.1. Assay Reagents

MMP-1, MMP-2, MMP-8, MMP-9, MMP-10, MMP-13, MMP-14, ADAM10, ADAM17 and Mca-KPLGL-Dpa-AR-NH$_2$ fluorogenic peptide substrate were purchased from R&D Systems (cat # 901-MP, 902-MP, 908-MP, 911-MP, 910-MP, 511-MM, 918-MP, 936-AD, 930-ADB, and ES010, respectively). All common chemicals were purchased from Sigma. NFF449 was purchased from Tocris (cat# 1391) and actinonin was from Sigma-Aldrich, St. Louis, MO, USA (cat# 01809).

4.2. Meprin α and Meprin β Substrate Synthesis

Meprin α and meprin β substrates (Mca-YVADAPK-(K-ε-Dnp) and Mca-EDEDED-(K-ε-Dnp), respectively) [12] were synthesized according to Fmoc solid-phase methodology on a peptide synthesizer. All peptides were synthesized as C-terminal amides to prevent diketopiperazine formation [13]. Cleavage and sidechain deprotection of peptide-resins was for at least 2 h using thioanisole-water-TFA (5:5:90). The substrates were purified and characterized by preparative RP HPLC and characterized by MALDI-TOF MS and analytical RP HPLC.

4.3. Meprins Expression Protocol

Recombinant human meprin α and meprin β were expressed using the Bac-to-Bac expression system (Gibco Life Technologies, Paisley, UK) as described before [5,14]. Media and supplements were obtained from Gibco Life Technologies. Recombinant Baculoviruses were amplified in adherently growing *Spodoptera frugiperda* (*Sf*)9 insect cells at 27 °C in Grace's insect medium supplemented with 10% fetal bovine serum, 50 units/mL penicillin, and 50 µg/mL streptomycin. Protein expression was performed in 500 mL suspension cultures of BTI-TN-5B1-4 insect cells growing in Express Five SFM supplemented with 4 mM glutamine, 50 units/mL penicillin, and 50 µg/mL streptomycin in Fernbach-flasks using a Multitron orbital shaker (INFORS AG, Bottmingen, Switzerland). Cells were infected at a density of 2 × 10^6 cells/mL with an amplified viral stock at a MOI of ~10. Protein expression was stopped after 72 h, and recombinant meprins were further purified from the media by ammonium sulfate precipitation (60% saturation) and affinity chromatography (Strep-tactin for Strep-tagged meprin α and Ni-NTA for His-tagged meprin β). Meprins were activated by trypsin, which was removed afterwards by affinity chromatography using a column containing immobilized chicken ovomucoid, a trypsin inhibitor.

4.4. Meprin α and Meprin β Assays in a 384-Well Plate

Both assays followed the same general protocol [15]. 5 µL of 2× enzyme solution (2.6 and 0.1 nM for meprin α and meprin β, respectively) in assay buffer (50 mM HEPES, 0.01% Brij-35, pH 7.5) were added to solid bottom black 384 low-volume plates (Nunc, cat# 264705). Next, 75 nL of test compounds or pharmacological control (actinonin) were added to corresponding wells using a 384-pin tool device (V&P Scientific, San Diego, CA, USA). After 30 min incubation at RT, the reactions were started by addition of 5 µL of 2× solutions of substrates (20 µM, meprin α Mca-YVADAPK-K(Dnp), and for meprin β Mca-EDEDED-K(Dnp)). Reactions were incubated at RT for 1 h, after which the fluorescence was measured using the Synergy H4 multimode microplate reader (Biotek Instruments) ($\lambda_{\text{excitation}}$ = 324 nm, $\lambda_{\text{emission}}$ = 390 nm).

Three parameters were calculated on a per-plate basis: (a) the signal-to-background ratio (S/B), (b) the coefficient for variation (CV; CV = (standard deviation/mean) × 100) for all compound test wells; and (c) the Z- or Z'-factor [16]. Z takes into account the effect of test compounds on the assay window, while Z' is based on controls.

4.5. Determination of Kinetic Parameters of Meprin α Mediated Proteolysis in the Presence of Potential Probe SR24717

Substrate stock solutions were prepared at various concentrations in HTS assay buffer (50 mM HEPES, 0.01% Brij-35, pH 7.5). Assays were conducted by incubating a range of substrate (2–50 μM) and SR24717 concentrations (0–1000 nM) with 1.3 nM meprin α at 25 °C. Fluorescence was measured on a multimode microplate reader Synergy H1 (Biotek Instruments, Winooski, VT, USA) using $\lambda_{excitation}$ = 324 nm and $\lambda_{emission}$ = 393 nm. Rates of hydrolysis were obtained from plots of fluorescence versus time, using data points from only the linear portion of the hydrolysis curve. The slope from these plots was divided by the fluorescence change corresponding to complete hydrolysis and then multiplied by the substrate concentration to obtain rates of hydrolysis in units of μM/s. Kinetic parameters were calculated by non-linear regression analysis using the GraphPad Prism 8.0 suite of programs.

4.6. ADAM10 and ADAM17 Assays

Both assays followed the same general protocol. 2.5 μL of 2× enzyme solution (20 nM) in assay buffer (10 mM HEPES, 0.001% Brij-35, pH 7.5) were added to solid bottom black 384 plates (Greiner, cat# 789075). Next, test compounds and pharmacological controls were added to corresponding wells using a 384-pin tool device (V&P Scientific, San Diego). After 30 min incubation at RT, the reactions were started by addition of 2.5 μL of 2× solutions of substrate (R&D Systems cat#: ES010, Mca-KPLGL-Dpa-AR-NH$_2$, 20 μM). Reactions were incubated at RT for 2 h, after which the fluorescence was measured using a Perkin Elmer Viewlux multimode microplate imager ($\lambda_{excitation}$ = 324 nm, $\lambda_{emission}$ = 393 nm). All compounds were tested in 10-point, 1:3 serial dilutions dose-response format starting with highest concentration of 100 μM. IC$_{50}$ values were determined using non-linear regression analysis using the GraphPad Prism 8.0 suite of programs.

4.7. MMP Assays

All assays followed the same general protocol. 5 μL of 2× enzyme solution (5 nM) in assay buffer (50 mM Tricine, 50 mM NaCl, 10 mM CaCl$_2$, 0.05% Brij-35, pH 7.5) were added to solid-bottom black 384 plates (Nunc, cat# 264705). Next, test compounds and pharmacological controls were added to corresponding wells using a 384-pin tool device (V&P Scientific, San Diego). After 30 min incubation at RT, the reactions were started by addition of 5 μL of 2× solutions of MMP substrate (R&D Systems cat#: ES010, 20 μM). Reactions were incubated at RT for 1 h, after which the fluorescence was measured using the Synergy H4 multimode microplate reader (Biotek Instruments) ($\lambda_{excitation}$ = 324 nm, $\lambda_{emission}$ = 390 nm). All compounds were tested in 10-point, 1:3 serial dilutions dose-response format starting with highest concentration of 100 μM. IC$_{50}$ values were determined using non-linear regression analysis using the GraphPad Prism 8.0 suite of programs.

4.8. Cell Toxicity Studies

Test compounds were solubilized in 100% DMSO and added to polypropylene 384-well plates (Greiner cat# 781280). 1250 of BJ skin fibroblasts or primary melanocytes were plated in 384-well plates in 8 μL of serum-free media (HybriCare for BT474, EMEM for HEK293). Test compounds and pharmacological assay control (lapatinib) were prepared as 10-point, 1:3 serial dilutions starting at 10 mM, then added to the cells using the pin tool mounted on Integra 384. Plates were incubated for 72 h at 37 °C, 5% CO$_2$ and 95% RH. After incubation, 8 μL of CellTiter-Glo® (Promega cat# G7570) was added to each well and incubated for 15 min at room temperature. Luminescence was recorded using a Biotek Synergy H1 multimode microplate reader. Viability was expressed as a percentage relative to wells containing media only (0%) and wells containing cells treated with DMSO only (100%). Three parameters were calculated on a per-plate basis: (a) the signal-to-background ratio (S/B), (b) the coefficient for variation (CV; CV = (standard deviation/mean) × 100) for all compound test wells, and (c) the Z′-factor. IC$_{50}$ values were calculated by fitting normalized data to sigmoidal log vs. response equation utilizing non-linear regression analysis from GraphPad Prism 8.

4.9. Molecular Modeling Studies

The crystal structure of human meprin β (PDB ID 4GWN) was used as a template for constructing a model of human meprin α using Prime (Schrödinger, LLC, New York, NY, USA), with the option of a single template to build a single chain and including the Zn ion. The homology model and the coordinates, 4GWN, were prepared using the protein preparation wizard in Maestro v12.2 (Schrodinger, LLC, NY, USA). Docking studies were performed using Glide SP v8.7 (Schrodinger, LLC, NY, USA) with no constraint. The docking grid was generated around the Zn ion with a box size of 18 × 18 × 18 Å3. SR19855 was prepared for Glide docking with LigPrep (Schrodinger, LLC, NY, USA) to include different conformational states. The docking pose with the highest docking score for each compound was then merged to the docked structures for energy minimization using the OPLS3e force field (Schrodinger, LLC, NY, USA) and the results were analyzed in pymol.

4.10. Compound Synthesis and Characterization

Synthesis of all test compounds followed the 6-step route summarized in Figure 3, which was adapted from the literature [10,17,18]. Details for all steps in the synthesis of SR24717 are shown, with supporting data. Other compounds followed the same synthetic methods, and data follows for all analogs reported herein.

Step 1, Ethyl 2-[(aryl)thio] acetate derivatives (**2**). Mercaptopyrimidine **1** (1 equiv.), ethyl 2-chloroacetate (2 equiv.) and potassium carbonate (1.2 equiv.) were heated at 80 °C in acetone. Reaction progress was monitored by LC/MS and after no more than 2 h, the mixture was cooled, filtered, and concentrated in vacuo to give the crude product **2**, which was dissolved in minimal dichloromethane, precipitated by addition of diethyl ether, and the precipitate was collected by filtration and dried under high vacuum. The product was isolated in 75% yield, with >95% purity by LC/MS, with no chromatography necessary.

Step 2, 2-[(Aryl)thio]acetohydrazide derivatives (**3**). A mixture of ethyl 2-[(aryl)thio]acetate (**2**) (1 equiv.) and hydrazine hydrate (2 equiv.) in ethanol was stirred for 1–2 h at room temperature. The colorless hydrazide **3** precipitated out of the solution and was collected by filtration. The precipitate was washed with water, dried under high vacuum, and was taken to the next step without purification. It was >95% pure by LC/MS analysis.

Step 3, Thiosemicarbazide derivatives (**5**). A solution of the hydrazide (**3**) (1 equiv.) and an aryl isothiocyanate **4**, typically a substituted phenyl isothiocyanate (1 equiv.), in hot ethanol was refluxed for 1 h. The colorless thiosemicarbazide **5** precipitated out of the solution and, after cooling, it was collected by filtration. The precipitate was washed with water, ether, dried under high vacuum, and was taken to the next step without purification. It was >95% pure by LC/MS analysis.

Step 4, Mercaptotriazole derivatives (**6**). A suspension of thiosemicarbazide (**5**) in aq. 2% NaOH was heated at 100 °C, and the solution became homogeneous. Reaction progress was monitored by LC/MS and cyclization was complete within 2 h. The solution was then cooled to room temperature. The pH was adjusted to 6–7 by the addition of 3 M HCl, at which point a colorless precipitate formed. The precipitate collected by filtration, washed with water, and dried under high vacuum. It was >95% pure by LC/MS analysis and was obtained in 70% isolated yield for steps 2–4, from compound **2**. This intermediate was potentially sensitive to air oxidation, so it was kept under inert atmosphere and then carried without delay to the next step.

Step 5, Mercaptotriazole esters (**7**). A mixture of the mercaptotriazole (**6**) (1 equiv.), ethyl 2-chloroacetate (1.12 equiv.) and potassium carbonate (1.5 equiv.) in acetone was heated at 80 °C. Reaction progress was monitored by LC/MS and the alkylation was complete within 2 h. Upon cooling, a colorless precipitate formed and it was collected by filtration. The precipitate washed with water and was dried under high vacuum. The precursor to the probe candidate was >95% pure by LC/MS analysis and was obtained in

85% isolated yield. Certain mercaptotriazole esters analogs were less crystalline and could be purified by column chromatography on silica gel at this stage, using a CH_2Cl_2 to 5% MeOH in CH_2Cl_2 gradient. Intermediate ester **7** was stable and could be stored cold under inert atmosphere for later use in the final step.

Step 6, final hydroxamic acids (**8**). A solution of mercaptotriazole ester (**7**) (1 equiv.) in MeOH/DCM (3:1) was cooled to 0 °C and was treated with saturated hydroxylamine hydrochloride solution (6 equiv.) followed by the addition of sat. aq. NaOH (12 equiv.). After reaction completion (typically~10 min) the mixture was concentrated in vacuo, water was added, and the pH was adjusted to 7.0 by the addition of 2 M HCl. A colorless precipitate formed (25–30% yield, >95% analytical purity by HPLC) and it was collected by filtration. The product was kept cold and under inert atmosphere to prevent slow hydrolysis of the hydroxamic acid to a carboxylate, presumably from atmospheric moisture. Such hydrolysis was also seen upon prep HPLC purification, so the final product was best isolated by precipitation.

Characterization data for the potential meprin α inhibitor SR24717: ^{1}H NMR (600 MHz, DMSO-d_6): δ ppm 10.75 (s, 1H), 9.01 (s, 1H), 8.58 (s, 1H), 7.16 (d, J = 8.8 Hz, 2H), 6.68 (d, J = 8.8 Hz, 2H), 5.89 (s, 1H), 6.94 (s, 1H), 4.48 (s, 2H), 3.81 (s, 6H), 3.78 (s, 2H), 2.94 (s, 6H). ^{13}C NMR (DMSO-d_6): δ ppm 170.9, 168.2, 164.2, 153.7, 151.9, 151.2, 128.4, 120.3, 112.4, 86.0, 60.2, 33.4, 24.4. MS (ESI, M + H) calcd for ($C_{19}H_{23}N_7O_4S_2$ + H): 478.13, found 477.86, purity by analytical HPLC >95%.

4.11. Data for Other Test Compounds

4.11.1. SR19855, *N*-hydroxy-2-((4-(4-methoxyphenyl)-5-((pyrimidin-2-ylthio)methyl)-4*H*-1,2,4-triazol-3-yl)thio)acetamide

Colorless solid, 42% yield for the final step. Other steps were comparable in yield to that for SR24717. Silica gel chromatography used MeOH/DCM 0–5% gradient, on the penultimate step. ^{1}H NMR (400 MHz, DMSO-d_6): δ ppm 10.76 (s, 1H), 9.02 (s, 1H), 8.54 (d, J = 4.88 Hz, 2H), 7.37 (d, J = 8.92 Hz, 2H), 7.19 (t, J = 4.88 Hz, 1H), 7.04 (d, J = 8.92 Hz, 2H), 4.42 (s, 2H), 3.80 (s, 6H), 3.77 (s, 2H). ^{13}C NMR (600 MHz, DMSO-d_6): δ ppm 170.0, 163.7, 160.2, 157.8, 153.0, 151.0, 128.9, 124.9, 117.5, 114.9, 55.5, 33.1, 23.9. MS (ESI, M + H) calcd for ($C_{16}H_{16}N_6O_3S_2$ + H): 405.07, found 405.01, purity by analytical HPLC > 95%.

4.11.2. SR24139, 2-((5-(((4,6-dimethylpyrimidin-2-yl)thio)methyl)-4-(4-methoxyphenyl)-4*H*-1,2,4-triazol-3-yl)thio)-*N*-hydroxyacetamide

Colorless solid, 45% yield for final step. Other steps were comparable in yield to that for SR24717. ^{1}H NMR (400 MHz, DMSO-d_6): δ ppm 10.75 (d, J = 1.2 Hz, 1H), 9.02 (d, J = 1.2 Hz, 1H), 7.35 (d, J = 8.9 Hz, 2H), 7.00 (d, J = 8.9 Hz, 2H), 6.93 (s, 1H), 4.47 (s, 2H), 3.79 (s, 6H), 3.78 (s, 2H), 2.28 (s, 6H). ^{13}C NMR (600 MHz, DMSO-d_6): δ ppm 167.7,167.0, 163.7, 160.01, 153.3, 151.0, 128.8, 124.8, 116.2, 114.7, 55.5, 33.1, 23.5, 23.2. MS (ESI, M + H) calcd for ($C_{18}H_{20}N_6O_3S_2$ + H): 433.10, found 433.05, purity by analytical HPLC > 97%.

4.11.3. SR24144, 2-((5-(((4,6-dimethoxypyrimidin-2-yl)thio)methyl)-4-(4-methoxyphenyl)-4*H*-1,2,4-triazol-3-yl)thio)-*N*-hydroxyacetamide

Colorless solid, 45% yield for the final step, recrystallized from EtOAc. Other steps were comparable in yield to that for SR24717. ^{1}H NMR (400 MHz, DMSO-d_6): δ ppm 10.76 (d, J = 0.88 Hz, 1H), 9.01 (d, J = 0.88 Hz, 1H), 8.37 (d, J = 8.92 Hz, 2H), 7.01 (d, J = 8.92 Hz, 2H), 5.91 (s, 1H), 4.49 (s, 2H), 3.80 (s, 6H), 3.79 (s, 3H), 3.77 (s, 2H). ^{13}C NMR (600 MHz, DMSO-d_6): δ ppm 170.5, 167.7, 163.7, 160.2, 153.0, 151.1, 128.9, 124.8, 114.8, 85.5, 55.5, 54.3, 33.1, 24.0. MS (ESI, M + H) calcd for ($C_{18}H_{20}N_6O_5S_2$ + H): 465.09, found 465.01, purity by analytical HPLC > 95%.

4.11.4. SR24460, 2-((5-(((4,6-dimethoxypyrimidin-2-yl)thio)methyl)-4-(3-methoxyphenyl)-4*H*-1,2,4-triazol-3-yl)thio)-*N*-hydroxyacetamide

Colorless solid, 38% yield for the final step. Other steps were comparable in yield to that for SR24717. ^{1}H NMR (400 MHz, DMSO-d_6): δ ppm 10.77 (d, J = 1.04 Hz, 1H), 9.03 (d, J = 1.48 Hz, 1H), 7.42 (dt, J = 7.88, 0.80 Hz, 1H), 7.04 (q, J = 7.08 Hz, 3H), 5.91 (s, 1H), 4.55 (s, 2H), 3.80 (s, 6H), 3.75 (s, 2H). ^{13}C NMR (600 MHz, DMSO-d_6): δ ppm 170.5, 167.7, 163.6, 159.8, 152.7, 150.6, 133.44, 130.5, 119.3, 115.6, 113.2, 85.5, 55.5, 54.2, 33.2, 24.0. (ESI, M + H) calcd for ($C_{18}H_{20}N_6O_5S_2$ + H): 465.09, found 465.02, purity by analytical HPLC > 98%.

4.11.5. SR24459, 2-((5-(((4,6-dimethoxypyrimidin-2-yl)thio)methyl)-4-(2-methoxyphenyl)-4*H*-1,2,4-triazol-3-yl)thio)-*N*-hydroxyacetamide

Colorless solid, 40% yield for the final step. Other steps were comparable in yield to that for SR24717.^{1}H NMR (400 MHz, DMSO-d_6): δ ppm 10.77 (s, 1H), 9.03 (s, 1H), 7.50 (dt, J = 6.88, 1.56 Hz, 1H), 7.39 (dd, J = 7.8, 1.56 Hz, 1H), 7.21 (d, J = 8.4 Hz,1 H), 7.05 (t, J = 7.64 Hz, 1H), 5.91 (s, 1H), 4.48 (d, J = 15.0 Hz, 1H), 4.37 (d, J = 15.0 Hz, 1H), 3.81 (s, 6H), 3.75 (s, 2H). ^{13}C NMR (600 MHz, DMSO-d_6): δ ppm 170.5, 167.8, 163.7, 154.3, 153.0, 151.2, 132.0, 129.1, 120.8, 120.5, 112.8, 85.5, 55.9, 54.2, 33.2, 23.8. (ESI, M + H) calcd for ($C_{18}H_{20}N_6O_5S_2$ + H): 465.09, found 464.98, purity by analytical HPLC > 95%.

4.11.6. SR24718, 2-((5-(((4,6-dimethoxypyrimidin-2-yl)thio)methyl)-4-(pyridin-3-yl)-4*H*-1,2,4-triazol-3-yl)thio)-*N*-hydroxyacetamide

Colorless solid, 42% yield for the final step. Other steps were comparable in yield to that for SR24717. ^{1}H NMR (400 MHz, DMSO-d_6): δ ppm 10.82 (s, 1H), 9.09 (s, 1H), 8.80 (d, J = 2.36 Hz, 1H), 8.75 (dd, J = 4.8, 1.4 Hz, 1H), 8.07 (td, J = 8.24, 2.48 Hz, 1H), 7.62 (dd, J = 8.24, 4.84 Hz, 1H), 5.98 (s, 1H), 4.63 (s, 2H), 3.87 (s, 6H), 3.84 (s, 2H). ^{13}C NMR (600 MHz, DMSO-d_6): δ ppm 170.5, 167.5, 163.5, 152.9, 151.0, 150.9, 148.2, 135.6, 129.6, 124.4, 85.6, 54.3, 33.7, 24.0. MS (ESI, M + H) calcd for ($C_{16}H_{17}N_7O_4S_2$ + H): 436.08, found 435.91, purity by analytical HPLC > 95%.

4.11.7. SR24003, 2-((5-(((4,6-dimethoxypyrimidin-2-yl)thio)methyl)-4-(4-(trifluoromethyl)phenyl)-4*H*-1,2,4-triazol-3-yl)thio)-*N*-hydroxyacetamide

Colorless solid, 35% yield for final step. Other steps were comparable in yield to that for SR24717. ^{1}H NMR (400 MHz, DMSO-d_6): δ ppm 10.78 (d, 1H), 9.02 (d, 1H), 7.88 (d, J = 8.31 Hz, 2H), 7.76 (d, J = 8.24 Hz, 2H), 5.88 (s, 1H), 4.58 (s, 2H), 3.79 (s, 9H). ^{13}C NMR (600 MHz, DMSO-d_6): δ ppm 170.5, 167.5, 163.5, 152.8, 150.5, 136.1, 130.4, 128.8, 126.8, 124.5, 122.7, 85.5, 54.3, 33.6, 24.0. MS (ESI, M + H) calcd for ($C_{18}H_{17}F_3N_6O_4S_2$ + H): 503.07, found 503.01, purity by analytical HPLC > 97%.

4.11.8. SR24462, 2-((4-(4-bromophenyl)-5-(((4,6-dimethoxypyrimidin-2-yl)thio)methyl)-4*H*-1,2,4 triazol-3-yl)thio)-*N*-hydroxyacetamide

Colorless solid, 35% yield for the final step. Other steps were comparable in yield to that for SR24717. ^{1}H NMR (400 MHz, DMSO-d_6): δ ppm 10.76 (s, 1H), 9.03 (s, 1H), 7.70 (d, J = 8.7 Hz, 2H), 7.45 (d, J = 8.7 Hz, 2H), 5.93 (s, 1H), 4.55 (s, 2H), 3.82 (s, 6H), 3.78 (s, 2H). ^{13}C NMR (600 MHz DMSO-d_6): δ ppm 170.5, 167.51, 163.6, 152.7, 150.6, 132.7, 131.8, 129.7, 123.6, 85.5, 54.3, 33.4, 24.0 MS (ESI, M + H) calcd for ($C_{17}H_{17}BrN_6O_4S_2$ + H): 513.0, 515.0, found 512.99, 514.95, purity by analytical HPLC > 95%.

4.11.9. SR24463, 2-((5-(((4,6-dimethoxypyrimidin-2-yl)thio)methyl)-4-(4-fluorophenyl)-4*H*-1,2,4-triazol-3-yl)thio)-*N*-hydroxyacetamide

Colorless solid, 35% yield for the final step. Other steps were comparable in yield to that for SR24717. ^{1}H NMR (400 MHz, DMSO-d_6): δ ppm 10.77 (s, 1H), 9.03 (s, 1H), 7.55–7.58 (m, 2H), 7.37 (t, J = 8.8 Hz, 2H), 5.93 (s, 1H), 4.53 (s, 2H), 3.82 (s, 6H), 3.79 (s, 2H). ^{13}C NMR (600 MHz, DMSO-d_6): δ ppm 170.5, 167.6, 163.4, 161.7, 152.8, 150.8, 130.1, 128.8,

116.8, 85.5, 54.3, 33.3, 24.0. MS (ESI, M + H) calcd for ($C_{17}H_{17}FN_6O_4S_2$ + H): 453.07, found 452.59, purity by analytical HPLC > 96%.

4.11.10. SR24467, 2-((5-(((4,6-dimethoxypyrimidin-2-yl)thio)methyl)-4-(m-tolyl)-4H-1,2,4-triazol-3-yl)thio)-N-hydroxyacetamide

Colorless solid, 42% yield for the final step. Other steps were comparable in yield to that for SR24717. ^{1}H NMR (400 MHz, DMSO-d_6): δ ppm 10.76 (s, 1H), 9.02 (s, 1H), 7.50 (t, *J* = 7.59 Hz, 1H), 7.31 (d, *J* = 7.47 Hz, 1H), 7.24 (d, *J* = 6.75 Hz, 2H), 5.90 (s, 1H), 4.51 (s, 2H), 3.80 (s, 9H), 2.29 (s, 3H). ^{13}C NMR (600 MHz, DMSO-d_6): δ ppm 170.5, 167.7, 163.6, 152.8, 150.7, 139.6, 132.4, 130.4, 129.5, 127.7, 124.5, 85.5, 54.3, 33.2, 24.0, 20.6. (ESI, M + H) calcd for ($C_{18}H_{20}N_6O_4S_2$ + H): 449.10, found 448.94, purity by analytical HPLC > 95%.

4.11.11. SR26467, 2-((5-(((4,6-dimethoxypyrimidin-2-yl)thio)methyl)-4-(naphthalen-1-yl)-4H-1,2,4-triazol-3-yl)thio)-N-hydroxyacetamide

Colorless solid, 32% yield for the final step. Other steps were comparable in yield to that for SR24717. Silica gel chromatography used MeOH/DCM 0–5% gradient, on the penultimate step. ^{1}H NMR (400 MHz, DMSO-d_6): δ ppm 10.76 (s, 1H), 9.02 (s, 1H), 8.06 (d, *J* = 8.32 Hz, 1H), 8.03 (d, *J* = 7.44 Hz, 1H), 7.67 (d, *J* = 7.36 Hz, 1H), 7.62–7.54 (m, 3H), 7.13 (d, *J* = 8.16 Hz, 1H), 5.74 (s, 1H), 4.41 (d, *J* = 15.1 Hz, 1H), 4.38 (d, *J* = 15.2 Hz, 1H), 3.80 (s, 2H), 3.59 (s, 6H). ^{13}C NMR (600 MHz, DMSO-d_6): δ ppm 170.1, 167.2, 163.6, 153.9, 151.8, 133.7, 130.9, 128.8, 128.4, 128.3, 128.1, 127.1, 127.0, 125.5, 121.4, 85.4, 54.0, 33.1, 23.7. MS (ESI, M + H) calcd for ($C_{21}H_{20}N_6O_4S_2$ + H): 485.10, found 484.91, purity by analytical HPLC > 95%.

4.11.12. SR26466, 2-((5-(((4,6-dimethoxypyrimidin-2-yl)thio)methyl)-4-(4-morpholinophenyl)-4H-1,2,4-triazol-3-yl)thio)-N-hydroxyacetamide

Colorless solid, 41% yield for the final step. Other steps were comparable in yield to that for SR24717. ^{1}H NMR (400 MHz, DMSO-d_6): δ ppm 10.76 (s, 1H), 9.02 (s, 1H), 7.24 (d, *J* = 8.96 Hz, 2H), 6.97 (d, *J* = 9.04 Hz, 2H), 5.91 (s, 1H), 4.50 (s, 2H), 3.81 (s, 6H), 3.78 (s, 2H), 3.75 (t, *J* = 4.88 Hz, 4H), 3.17 (t, *J* = 4.76 Hz, 4H). ^{13}C NMR (600 MHz, DMSO-d_6): δ ppm 170.5, 167.7, 163.7, 153.1, 151.6, 151.3, 128.1, 122.6, 114.7, 85.6, 66.0, 54.3, 47.4, 33.0, 24.0. MS (ESI, M + H) calcd for ($C_{21}H_{26}N_7O_5S_2$ + H): 520.14, found 519.95, purity by analytical HPLC > 95%.

4.11.13. SR26465, 2-((4-(4-(diethylamino)phenyl)-5-(((4,6-dimethoxypyrimidin-2-yl)thio)methyl)-4H-1,2,4-triazol-3-yl)thio)-N-hydroxyacetamide

Colorless solid, 41% yield for the final step. Other steps were comparable in yield to that for SR24717. ^{1}H NMR (400 MHz, CD3CN-d_6): δ ppm 7.06 (d, *J* = 8.88 Hz, 2H), 6.59 (d, *J* = 9.04 Hz, 2H), 5.71 (s, 1H), 4.51 (s, 2H), 3.80 (s, 6H), 3.69 (s, 2H), 3.33 (q, *J* = 7.04 Hz, 4H), 1.11 (t, *J* = 7.08 Hz, 6H). ^{13}C NMR (600 MHz, DMSO-d_6): δ ppm 170.4, 167.7, 163.8, 153.3, 151.6, 148.1, 128.2, 118.8, 111.0, 85.6, 54.2, 43.7, 32.8, 23.8, 12.3. MS (ESI, M + H) calcd for ($C_{21}H_{27}N_7O_4S_2$ + H): 506.16, found 505.96, purity by anal. HPLC > 95%.

5. Conclusions

The lead molecule emerging from this study, SR24717, optimized from a high through-put screening hit, shows a promising profile with selectivity for meprin α. Three interactions, a π-π interaction and two cation-π interactions, appear to be responsible for this selectivity vs. meprin β and other metzincins. Accordingly, modification of substituent effects on the aryl and heteroaryl rings impact affinity and selectivity. Compounds in this chemical series may prove useful for the study of the efficacy and safety of meprin α-selective inhibition in animal models of human disease.

Author Contributions: Conceptualization, D.M., T.P.S., L.D.S. and T.D.B.; chemistry methodology, C.W. and T.D.B.; biochemical assays development, D.M.; high throughput screening implementation, T.P.S. and L.D.S.; cytotoxicity studies, J.D.; molecular modeling studies, H.P.; compound management, T.P.S. and L.D.S.; preparation, purification, and characterization of meprin substrates, G.B.F.; expression of meprins, C.B.-P.; writing—original draft preparation, C.W., D.M. and T.D.B.; writing—review and editing, all authors.; project administration, D.M.; funding acquisition, C.B.-P. and D.M. All authors have read and agreed to the published version of the manuscript.

Funding: This work was supported by the National Institutes of Health (AR066676 and CA249788 to D.M.). This work was also supported by the Deutsche Forschungsgemeinschaft (DFG) grant SFB877 "Proteolysis as a Regulatory Event in Pathophysiology" (project A9 and A15) (both to C.B.-P.).

Institutional Review Board Statement: Not applicable.

Informed Consent Statement: Not applicable.

Data Availability Statement: The data presented in this study are available on request from the corresponding author.

Acknowledgments: We thank Pierre Baillargeon and Lina DeLuca (Lead Identification, Scripps Florida) for compound management support.

Conflicts of Interest: The authors declare no conflict of interest.

Abbreviations

MMP	matrix metalloprotease
ADAM	a disintegrin and metalloprotease

References

1. Prox, J.; Arnold, P.; Becker-Pauly, C. Meprin alpha and meprin beta: Procollagen proteinases in health and disease. *Matrix Biol.* **2015**, *44–46*, 7–13. [CrossRef] [PubMed]
2. Broder, C.; Arnold, P.; Vadon-Le Goff, S.; Konerding, M.A.; Bahr, K.; Müller, S.; Overall, C.M.; Bond, J.S.; Koudelka, T.; Tholey, A.; et al. Metalloproteases meprin alpha and meprin beta are C- and N-procollagen proteinases important for collagen assembly and tensile strength. *Proc. Natl. Acad. Sci. USA* **2013**, *110*, 14219–14224. [CrossRef] [PubMed]
3. Peters, F.; Becker-Pauly, C. Role of meprin metalloproteases in metastasis and tumor microenvironment. *Cancer Metastasis Rev.* **2019**, *38*, 347–356. [CrossRef] [PubMed]
4. Kruse, M.N.; Becker, C.; Lottaz, D.; Köhler, D.; Yiallouros, I.; Krell, H.W.; Sterchi, E.E.; Stöcker, W. Human meprin alpha and beta homo-oligomers: Cleavage of basement membrane proteins and sensitivity to metalloprotease inhibitors. *Biochem. J.* **2004**, *378 Pt 2*, 383–389. [CrossRef]
5. Madoux, F.; Tredup, C.; Spicer, T.P.; Scampavia, L.; Chase, P.S.; Hodder, P.S.; Fields, G.B.; Becker-Pauly, C.; Minond, D. Development of high throughput screening assays and pilot screen for inhibitors of metalloproteases meprin alpha and beta. *Biopolymers* **2014**, *102*, 396–406. [CrossRef] [PubMed]
6. Ramsbeck, D.; Hamann, A.; Schlenzig, D.; Schilling, S.; Buchholz, M. First insight into structure-activity relationships of selective meprin beta inhibitors. *Bioorg. Med. Chem. Lett.* **2017**, *27*, 2428–2431. [CrossRef] [PubMed]
7. Ramsbeck, D.; Hamann, A.; Richter, G.; Schlenzig, D.; Geissler, S.; Nykiel, V.; Cynis, H.; Schilling, S.; Buchholz, M. Structure-Guided Design, Synthesis, and Characterization of Next-Generation Meprin beta Inhibitors. *J. Med. Chem.* **2018**, *61*, 4578–4592. [CrossRef] [PubMed]
8. Tan, K.; Jäger, C.; Schlenzig, D.; Schilling, S.; Buchholz, M.; Ramsbeck, D. Tertiary-Amine-Based Inhibitors of the Astacin Protease Meprin alpha. *ChemMedChem* **2018**, *13*, 1619–1624. [CrossRef] [PubMed]
9. Hou, S.; Diez, J.; Wang, C.; Becker-Pauly, C.; Fields, G.B.; Bannister, T.D.; Spicer, T.P.; Scampavia, L.D.; Minond, D.A. Discovery and Optimization of Selective Inhibitors of Meprin α (Part I). *Pharmaceuticals* **2021**. Accepted.
10. Altintop, M.D.; Kaplancikli, Z.A.; Çiftçi, G.A.; Demirel, R. Synthesis and biological evaluation of thiazoline derivatives as newantimicrobial and anticancer agents. *Eur. J. Med. Chem.* **2014**, *74*, 264–277. [CrossRef] [PubMed]
11. Tan, K.; Jäger, C.; Körschgen, H.; Geissler, S.; Schlenzig, D.; Buchholz, M.; Stöcker, W.; Ramsbeck, D. Heteroaromatic Inhibitors of the Astacin Proteinases Meprin a, Meprin b and Ovastacin Discovered by a Scaffold-Hopping Approach. *ChemMedChem* **2020**. [CrossRef]
12. Broder, C.; Becker-Pauly, C. The metalloproteases meprin alpha and meprin beta: Unique enzymes in inflammation, neurodegeneration, cancer and fibrosis. *Biochem. J.* **2013**, *450*, 253–264. [CrossRef] [PubMed]
13. Fields, G.B.; Lauer-Fields, J.L.; Liu, R.-Q.; Barany, G. Principles and Practice of Solid-Phase Peptide Synthesis. In *Synthetic Peptides: A User's Guide*, 2nd ed.; Grant, G.A., Ed.; W.H. Freeman & Co.: New York, NY, USA, 2001; pp. 93–219.

14. de Jong, G.I.; Buwalda, B.; Schuurman, T.; Luiten, P.G. Synaptic plasticity in the dentate gyrus of aged rats is altered after chronic nimodipine application. *Brain Res.* **1992**, *596*, 345–348. [CrossRef]
15. Becker-Pauly, C.; Höwel, M.; Walker, T.; Vlad, A.; Aufenvenne, K.; Oji, V.; Lottaz, D.; Sterchi, E.E.; Debela, M.; Magdolen, V.; et al. The alpha and beta subunits of the metalloprotease meprin are expressed in separate layers of human epidermis, revealing different functions in keratinocyte proliferation and differentiation. *J. Investig. Dermatol.* **2007**, *127*, 1115–1125. [CrossRef] [PubMed]
16. Zhang, J.H.; Chung, T.D.; Oldenburg, K.R. A Simple Statistical Parameter for Use in Evaluation and Validation of High Throughput Screening Assays. *J. Biomol. Screen.* **1999**, *4*, 67–73. [CrossRef] [PubMed]
17. Kaplancıklıa, Z.A.; Yurttaas, L.; Turan-Zitouni, G.; Ã-zdemir, A.; Göger, G.; Demirci, F.; Mohsen, U. Synthesis and Antimicrobial Activity of New Pyrimidine-Hydrazones. *Lett. Drug Des. Discov.* **2014**, *11*, 76–81. [CrossRef]
18. Lokwani, D.; Azad, R.; Sarkate, A.; Reddanna, P.; Shinde, D. Structure Based Library Design (SBLD) for new 1,4-dihydropyrimidine scaffold as simultaneous COX-1/COX-2 and 5-LOX inhibitors. *Bioorg. Med. Chem.* **2015**, *23*, 4533–4543. [CrossRef] [PubMed]

 *pharmaceuticals*

Review

Developments in Carbohydrate-Based Metzincin Inhibitors

Doretta Cuffaro, Elisa Nuti ***, Felicia D'Andrea** **and Armando Rossello**

Department of Pharmacy, University of Pisa, via Bonanno 6, 56126 Pisa, Italy;
doretta.cuffaro@farm.unipi.it (D.C.); felicia.dandrea@farm.unipi.it (F.D.); armando.rossello@farm.unipi.it (A.R.)
* Correspondence: elisa.nuti@farm.unipi.it

Received: 5 October 2020; Accepted: 6 November 2020; Published: 10 November 2020

Abstract: Matrix metalloproteinases (MMPs) and A disintegrin and Metalloproteinase (ADAMs) are zinc-dependent endopeptidases belonging to the metzincin superfamily. Upregulation of metzincin activity is a major feature in many serious pathologies such as cancer, inflammations, and infections. In the last decades, many classes of small molecules have been developed directed to inhibit these enzymes. The principal shortcomings that have hindered clinical development of metzincin inhibitors are low selectivity for the target enzyme, poor water solubility, and long-term toxicity. Over the last 15 years, a novel approach to improve solubility and bioavailability of metzincin inhibitors has been the synthesis of carbohydrate-based compounds. This strategy consists of linking a hydrophilic sugar moiety to an aromatic lipophilic scaffold. This review aims to describe the development of sugar-based and azasugar-based derivatives as metzincin inhibitors and their activity in several pathological models.

Keywords: carbohydrates; glycoconjugates; MMPs; ADAMs; iminosugars; metzincin inhibitors

---

1. Introduction

1.1. Metzincins

Tissue remodeling is a crucial process in various pathological and physiological events in living organisms. In general, protein degradation during tissue turnover is regulated by a multitude of proteases, among which metzincins play an important role. Metzincin superfamily includes metalloproteinases which share a similar catalytic site constituted by a zinc ion, a zinc binding consensus motif, and a specific conserved methionine residue [1]. In the active form of the enzyme, the zinc ion is disposed in a highly conserved motif, tetrahedrally complexed with three histidines, and with a fourth coordination place available for the substrate and/or water. The consensus motif (HEXXHXXG/NXXH/D) is followed by a structurally conserved methionine which participates in the structural integrity of the catalytic domain [2]. Metzincins are classified in five different subfamilies on the basis of structural similarities: the matrixins (matrix metalloproteinases, MMPs), the serralysins (large bacterial proteinases), the astacins, the pappalysins and the adamalysins (ADAMs (A Disintegrin And Metalloproteinases), and ADAMTSs (ADAMs with thrombospondin motifs) [3].

1.2. Matrix Metalloproteinases

Matrix metalloproteinases (MMPs), or matrixins, constitute a family of 24 zinc-dependent endopeptidases homogeneous for structure, function, and localization which are critical for the degradation of the extracellular matrix (ECM) [4]. In addition to their functions as tissue-remodeling enzymes, MMPs are also involved in the selective cleavage of many non-ECM targets, such as cytokines, cell surface receptors, chemokines, and cell–cell adhesion molecules [3]. The widely used classification of MMPs (Figure 1) is based on substrate specificity, sequence similarity, domain organization, and partially on their cellular organization [2,3]:

- Archetypal MMPs have a similar structure and are divided into:

    - Collagenases (MMP-1, MMP-8, MMP-13): specifically cleave the collagen triple helix. Moreover, their collagenolytic activity, due to a cooperation between the hemopexin and the catalytic domains, acts on a various number of ECM and non-ECM molecules.
    - Stromelysins (MMP-3, MMP-10): structurally similar to collagenases, show wide substrate specificity but are unable to degrade native collagen.
    - Others (MMP-12, MMP-19, MMP-20, and MMP-27): not classified in the previous categories. Metalloelastase (MMP-12) is mainly expressed in macrophages and digests elastin and other proteins.

- Gelatinases (MMP-2 or Gelatinase A and MMP-9 or Gelatinase B): owing to the presence of three fibronectin type II repeats inside the catalytic domain, they readily digest type IV collagen, gelatin, and a number of ECM molecules including laminin, fibronectin, and aggrecan core proteins.
- Matrilysins (MMP-7 and MMP-26): characterized by the lack of the hinge region and of hemopexin domain.
- Furin-activatable MMPs:

    - Secreted (MMP-11, MMP-21, and MMP-28): activated by furin-like proteases before secretion.
    - Membrane-type MMPs (MT-MMPs): classified in type I transmembrane proteins (MMP-14, MMP-15, MMP-16, and MMP-24) and type II transmembrane protein (MMP-23).
    - Glycosylphosphatidylinositol (GPI)-anchored proteins (MMP-17 and MMP-25).

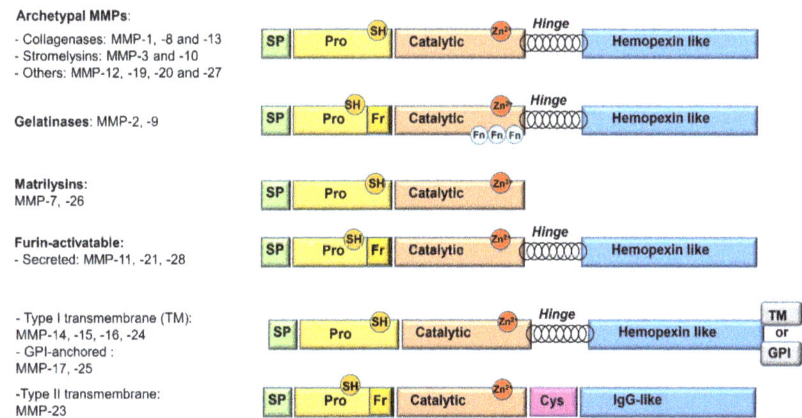

Figure 1. Schematic overview of the matrix metalloproteinase (MMP) domains and family member classification. SP: signal peptide; Pro: propeptide; TM: transmembrane domain; GPI: glycosylphosphatidylinositol; Cys: cysteine array; Fn: fibronectin repeat; Fr: furin-cleavage site; SH: thiol group.

Currently, the 3D structures of a number of MMPs have been determined by X-ray crystallography and NMR methods [5–8]. From a structural point of view, the family of MMPs mostly shares a common three domain-based structure that consists of a pro-peptide domain (about 80 residues) at the amino-terminal, a central catalytic domain (about 160–170 residues), and a hemopexin domain at the carboxylic-terminal. The pro-peptide domain confers latency to the enzyme by occupying the active site zinc, making the catalytic domain inaccessible to substrates. The pro-domain contains a sequence PRCGXPD with a conserved cysteine, fundamental for the activation process. The *catalytic domain* is strictly conserved among MMPs and also shows similarity for all metalloproteinases, leading to difficulties in site-targeted selective inhibitor design. It presents a specific consensus motif (His-Glu-X-X-His-X-X-Gly-X-X-His) typical of all MMPs. In addition, two structural zinc ions and at least one calcium ion are included. A hinge domain (about 60–70 residues) connects the catalytic domain to the hemopexin one [9].

The catalytic mechanism starts with the enzyme activation due to proteolytic cleavage of the *N*-terminal pro-domain. The removal of the pro-peptide disrupts the interaction between the conserved cysteine and the zinc, allowing the thiol group to be replaced by a water molecule. This so called "cysteine-switch mechanism" enables the partially activated enzyme to further hydrolyze the pro-peptide, resulting in its removal [4].

The substrate binding involves principally the zinc ion. In the active form of the enzyme, the zinc is tetrahedrally coordinated by three histidines (His 228, His 222, and His 218 in full MMP-3 numbering) and a water molecule. The glutamate residue, coordinating the water molecule, makes it a good nucleophile to attack the substrate scissile peptide bond (Figure 2).

Figure 2. General representation of amide bond hydrolysis in the catalytic site of MMPs.

MMP Catalytic Binding Site

The catalytic binding site as well as the substrates differ significantly among MMPs. Therefore, Schechter and Berger reported [10] a nomenclature system to identify amino acids in the substrate as well as the recognizing areas on the surface of the enzyme (Figure 3).

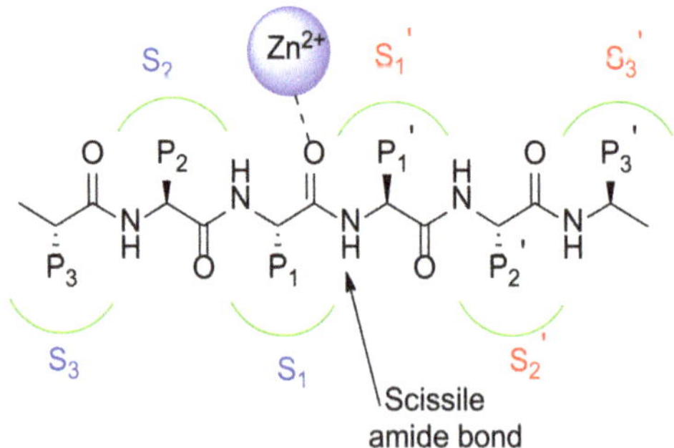

Figure 3. Schematic representation of MMP catalytic binding site.

This generally accepted nomenclature identifies all of the subsites with S1, S2, and S3 for the sites on the left side of the zinc ion, while S1′, S2′, and S3′ are used to recognize the subsites on the right side of the zinc ion. The amino acid positions (P) of the substrate are countered from the scissile amide bond and have the same numbering to that of the subsites they occupy on the enzyme surface.

Basically, the residues on the *N*-terminal side of the substrate are named P1, P2, and P3 whereas the residues on the *C*-terminal side are termed P1′, P2′, and P3′.

The MMP substrate binding cleft is mainly represented by a hydrophobic pocket, namely S1′ [11]. The S1′ pocket, also called the specificity pocket, confers the substrate-recognition varying among the different isoforms of MMPs in its aminoacidic sequence and depth. Although S2′ and S3′ pockets participate in substrate binding together with S1′, they are shallower pockets more solvent exposed than S1′. For these characteristics they result more difficult to target using synthetic inhibitors.

MMPs are classified on the basis of S1′ depth in shallow (MMP-1, MMP-7), medium (MMP-2, MMP-8, MMP-9), and deep (MMP-3, MMP-11, MMP-12, MMP-13, MMP-14) pocket MMPs. The S1′ pocket accommodates the side chain of the substrate that will become the new *N*-terminus [12].

1.3. ADAM Metalloproteinases

The vast majority of ADAMs are type I transmembrane proteins approximately constituted by 750 amino acids which belong to the metzincin superfamily [13,14].

ADAM structure is constituted by a metalloprotease domain, similarly to MMPs, and a unique disintegrin domain responsible for binding to integrins, which explains their name: ADAM (A Disintegrin and A Metalloproteinase). The main function of ADAMs is the cleavage of cell surface proteins among which are growth factors, cytokines, receptors, and adhesion molecules [15]. Their domain structure consists of a pro-domain, a metalloprotease domain, a disintegrin domain, a cysteine-rich domain, an EGF-like domain, a transmembrane domain, and a cytoplasmic tail. In the last years, 3D structures of the most important ADAMs have been published disclosing important structural information and specificities of the various domains [16–19].

The catalytic metalloproteinase domain is quite similar to the one of MMPs. The catalytic domain is divided into two subdomains, and the active site cleft is positioned between the two [2,20].

Similarly to MMPs, the groove between the two subdomains has six subsites (S3, S2, S1, S1′, S2′, S3′) determining specificity for particular amino acid sequences (P3, P2, P1, P1′, P2′, P3′) in the substrate. Particularly interesting are the subsites P1 and P1′, where the proteolytic cleavage occurs.

ADAM Ectodomain Shedding

The ADAMs have long been studied for their ability to cleave a large number of cell surface proteins, releasing a soluble fragment from its trans-membrane precursor. This mechanism is known as ectodomain shedding [21]. The cleaved transmembrane protein can be the substrate for cytoplasmic proteases or intramembrane proteolytic complexes, capable of releasing intracellular molecules responsible for gene transcription regulation. Alternatively, the ectodomain shedding, together with the protein internalization, can control the half-life of transmembrane proteins on the cell surface.

The ADAM mediated proteolysis is not only essential for providing extracellular signals, but also represents a prerequisite for intracellular signaling by regulated intramembrane proteolysis (RIP). RIP can be defined as a proteolytic cascade, initiated by an ectodomain sheddase action on a transmembrane protein which leads to intracellular release of soluble fragments [22].

1.4. Metzincin Inhibitors

Misregulation of the metzincin activity is a major feature in many serious pathologies [23]. In particular, it is well known that overexpression of these enzymes can lead to massive tissue degradation, dangerous for the promotion of cellular events such as tumor invasiveness [24] and inflammation spreading [25,26]. Moreover, an imbalanced expression of MMPs due to an inadequate regulation of TIMPs, is a hallmark of different bacterial, viral, or combined systemic infection such as sepsis [27,28]. Furthermore, recent findings about metzincin involvement in central nervous system (CNS) disorders, and in particular in Alzheimer's disease (AD), make them modulators for promising therapeutic strategies [29–31]. The strong involvement of metzincins in these critical diseases has been repeatedly confirmed by gene-direct studies and animal models so that many resources have been invested in the last 30 years by pharmaceutical companies in the development of metzincin inhibitors [32].

A classical metzincin inhibitor consists of a "backbone" and a "zinc-binding group" (ZBG) [33]. The backbone is a classic drug-like structure designed to establish noncovalent bonds with the protein, such as hydrophobic and electrostatic interactions and hydrogen bonds. The zinc-binding group is a chelating group for the catalytic zinc atom that is also able to give hydrogen bonds with the enzyme. The hydroxamate group is the most used ZBG due to its excellent chelating properties [34]. The zinc chelation mediated by hydroxamates proceeds with a bidentate structure, in which each oxygen is located at an optimal distance to the catalytic zinc. Additionally, hydroxamate efficacy is potentiated by hydrogen bond formation between its heteroatoms and some residues conserved in all MMPs (the reason why this moiety suffers from poor selectivity).

Initially, three main generations of zinc-binding metzincin inhibitors were developed, exploiting the growing knowledge of these proteinases. Small molecules such as the peptidomimetic Marimastat, and Ilomastat (or Galardin) [35] (first generation, Figure 4), the sulfonamido-based CGS-27023A [36] and Prinomastat (second generation, Figure 4), NNGH [37] and ARP100 [38] (third generation, Figure 4), were identified as progenitors of many potent metzincin inhibitors [39]. Many MMP inhibitors showed high potency in preclinical studies but the selectivity goal was not achieved, and they eventually failed in clinical trials. The most important syndrome derived from the use of broad-spectrum metzincin inhibitors is a Musculoskeletal Syndrome (MSS) and has nowadays been ascribed to MMP-1 and MMP-14 inhibition [40].

In the last 15 years new strategies have been developed in order to selectively target MMPs such as using non-zinc binding inhibitors [41–43], or exosite-binding inhibitors [44–46].

The first synthetic ADAM inhibitors relied on Zn^{2+} binding groups to interact with the catalytic active site. One of the most studied ADAMs is ADAM17 (or TACE, TNF-α converting enzyme), a protease involved in the shedding of transmembrane protein ectodomains [47,48]. Recently, selective ADAM17 inhibitors able to target the catalytic binding site [49–51] or the enzyme exosites have been reported [52,53].

Although several small molecule MMP [54] or ADAM [55] inhibitors with high in vitro activity have been described, the principal obstacles that have hindered clinical development of metzincin inhibitors are the inadequate selectivity for the target enzyme, the poor water solubility and long-term toxicity. In fact, the majority of the metzincin inhibitors reported so far presented a hydrophobic structure which gave high affinity for the target enzyme due to hydrophobic interactions. However, the high lipophilicity of these scaffolds negatively affected their bioavailability, increasing the human serum albumin retention (HSA) [56] and hindering the target reach. A good water solubility would confer a good bioavailability in physiological media, representing a highly desirable property for drug-like compounds.

Therefore, in the last 15 years, glycoconjugation has been investigated as an approach to improve the hydrophilicity of metzincin inhibitors and thus increase their oral bioavailability, avoiding a detrimental effect on their inhibitory activity. In the next paragraph we will describe and fully detail all the carbohydrate-based inhibitors of MMPs and ADAMs reported in literature so far.

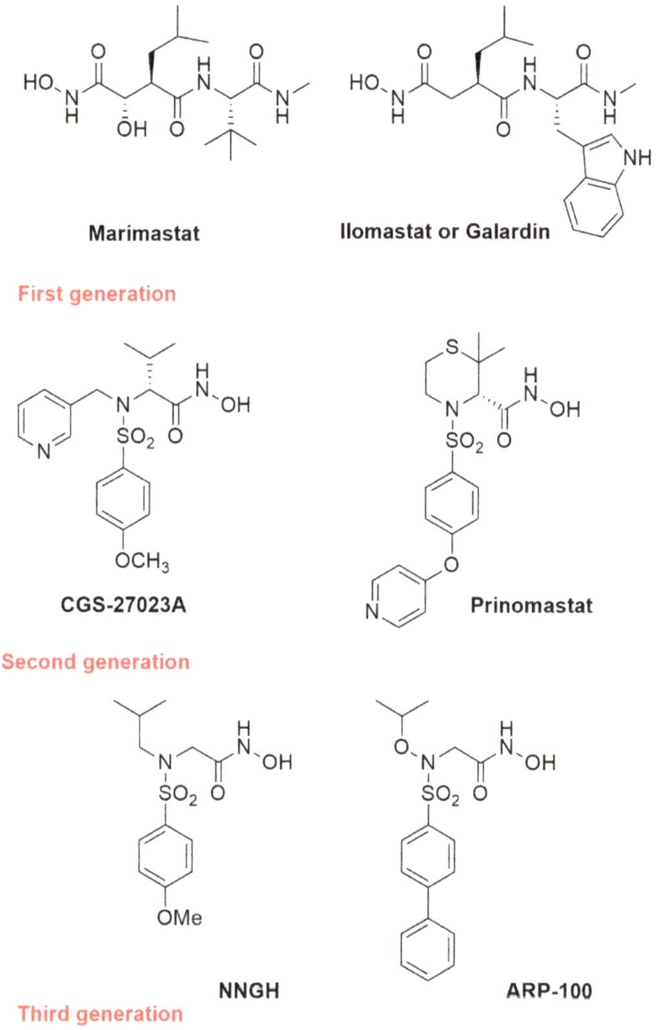

Figure 4. Chemical structure of the most representative zinc-chelating metzincin inhibitors.

2. Carbohydrate-Based Metzincin Inhibitors

In this review carbohydrate-based metzincin inhibitors will be described and classified on the basis of the sugar moiety in sugar-based inhibitors and azasugar-based inhibitors.

2.1. Sugar-Based MMPIs

Nowadays, it is well known how a wide range of glycosylated derivatives such as glycoproteins or glycosylated natural products, are crucial for physiological [57] and pathological processes [58]. Particularly interesting is the protein-carbohydrate recognition, a phenomenon fundamental in many cellular mechanisms, such as migration, adhesion, cell-differentiation, tumor progression, infections (by viruses or bacteria), and immune response among others [59–63]. In the last years, the ubiquity but also the biological properties of glycoproteins or glycosylated natural products have driven research towards the synthesis of glycoconjugates for the development of drug candidates with anti-infectious, anti-inflammatory, anticancer, or vaccine activity [64–66].

Due to their polyhydroxylated nature, carbohydrates are often employed as biocompatible enhancers of the hydrophilic profile of water insoluble drugs. In the design and development of potential drugs, the study of physico-chemical properties (stability, lipophilicity, aqueous solubility, and bioavailability) is essential to obtain compounds with optimal potency and selectivity [67–70]. For this reason, the insertion of a sugar moiety could be crucial to improve the bioavailability profile but also to recognize specific cellular entities, thus increasing the activity of the parent non-glycoconjugated compound.

In the last years some water-soluble MMPIs have been reported in literature by different research groups [71]. The first two sugar-based MMPIs (**1** and **2**, Figure 5) were reported by Fragai et al., in 2005 [72]. These two compounds present a sulfur-containing constrained structure that can be synthesized in a diastereomerically pure α-*O*-glyco form through a totally chemo-, regio-, and stereoselective Diels–Alder reaction.

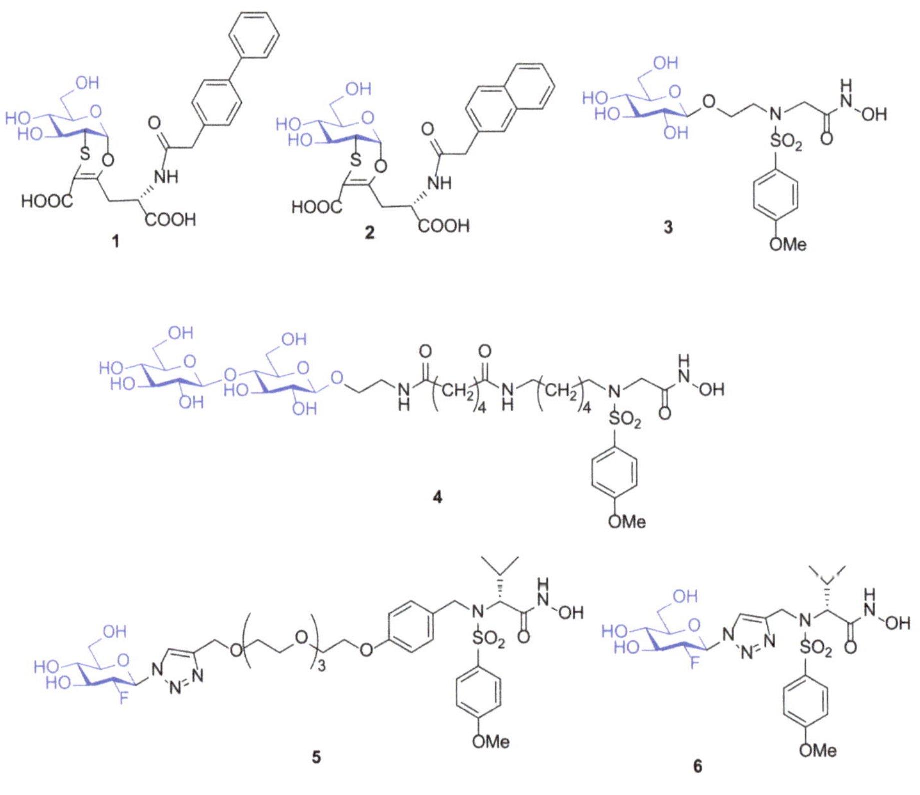

Figure 5. Chemical structure of glycoconjugated MMPIs **1–6**.

A virtual screening of various molecules has been carried out by docking analysis in MMP-12 catalytic domain. Compounds **1** and **2**, which are characterized by a side chain with a biphenyl group (**1**) or a naphthyl group (**2**) linked to the homoglutamic nitrogen and presenting as ZBG a carboxylic acid, afforded the best activity results. A NMR study on ligand–protein interactions was performed to confirm the in silico results. Binding of **1** and **2** to ^{15}N-enriched MMP-12 catalytic domain was monitored by ^{1}H,^{15}N HSQC NMR spectroscopy and pointed out the lack of interactions between the carboxylic acid and the protein and confirmed the good interaction of the lipophilic moieties with the S1′ pocket of the enzyme. These compounds were poor MMP-12 inhibitors with IC$_{50}$ in the high micromolar range but represent the first carbohydrate-based MMP-12 inhibitors reported in literature.

In 2006 the sulfonamido-based MMPI **3** (Figure 5) was published by Calderone et al. [73] as a more soluble derivative of NNGH [37] in which the isopropyl group was replaced by a glycosylated *N*-hydroxyethyl chain. The β-*O*-glucopyranoside derivative **3** showed a higher water solubility (>30 mM) compared to NNGH and presented a nanomolar activity (K_i) for MMP-1 (286 nM), MMP-8 (9 nM), MMP-12 (14 nM), and MMP-13 (1.7 nM). The X-ray analysis of compound **3**-MMP-12 complex (Figure 6) evidenced a lack of significant participation in binding of the glucose ring. The MMP-12-compound **3** complex showed the classical coordination of the hydroxamate moiety with the catalytic zinc ion, the good fitting of the hydrophobic moiety in the S1′ pocket and the protrusion of the glucose portion out of the protein towards the solvent region. Although the glucose interaction with the protein could be marginal, it seemed to have an effective role on the selectivity profile for MMP-13 compared to other MMPs. Moreover, the interaction of compound **3** and NNGH with human serum albumin (HSA) was reported. While NNGH is a strong binder, inhibitor **3** did not appreciably interact with HSA, probably due to the glucose moiety.

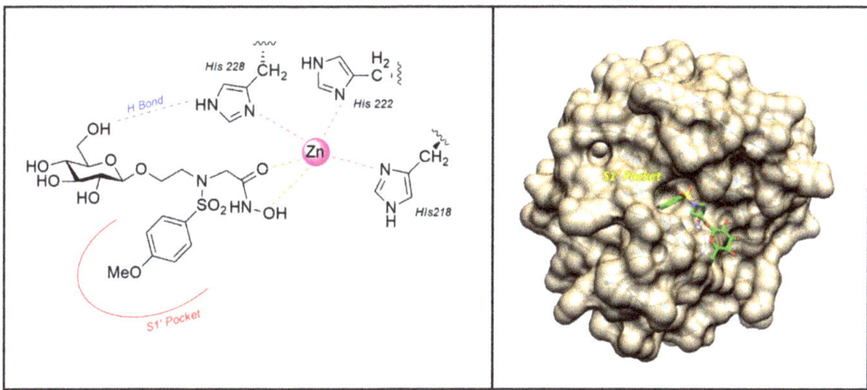

Figure 6. The binding mode of **3** [73] in the MMP-12 catalytic domain (PDB: 3N2U). Image generated with Chimera, version 1.13.

Afterwards, the same research group reported the synthesis of a bifunctional MMPI, **4** (Figure 3), which was structurally related to compound **3**. The glucose unit of **3** was replaced by a lactose moiety and the spacer length was increased [74]. The lactose was chosen in order to bind Galectin-3 (Gal-3), a member of galectin family strictly involved in cancer progression. The presence of Gal-3 in tumor invasion area correlates with MMP upregulation. For this reason, a bifunctional architecture, which combines MMP inhibitory activity with galectin targeting, was adopted as design strategy. The inhibitory potency of compound **4**, tested on a set of MMPs, revealed that the lactose functionalization did not impair the activity of the ligand (K_is in the nanomolar range) and did not improve the selectivity. The ability of **4** to bind MMPs and Gal-3 simultaneously was demonstrated by a NMR study in MMP-12 catalytic domain. A ternary complex Gal-3-compound **4**-MMP-12 was formed in solution and the ligand **4** was able to gather Gal-3 and MMP-12 side by side.

In 2012, Hugenberg et al. [75] described a MMPI series of fluorinated triazole-substituted hydroxamates. Among them, two 2-fluoro-glucopyranoside derivatives, **5** and **6** (Figure 5), were synthesized by click chemistry reaction. These compounds differ only in the linker connection that is a triazole methyl chain in **6** and a triazole PEG chain in **5**. The conjugation was achieved through copper-catalyzed click chemistry reaction between 2-fluoro glucose and the alkynyl sulfonamide scaffold. The glucose derivatives reported high activity (K_i ranging picomolar for **5** and nanomolar for **6**) for MMP-2, MMP-8, MMP-9, and MMP-13 but no selectivity, and an excellent cLogD value. These compounds

Table 1. Inhibitory activity (IC$_{50}$ nM) of sugar-based MMP inhibitors against MMP-1, -2, -9, -12.

Compound	MMP-1	MMP-2	MMP-9	MMP-12	Ref
1	- [a]	-	-	490,000	[72]
2	-	-	-	790,000	[72]
3	286	-	-	14.3	[73]
4	-	-	80 (K_i)	17 (K_i)	[74]
5	-	30	43	-	[75]
6	-	0.2	0.6	0.5	[75]
8	-	-	6800	63	[76]
9	-	-	9400	73	[76]
10	-	-	4400	72	[76]
11	-	-	3100	53	[76]
12	-	-	1220	36	[76]
13	-	-	1200	28	[76]
14	19,000	330	1200	18	[76]
15	40,000	320	5400	40	[76]
16	50,000	100	860	12	[76]
17	-	-	1470	42	[76]
18	-	-	880	19	[77]
19	-	-	1120	98	[77]
20	-	-	520	65	[77]
21	-	-	1200	18	[77]
22	-	-	>50,000	1300	[77]
23	-	-	1235	83	[77]
24	-	-	890	41	[77]
25	-	-	2200	51	[77]

[a] "-": not tested or unknown from the corresponding original reference.

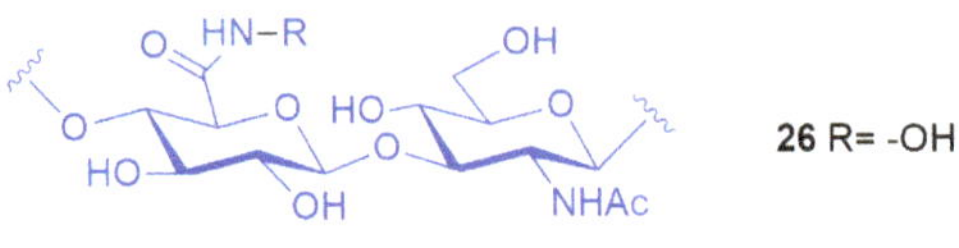

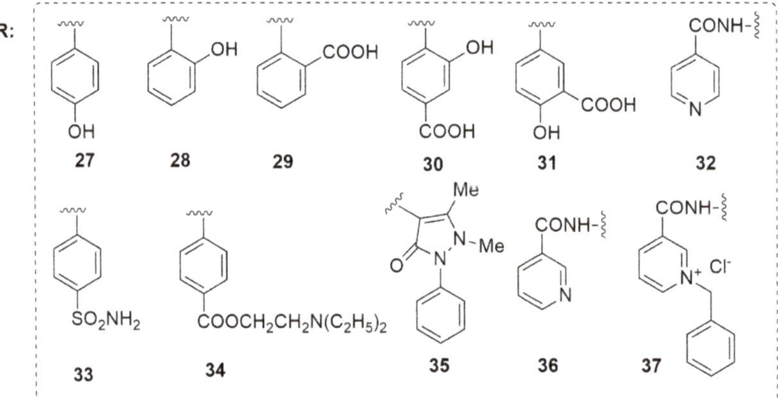

Figure 9. Chemical structure of hyaluronic acid-based MMP-2 inhibitors **26–37**.

2.2. Azasugar-Based Inhibitors

2.2.1. Azasugar Biological Background

Azasugars or iminosugars, whether of natural or synthetic origin, are hydroxylated carbohydrate mimics, where a basic nitrogen substitutes the classic endocyclic oxygen [85]. This modification appears simple and not significant but provides important biological properties and also gives rise to many synthetic challenges [86–92]. Historically, iminosugars are known as inhibitors of glycosidases, enzymes involved in the glycosidic bond hydrolysis in different biological processes, such as intestinal digestion, post-translational processing of the sugar chain of glycoproteins, and lysosomal catabolism of glycoconjugates [93].

Despite being a structurally diversified class of molecules (polyhydroxylated piperidines and pyrrolidines and their derivatives), iminosugars attracted considerable interest due to their biomedical relevant activities. In fact, this class of compounds displayed different biological properties: as antiviral agents (against HIV-1, herpes simplex virus, bovine viral diarrhea virus (BVDV), and hepatitis C virus (HCV)), as antidiabetics, as agents for the treatment of lysosomal storage disorders such as Gaucher's and Niemann–Pick type C, in immune modulation and as anticancer agents [94].

Among these polyhydroxylated alkaloids, 1,5-dideoxy-1,5-imino-D-glucitol or 1-deoxy-D -nojirimycin [95] (DNJ, Figure 10) and its *N*-alkylated derivatives have prime importance. In fact, most of the commercially available azasugar drugs are based on DNJ structure, such as Glyset (*N*-hydroxyethyl -1-deoxy-D-nojirimycin, commonly known as Miglitol, Figure 10) [96–98], a well-known human alpha-glucosidase inhibitor used to treat type II diabetes and Zavesca (*N*-butyl-1-deoxy-D-nojirimycin commonly known as Miglustat, Figure 10) [99], which is an approved prominent inhibitor of glycosphingolipids for Gaucher disease. Recently DNJ has been studied also as MMPI, with not completely clear results. In 2010 Wang et al. [100] reported the DNJ effect on reducing B16F10 cell metastatic capability by suppressing the activity and the expression of MMP-2 and -9, but simultaneously imbalancing MMP-2 and TIMP-2 activity. Moreover, in 2013 the mulberry DNJ pleiotropic effect on the development of atherosclerosis was reported [101]. Mulberry DNJ is able to inhibit migration of A7r5 vascular smooth cells in a dose dependent manner and both the MMP-2 and MMP-9 activities in these cells. In this case the DNJ-inhibition of VSMC migration is a synergistic effect due to two different pathways: AMPK activation and the inhibition of F-actin activity. MMP-2 and MMP-9 inhibition is a key step in RhoB activation in the AMPK activation path.

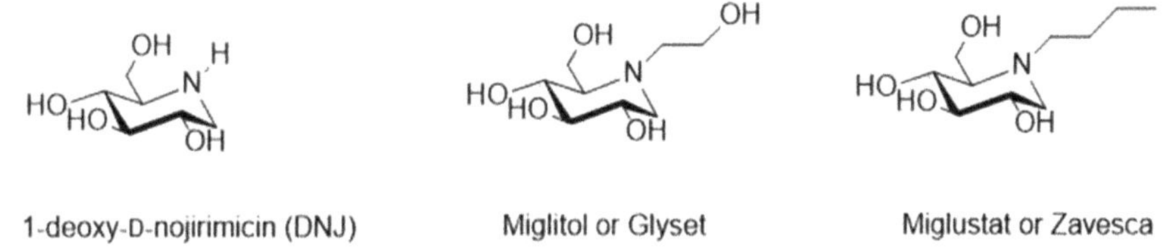

Figure 10. Chemical structure of DNJ, Miglitol and Zavesca.

2.2.2. Azasugar-Based ADAM/MMP Inhibitors

Based on a classical MMP inhibitor scaffold, Moriyama et al., reported [102–104] the design, synthesis, and SAR study of azasugar-based MMP/ADAM inhibitors (**38–53**, Series 1–4 Figure 11). This class of compounds is based on the structure of known MMP inhibitors constituted by a sulfonamide scaffold as structural motif inserted in an azasugar having different configuration (L-*altro*, L-*ido*, L-*gluco*, L-*manno*, and L-*gulo*). The first series (Series 1 Figure 11) [102] was prepared to investigate the stereochemistry and the different protection of hydroxyl groups attached to a *p*-methoxysulfonamido scaffold. The inhibition data on MMP-1, MMP-3, MMP-9, and ADAM17 (also known as TACE: *tumor necrosis factor-α-converting enzyme*) led to identify compound **40** (L-*altro*) with a nanomolar activity for all the tested enzymes. In a further work [103], a SAR analysis on compound **40** has been carried out, changing the stereochemistry configuration of azapyranose unit and the type of aromatic ring (Series 2 and Series 3, Figure 11). All these compounds were tested on MMP-1, -3, -9, and TACE and the phenoxyphenyl derivative **46** (L-*ido*) was selected as the best one with an inhibitory activity in the low nanomolar range (K_i MMP-1: 8.0 nM; K_i MMP-3: 0.5 nM; K_i MMP-9: 0.06 nM; K_i TACE: 2.3 nM). Compound **40** was used as starting point to develop the fourth series [104] of azasugar MMP/ADAM inhibitors, changing its stereochemistry (compound **49**, L-*gulo*) and analyzing different groups in the P1′ position (**50–53**, L-*gulo*). The phenoxyphenyl derivatives **46** and **49** presented an excellent stability in aqueous solution and the best inhibitory activity for MMP-1, MMP-3, MMP-9, and TACE (K_i in the low nanomolar range). Moreover, azasugar derivative **49** was effective on a mouse TPA-induced epidermal hyperplasia model used to evaluate its efficacy as antipsoriatic agent. The binding mode of compound **49** in MMP-3 catalytic domain determined by crystallographic analysis showed that the stereochemistry of C-2 hydroxyl group was not fundamental for the interaction with the enzyme.

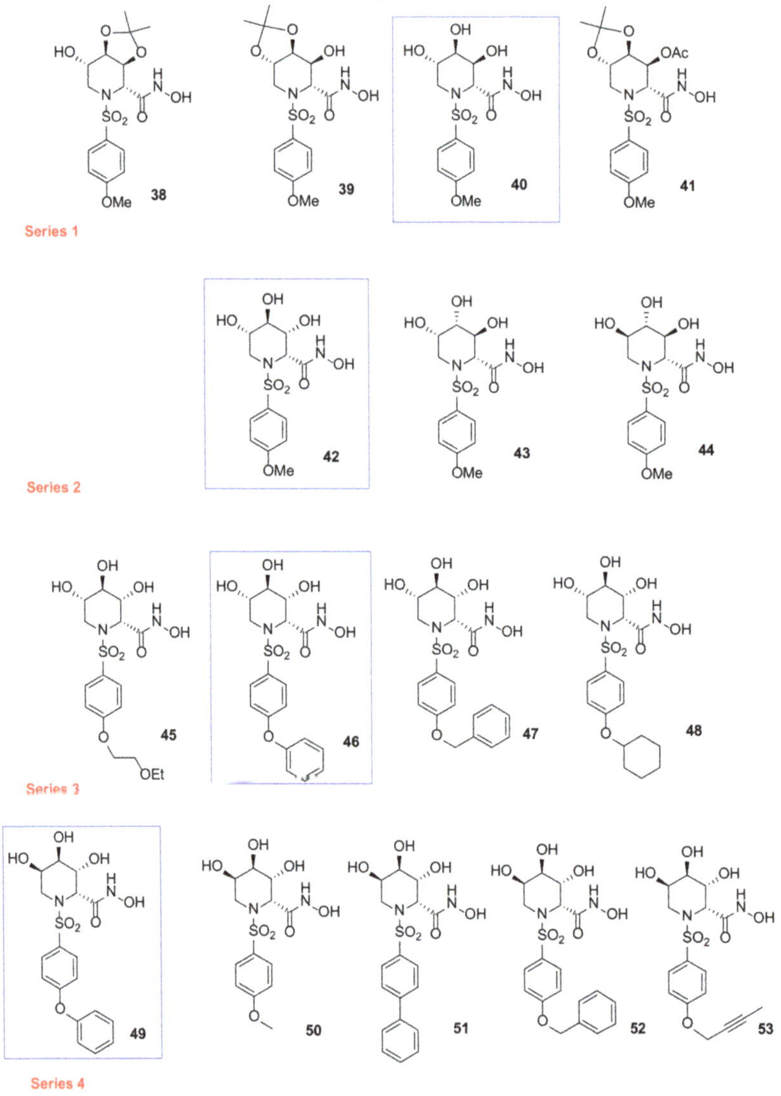

Figure 11. Chemical structure of azasugar-based MMP/ADAM inhibitors (Series 1–4).

Compound **49** was also tested by other research groups. In 2009, Chikaraishi et al. [105] studied the antiangiogenic activity of **49** by analyzing its suppressing activity against vascular endothelial cell tube formation by an in vitro HUVEC and fibroblast coculture assay and by in vivo retinal neovascularization in a murine ischemia-induced proliferative retinopathy model. In the in vitro angiogenesis model, **49** was able to decrease VEGF-induced HUVEC tube formation. Furthermore, in the in vivo angiogenesis model, administration of **49** reduced retinal neovascularization avoiding side effects on physiological revascularization to the oxygen-induced obliteration area.

In 2018, Sylte et al. [106] reported a study on the binding strengths of compound **49** (Figure 11) and Galardin (Figure 4) for the bacterial MMPs, thermolysin, pseudolysin, and auerolysin, compared to human MMP-9 and MMP-14. The obtained K_i values of **49** on MMP-9 were approximately 10-fold lower than the previously reported ones. Enzymatic data revealed a stronger inhibition of MMP-9 and MMP-14 by azasugar **49** compared to Galardin probably due to a more favorable interaction of diphenyl ether moiety with the S1' pocket, as confirmed by a docking study. The opposite effect was found on bacterial MMPs, where Galardin was highly effective while compound **49** was inactive.

The azasugar-based MMP/ADAM inhibitors described so far showed nanomolar inhibition of MMPs and ADAMs but none of these structures presented a good selectivity profile. Based on the SAR analysis of the previous papers, a TACE selective azasugar-based inhibitor series was reported (**54–58**, Figure 12) [107]. In this series the iminosugar stereochemistry was analogous to the one of compounds **49–53** (L-*gulo*) because this was the most suitable in order to inhibit TACE, as described in Moryiama's papers [102–104]. The azasugar was combined with a butyn-2-yloxy aromatic scaffold reported as suitable P1' substituent in order to achieve TACE selectivity. The new inhibitors were tested by enzymatic assay on human MMPs and TACE, revealing excellent activity for TACE (K_i 0.53–1.85 nM), high selectivity over MMP-1 and moderate selectivity over MMP-3 and MMP-9. In particular, compound **56** with a 2,3-*O*-acetonide group had a very potent TACE inhibitory activity (K_i 0.57 nM) and a good selectivity profile with a 136-fold selectivity for TACE over MMP-9, 52-fold over MMP-3, and 1500-fold over MMP-1.

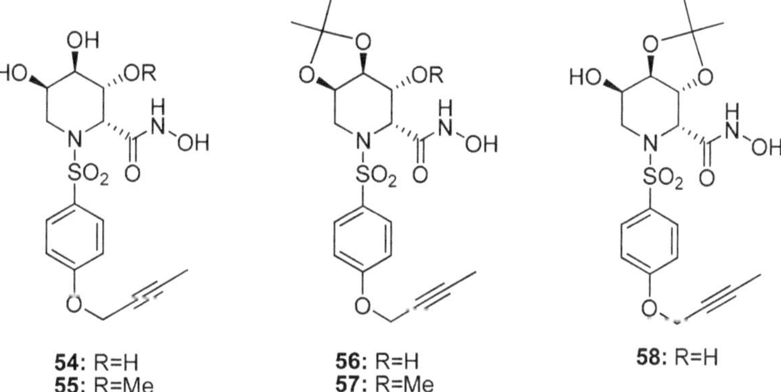

Figure 12. Chemical structure of azasugar-based TACE inhibitors (**54–58**).

Biological results for azasugar-based metzincin inhibitors are summarized in Table 2.

Table 2. Inhibitory activity (IC$_{50}$ nM) of azasugar-based metzincin inhibitors against ADAM17 and MMP-1, -3, -9.

Compound	ADAM17	MMP-1	MMP-3	MMP-9	Ref
38	22	554	44	310	[102]
39	40	16	3.7	21	[102]
40	71	84	1.7	157	[102]
41	21	50	50	47	[102]
42	12	25	7.7	4.8	[103]
43	510	>850	490	780	[103]
44	340	450	85	82	[103]
45	8.7	>850	42	64	[103]
46	2.3	8.0	0.5	0.06	[103]
47	1.6	850	2.6	6.1	[103]
48	67	100	1.8	0.9	[103]
49	6.2	5.3	0.35	0.097	[104]
50	15	26	2	2	[104]
51	21	162	50	47	[104]
52	1.7	>850	2.1	7.4	[104]
53	0.53	128	3.3	14	[104]
54	0.53	128	3.3	14.2	[107]
55	1.85	90	0.43	12	[107]
56	0.57	>850	29.6	77	[107]
57	0.84	552	11.5	98	[107]
58	1.85	>850	58	118	[107]

3. Conclusions

MMPs and ADAMs are zinc-dependent endopeptidases belonging to a larger family of proteases known as metzincins. Upregulation of metzincin activity is a major feature in many serious pathologies such as cancer, inflammatory disorders, neurological diseases, and infections. Several molecules have been discovered in the last years as MMP or ADAM inhibitors presenting very high activity in vitro but also inadequate selectivity for the target enzyme, poor water solubility, and long-term toxicity. For this reason, over the last 15 years, a novel approach to improve solubility and bioavailability of metzincin inhibitors has been the synthesis of carbohydrate-based inhibitors. This strategy consists of the conjugation of a hydrophilic carbohydrate moiety to an aromatic backbone containing the ZBG. In this review we described all the carbohydrate-based metzincin inhibitors reported in literature, classifying them on the basis of the carbohydrate (sugar-based and azasugar-based) and on the basis of the biological target (MMPs or ADAMs). Some promising molecules showing a nanomolar activity and a good selectivity profile for the target enzyme together with an improved water solubility have been discovered. Among sugar-based inhibitors, relevant inhibitory results have been achieved by the GlcNAc-derivative **15**, showing a nanomolar activity for MMP-12, an interesting specificity as well as a good water solubility. Among azasugar-based inhibitors, the phenoxyphenyl derivative **49** is one of the most promising compounds, with high activity and selectivity for MMP-1, MMP-3, and MMP-9 and good inhibitory results in different biological models used to evaluate its antiangiogenic activity. On the basis of the published results here presented, carbohydrate-based inhibitors resulted an emerging class of potent, selective, and water-soluble metzincin inhibitors. Further in vitro and in vivo studies should be performed to finally prove the potentialities of this class of compounds.

Author Contributions: Writing—original draft preparation, D.C. and F.D.; writing—review and editing, E.N.; funding acquisition, A.R. All authors have read and agreed to the published version of the manuscript.

Funding: This research was funded by University of Pisa (PRA_2018_20).

Conflicts of Interest: The authors declare no conflict of interest.

References

1. Bode, W.; Gomis-Rüth, F.-X.; Stöckler, W. Astacins, serralysins, snake venom and matrix metalloproteinases exhibit identical zinc-binding environments (HEXXHXXGXXH and Met-turn) and topologies and should be grouped into a common family, the 'metzincins'. *FEBS Lett.* **1993**, *331*, 134–140. [CrossRef]
2. Tallant, C.; García-Castellanos, R.; Baumann, U.; Gomis-Rüth, F.X. On the Relevance of the Met-turn Methionine in Metzincins. *J. Biol. Chem.* **2010**, *285*, 13951–13957. [CrossRef] [PubMed]
3. Vandenbroucke, R.E.; Libert, C. Is there new hope for therapeutic matrix metalloproteinase inhibition? *Nat. Rev. Drug Discov.* **2014**, *13*, 904–927. [CrossRef] [PubMed]
4. Murphy, G.; Nagase, H. Progress in matrix metalloproteinase research. *Mol. Asp. Med.* **2008**, *29*, 290–308. [CrossRef] [PubMed]
5. Maskos, K. Crystal structures of MMPs in complex with physiological and pharmacological inhibitors. *Biochimie* **2005**, *87*, 249–263. [CrossRef]
6. Ogata, H.; Decaneto, E.; Grossman, M.; Havenith, M.; Sagi, I.; Lubitz, W.; Knipp, M. Crystallization and preliminary X-ray crystallographic analysis of the catalytic domain of membrane type 1 matrix metalloproteinase. *Acta Crystallogr. Sect. F Struct. Biol. Commun.* **2014**, *70*, 232–235. [CrossRef]
7. Tallant, C.; Marrero, A.; Gomis-Rüth, F.X. Matrix metalloproteinases: Fold and function of their catalytic domains. *Biochim. Biophys. Acta Bioenerg.* **2010**, *1803*, 20–28. [CrossRef]
8. Bertini, I.; Calderone, V.; Cosenza, M.; Fragai, M.; Lee, Y.-M.; Luchinat, C.; Mangani, S.; Terni, B.; Turano, P. Conformational variability of matrix metalloproteinases: Beyond a single 3D structure. *Proc. Natl. Acad. Sci. USA* **2005**, *102*, 5334–5339. [CrossRef]
9. Page-McCaw, A.; Ewald, A.J.; Werb, Z. Matrix metalloproteinases and the regulation of tissue remodelling. *Nat. Rev. Mol. Cell Biol.* **2007**, *8*, 221–233. [CrossRef]
10. Schechter, E.; Berger, A. On the size of the active site in proteases: Pronase. *Biochem. Biophys. Res. Commun.* **1972**, *46*, 1956–1960. [CrossRef]
11. Gupta, S.P.; Patil, V.M. Specificity of Binding with Matrix Metalloproteinases. *Exp. Suppl.* **2012**, *103*, 35–56. [CrossRef] [PubMed]
12. Gimeno, A.; Beltrán-Debón, R.; Mulero, M.; Pujadas, G.; Garcia-Vallvé, S. Understanding the variability of the S1′ pocket to improve matrix metalloproteinase inhibitor selectivity profiles. *Drug Discov. Today* **2020**, *25*, 38–57. [CrossRef] [PubMed]
13. Edwards, D.R.; Handsley, M.M.; Pennington, C.J. The ADAM metalloproteinases. *Mol. Asp. Med.* **2008**, *29*, 258–289. [CrossRef]
14. Giebeler, N.; Zigrino, P. A Disintegrin and Metalloprotease (ADAM): Historical Overview of Their Functions. *Toxins* **2016**, *8*, 122. [CrossRef] [PubMed]
15. Seals, D.F.; Courtneidge, S.A. The ADAMs family of metalloproteases: Multidomain proteins with multiple functions. *Genes Dev.* **2003**, *17*, 7–30. [CrossRef] [PubMed]
16. Hall, T.; Shieh, H.-S.; Day, J.E.; Caspers, N.; Chrencik, J.E.; Williams, J.M.; Pegg, L.E.; Pauley, A.M., Moon, A.F.; Krahn, J.M.; et al. Structure of human ADAM-8 catalytic domain complexed with batimastat. *Acta Crystallogr. Sect. F Struct. Biol. Cryst. Commun.* **2012**, *68*, 616–621. [CrossRef]
17. Seegar, T.C.M.; Killingsworth, L.B.; Saha, N.; Meyer, P.A.; Patra, D.; Zimmerman, B.; Janes, P.W.; Rubinstein, E.; Nikolov, D.B.; Skiniotis, G.; et al. Structural Basis for Regulated Proteolysis by the α-Secretase ADAM10. *Cell* **2017**, *171*, 1638–1648. [CrossRef]
18. Maskos, K.; Fernandez-Catalan, C.; Huber, R.; Bourenkov, G.P.; Bartunik, H.; Ellestad, G.A.; Reddy, P.; Wolfson, M.F.; Rauch, C.T.; Castner, B.J.; et al. Crystal structure of the catalytic domain of human tumor necrosis factor-α-converting enzyme. *Proc. Natl. Acad. Sci. USA* **1998**, *95*, 3408–3412. [CrossRef]
19. Orth, P.; Reichert, P.; Wang, W.; Prosise, W.W.; Yarosh-Tomaine, T.; Hammond, G.; Ingram, R.N.; Xiao, L.; Mirza, U.A.; Zou, J.; et al. Crystal Structure of the Catalytic Domain of Human ADAM33. *J. Mol. Biol.* **2004**, *335*, 129–137. [CrossRef]
20. Stone, A.L.; Kroeger, M.; Sang, Q.X.A. Structure–Function Analysis of the ADAM Family of Disintegrin-Like and Metalloproteinase-Containing Proteins (Review). *J. Protein Chem.* **1999**, *18*, 447–465. [CrossRef]
21. Weber, S.; Saftig, P. Ectodomain shedding and ADAMs in development. *Development* **2012**, *139*, 3693–3709. [CrossRef] [PubMed]

22. Murphy, G. The ADAMs: Signalling scissors in the tumour microenvironment. *Nat. Rev. Cancer* **2008**, *8*, 929–941. [CrossRef] [PubMed]

23. Boucher, B.J. Matrix metalloproteinase protein inhibitors: Highlighting a new beginning for metalloproteinases in medicine. *Met. Med.* **2016**, *3*, 75–79. [CrossRef]

24. Winer, A.; Adams, S.; Mignatti, P. Matrix Metalloproteinase Inhibitors in Cancer Therapy: Turning Past Failures into Future Successes. *Mol. Cancer Ther.* **2018**, *17*, 1147–1155. [CrossRef]

25. Nissinen, L.; Kähäri, V.-M. Matrix metalloproteinases in inflammation. *Biochim. Biophys. Acta Gen. Subj.* **2014**, *1840*, 2571–2580. [CrossRef]

26. Hu, J.; Van den Steen, P.E.; Sang, Q.X.A.; Opdenakker, G. Matrix metalloproteinase inhibitors as therapy for inflammatory and vascular diseases. *Nat. Rev. Drug Discov.* **2007**, *6*, 480–498. [CrossRef]

27. Muri, L.; Leppert, D.; Grandgirard, D.; Leib, S.L. MMPs and ADAMs in neurological infectious diseases and multiple sclerosis. *Cell. Mol. Life Sci.* **2019**, *76*, 3097–3116. [CrossRef]

28. Mishra, H.K.; Ma, J.; Walcheck, B. Ectodomain Shedding by ADAM17: Its Role in Neutrophil Recruitment and the Impairment of This Process during Sepsis. *Front. Cell. Infect. Microbiol.* **2017**, *7*, 138. [CrossRef]

29. Rivera, S.; García-González, L.; Khrestchatisky, M.; Baranger, K. Metalloproteinases and their tissue inhibitors in Alzheimer's disease and other neurodegenerative disorders. *Cell. Mol. Life Sci.* **2019**, *76*, 3167–3191. [CrossRef]

30. Marcello, E.; Borroni, B.; Pelucchi, S.; Gardoni, F.; Di Luca, M. ADAM10 as a therapeutic target for brain diseases: From developmental disorders to Alzheimer's disease. *Expert Opin. Ther. Targets* **2017**, *21*, 1017–1026. [CrossRef]

31. Qian, M.; Shen, X.; Wang, H. The Distinct Role of ADAM17 in APP Proteolysis and Microglial Activation Related to Alzheimer's Disease. *Cell. Mol. Neurobiol.* **2016**, *36*, 471–482. [CrossRef] [PubMed]

32. Fischer, T.; Senn, N.; Riedl, R. Design and Structural Evolution of Matrix Metalloproteinase Inhibitors. *Chem. Eur. J.* **2019**, *25*, 7960–7980. [CrossRef] [PubMed]

33. Fisher, J.F.; Mobashery, S. Recent advances in MMP inhibitor design. *Cancer Metastasis Rev.* **2006**, *25*, 115–136. [CrossRef] [PubMed]

34. Verma, R.P. Hydroxamic Acids as Matrix Metalloproteinase Inhibitors. *Exp. Suppl.* **2012**, *103*, 137–176. [CrossRef]

35. Whittaker, M.; Floyd, C.D.; Brown, P.; Gearing, A.J.H. Design and Therapeutic Application of Matrix Metalloproteinase Inhibitors. *Chem. Rev.* **1999**, *99*, 2735–2776. [CrossRef]

36. MacPherson, L.J.; Bayburt, E.K.; Capparelli, M.P.; Carroll, B.J.; Goldstein, R.; Justice, M.R.; Zhu, L.; Hu, S.-I.; Melton, R.A.; Fryer, L.; et al. Discovery of CGS 27023A, a Non-Peptidic, Potent, and Orally Active Stromelysin Inhibitor That Blocks Cartilage Degradation in Rabbits. *J. Med. Chem.* **1997**, *40*, 2525–2532. [CrossRef]

37. Bertini, I.; Calderone, V.; Fragai, M.; Luchinat, C.; Mangani, S.; Terni, B. Crystal Structure of the Catalytic Domain of Human Matrix Metalloproteinase 10. *J. Mol. Biol.* **2004**, *336*, 707–716. [CrossRef]

38. Rossello, A.; Nuti, E.; Orlandini, E.; Carelli, P.; Rapposelli, S.; Macchia, M.; Minutolo, F.; Carbonaro, L.; Albini, A.; Benelli, R.; et al. New N-arylsulfonyl-N-alkoxyaminoacetohydroxamic acids as selective inhibitors of gelatinase A (MMP-2). *Bioorganic Med. Chem.* **2004**, *12*, 2441–2450. [CrossRef]

39. Levin, M.; Udi, Y.; Solomonov, I.; Sagi, I. Next generation matrix metalloproteinase inhibitors—Novel strategies bring new prospects. *Biochim. Biophys. Acta Bioenerg.* **2017**, *1864*, 1927–1939. [CrossRef]

40. Renkiewicz, R.; Qiu, L.; Lesch, C.; Sun, X.; Devalaraja, R.; Cody, T.; Kaldjian, E.; Welgus, H.; Baragi, V. Broad-spectrum matrix metalloproteinase inhibitor marimastat-induced musculoskeletal side effects in rats. *Arthritis Rheum.* **2003**, *48*, 1742–1749. [CrossRef]

41. Senn, N.; Ott, M.; Lanz, J.; Riedl, R. Targeted Polypharmacology: Discovery of a Highly Potent Non-Hydroxamate Dual Matrix Metalloproteinase (MMP)-10/-13 Inhibitor. *J. Med. Chem.* **2017**, *60*, 9585–9598. [CrossRef] [PubMed]

42. Nara, H.; Sato, K.; Naito, T.; Mototani, H.; Oki, H.; Yamamoto, Y.; Kuno, H.; Santou, T.; Kanzaki, N.; Terauchi, J.; et al. Discovery of Novel, Highly Potent, and Selective Quinazoline-2-carboxamide-Based Matrix Metalloproteinase (MMP)-13 Inhibitors without a Zinc Binding Group Using a Structure-Based Design Approach. *J. Med. Chem.* **2014**, *57*, 8886–8902. [CrossRef] [PubMed]

43. Choi, J.Y.; Fuerst, R.; Knapinska, A.M.; Taylor, A.B.; Smith, L.; Cao, X.; Hart, P.J.; Fields, G.B.; Roush, W.R. Structure-Based Design and Synthesis of Potent and Selective Matrix Metalloproteinase 13 Inhibitors. *J. Med. Chem.* **2017**, *60*, 5816–5825. [CrossRef] [PubMed]

44. Dufour, A.; Sampson, N.S.; Li, J.; Kuscu, C.; Rizzo, R.C.; DeLeon, J.L.; Zhi, J.; Jaber, N.; Liu, E.; Zucker, S.; et al. Small-Molecule Anticancer Compounds Selectively Target the Hemopexin Domain of Matrix Metalloproteinase-9. *Cancer Res.* **2011**, *71*, 4977–4988. [CrossRef] [PubMed]

45. Remacle, A.G.; Golubkov, V.S.; Shiryaev, S.A.; Dahl, R.; Stebbins, J.L.; Chernov, A.V.; Cheltsov, A.V.; Pellecchia, M.; Strongin, A.Y. Novel MT1-MMP Small-Molecule Inhibitors Based on Insights into Hemopexin Domain Function in Tumor Growth. *Cancer Res.* **2012**, *72*, 2339–2349. [CrossRef]

46. Spicer, T.P.; Jiang, J.; Taylor, A.B.; Choi, J.Y.; Hart, P.J.; Roush, W.R.; Fields, G.B.; Hodder, P.S.; Minond, D. Characterization of Selective Exosite-Binding Inhibitors of Matrix Metalloproteinase 13 That Prevent Articular Cartilage Degradation In Vitro. *J. Med. Chem.* **2014**, *57*, 9598–9611. [CrossRef]

47. Saftig, P.; Reiss, K. The "A Disintegrin and Metalloproteases" ADAM10 and ADAM17: Novel drug targets with therapeutic potential? *Eur. J. Cell Biol.* **2011**, *90*, 527–535. [CrossRef]

48. Zunke, F.; Rose-John, S. The shedding protease ADAM17: Physiology and pathophysiology. *Biochim. Biophys. Acta Bioenerg.* **2017**, *1864*, 2059–2070. [CrossRef]

49. Fridman, J.S.; Caulder, E.; Hansbury, M.; Liu, X.; Yang, G.; Wang, Q.; Lo, Y.; Zhou, B.-B.S.; Pan, M.; Thomas, S.M.; et al. Selective Inhibition of ADAM Metalloproteases as a Novel Approach for Modulating ErbB Pathways in Cancer. *Clin. Cancer Res.* **2007**, *13*, 1892–1902. [CrossRef]

50. Zhou, B.-B.S.; Peyton, M.; He, B.; Liu, C.; Girard, L.; Caudler, E.; Lo, Y.; Baribaud, F.; Mikami, I.; Reguart, N.; et al. Targeting ADAM-mediated ligand cleavage to inhibit HER3 and EGFR pathways in non-small cell lung cancer. *Cancer Cell* **2006**, *10*, 39–50. [CrossRef]

51. Nuti, E.; Casalini, F.; Santamaria, S.; Fabbi, M.; Carbotti, G.; Ferrini, S.; Marinelli, L.; La Pietra, V.; Novellino, E.; Camodeca, C.; et al. Selective Arylsulfonamide Inhibitors of ADAM-17: Hit Optimization and Activity in Ovarian Cancer Cell Models. *J. Med. Chem.* **2013**, *56*, 8089–8103. [CrossRef] [PubMed]

52. Minond, D.; Cudic, M.; Bionda, N.; Giulianotti, M.; Maida, L.; Houghten, R.A.; Fields, G.B. Discovery of Novel Inhibitors of a Disintegrin and Metalloprotease 17 (ADAM17) Using Glycosylated and Non-glycosylated Substrates. *J. Biol. Chem.* **2012**, *287*, 36473–36487. [CrossRef] [PubMed]

53. Knapinska, A.M.; Dreymueller, D.; Ludwig, A.; Smith, L.; Golubkov, V.; Sohail, A.; Fridman, R.; Giulianotti, M.; LaVoi, T.M.; Houghten, R.A.; et al. SAR Studies of Exosite-Binding Substrate-Selective Inhibitors of A Disintegrin And Metalloprotease 17 (ADAM17) and Application as Selective in Vitro Probes. *J. Med. Chem.* **2015**, *58*, 5808–5824. [CrossRef] [PubMed]

54. Nuti, E.; Tuccinardi, T.; Rossello, A. Matrix Metalloproteinase Inhibitors: New Challenges in the Era of Post Broad-Spectrum Inhibitors. *Curr. Pharm. Des.* **2007**, *13*, 2087–2100. [CrossRef] [PubMed]

55. Camodeca, C.; Cuffaro, D.; Nuti, E.; Rossello, A. ADAM Metalloproteinases as Potential Drug Targets. *Curr. Med. Chem.* **2019**, *26*, 2661–2689. [CrossRef]

56. Digilio, G.; Tuccinardi, T.; Casalini, F.; Cassino, C.; Dias, D.M.; Geraldes, C.F.G.C.; Catanzaro, V.; Maiocchi, A.; Rossello, A. Study of the binding interaction between fluorinated matrix metalloproteinase inhibitors and Human Serum Albumin. *Eur. J. Med. Chem.* **2014**, *79*, 13–23. [CrossRef]

57. Gabius, H.-J. *The Sugar Code. Fundamentals of Glycosciences*; Wiley-VCH: Weinheim, Germany, 2009.

58. Ernst, B.; Hart, G.W.; Sinaý, P. *Carbohydrates in Chemistry and Biology*; Wiley-VCH: Weinheim, Germany, 2000.

59. Mullapudi, S.S.; Mitra, D.; Li, M.; Kang, E.-T.; Chiong, E.; Neoh, K.G. Potentiating anti-cancer chemotherapeutics and antimicrobials via sugar-mediated strategies. *Mol. Syst. Des. Eng.* **2020**, *5*, 772–791. [CrossRef]

60. Micoli, F.; Costantino, P.; Adamo, R. Potential targets for next generation antimicrobial glycoconjugate vaccines. *FEMS Microbiol. Rev.* **2018**, *42*, 388–423. [CrossRef]

61. Khatun, F.; Toth, I.; Stephenson, R.J. Immunology of carbohydrate-based vaccines. *Adv. Drug Deliv. Rev.* **2020**. [CrossRef]

62. Mettu, R.; Chen, C.-Y.; Wu, C.-Y. Synthetic carbohydrate-based vaccines: Challenges and opportunities. *J. Biomed. Sci.* **2020**, *27*, 1–22. [CrossRef]

63. Gragnani, T.; Cuffaro, D.; Fallarini, S.; Lombardi, G.; D'Andrea, F.; Guazzelli, L. Selectively Charged and Zwitterionic Analogues of the Smallest Immunogenic Structure of Streptococcus Pneumoniae Type 14. *Molecules* **2019**, *24*, 3414. [CrossRef] [PubMed]

64. Hossain, F.; Andreana, P.R. Developments in Carbohydrate-Based Cancer Therapeutics. *Pharmaceuticals* **2019**, *12*, 84. [CrossRef] [PubMed]

65. D'Andrea, F.; Vagelli, G.; Granchi, C.; Guazzelli, L.; Tuccinardi, T.; Poli, G.; Iacopini, D.; Minutolo, F.; Di Bussolo, V. Synthesis and Biological Evaluation of New Glycoconjugated LDH Inhibitors as Anticancer Agents. *Molecules* **2019**, *24*, 3520. [CrossRef]

66. Drinnan, N.B.; Vari, F. Aspects of the stability and bioavailability of carbohydrates and carbohydrate derivatives. *Mini Revi. Med. Chem.* **2003**, *3*, 633–649. [CrossRef] [PubMed]

67. Macmillan, D.; Daines, A.M. Recent Developments in the Synthesis and Discovery of Oligosaccharides and Glycoconjugates for the Treatment of Disease. *Curr. Med. Chem.* **2003**, *10*, 2733–2773. [CrossRef] [PubMed]

68. Makuch, S.; Woźniak, M.; Krawczyk, M.; Pastuch-Gawołek, G.; Szeja, W.; Agrawal, S. Glycoconjugation as a Promising Treatment Strategy for Psoriasis. *J. Pharmacol. Exp. Ther.* **2020**, *373*, 204–212. [CrossRef]

69. Molejon, M.I.; Weiz, G.; Breccia, J.D.; Vaccaro, M.I. Glycoconjugation: An approach to cancer therapeutics. *World J. Clin. Oncol.* **2020**, *11*, 110–120. [CrossRef]

70. Cuffaro, D.; Nuti, E.; Rossello, A. An overview of carbohydrate-based Carbonic Anhydrase inhibitors. *J. Enzyme Inhib. Med. Chem.* **2020**. [CrossRef]

71. Nativi, C.; Richichi, B.; Roelens, S. Carbohydrate-containing matrix metalloproteinase inhibitors. In *Carbohydrates in Drug Design and Discovery*; Jimenez-Barbero, J., Canada, F.J., Martin-Santamaria, S., Eds.; RSC Drug Discovery: Cambridge, UK, 2015.

72. Fragai, M.; Nativi, C.; Richichi, B.; Venturi, C. Design In Silico, Synthesis and Binding Evaluation of a Carbohydrate-Based Scaffold for Structurally Novel Inhibitors of Matrix Metalloproteinases. *ChemBioChem* **2005**, *6*, 1345–1349. [CrossRef]

73. Calderone, V.; Fragai, M.; Luchinat, C.; Nativi, C.; Richichi, B.; Roelens, S. A High-Affinity Carbohydrate-Containing Inhibitor of Matrix Metalloproteinases. *ChemMedChem* **2006**, *1*, 598–601. [CrossRef]

74. Bartoloni, M.; Domínguez, B.E.; Dragoni, E.; Richichi, B.; Fragai, M.; André, S.; Gabius, H.-J.; Ardá, A.; Luchinat, C.; Jiménez-Barbero, J.; et al. Targeting Matrix Metalloproteinases: Design of a Bifunctional Inhibitor for Presentation by Tumour-Associated Galectins. *Chem. A Eur. J.* **2012**, *19*, 1896–1902. [CrossRef] [PubMed]

75. Hugenberg, V.; Breyholz, H.-J.; Riemann, B.; Hermann, S.; Schober, O.; Schäfers, M.; Gangadharmath, U.; Mocharla, V.; Kolb, H.; Walsh, J.; et al. A New Class of Highly Potent Matrix Metalloproteinase Inhibitors Based on Triazole-Substituted Hydroxamates: (Radio)Synthesis and in Vitro and First in Vivo Evaluation. *J. Med. Chem.* **2012**, *55*, 4714–4727. [CrossRef] [PubMed]

76. Nuti, E.; Cuffaro, D.; D'Andrea, F.; Rosalia, L.; Tepshi, L.; Fabbi, M.; Carbotti, G.; Ferrini, S.; Santamaria, S.; Camodeca, C.; et al. Sugar-Based Arylsulfonamide Carboxylates as Selective and Water-Soluble Matrix Metalloproteinase-12 Inhibitors. *ChemMedChem* **2016**, *11*, 1626–1637. [CrossRef] [PubMed]

77. Cuffaro, D.; Camodeca, C.; D'Andrea, F.; Piragine, E.; Testai, L.; Calderone, V.; Orlandini, E.; Nuti, E.; Rossello, A. Matrix metalloproteinase-12 inhibitors: Synthesis, structure-activity relationships and intestinal absorption of novel sugar-based biphenylsulfonamide carboxylates. *Bioorganic Med. Chem.* **2018**, *26*, 5804–5815. [CrossRef]

78. Santamaria, S.; Nuti, F.; Cercignani, G.; Marinelli, L.; La Pietra, V.; Novellino, E.; Rossello, A. N-O-Isopropyl sulfonamido-based hydroxamates: Kinetic characterisation of a series of MMP-12/MMP-13 dual target inhibitors. *Biochem. Pharmacol.* **2012**, *84*, 813–820. [CrossRef]

79. Nuti, E.; Cuffaro, D.; Bernardini, E.; Camodeca, C.; Panelli, L.; Chaves, S.; Ciccone, L.; Tepshi, L.; Vera, L.; Orlandini, E.; et al. Development of Thioaryl-Based Matrix Metalloproteinase-12 Inhibitors with Alternative Zinc-Binding Groups: Synthesis, Potentiometric, NMR, and Crystallographic Studies. *J. Med. Chem.* **2018**, *61*, 4421–4435. [CrossRef]

80. D'Andrea, F.; Nuti, E.; Becherini, S.; Cuffaro, D.; Husanu, E.; Camodeca, C.; De Vita, E.; Zocchi, M.R.; Poggi, A.; D'Arrigo, C.; et al. Design and Synthesis of Ionic Liquid-Based Matrix Metalloproteinase Inhibitors (MMPIs): A Simple Approach to Increase Hydrophilicity and to Develop MMPI-Coated Gold Nanoparticles. *ChemMedChem* **2019**, *14*, 686–698. [CrossRef]

81. Nuti, E.; Panelli, L.; Casalini, F.; Avramova, S.I.; Orlandini, E.; Santamaria, S.; Nencetti, S.; Tuccinardi, T.; Martinelli, A.; Cercignani, G.; et al. Design, Synthesis, Biological Evaluation, and NMR Studies of a New Series of Arylsulfones As Selective and Potent Matrix Metalloproteinase-12 Inhibitors. *J. Med. Chem.* **2009**, *52*, 6347–6361. [CrossRef]

82. Antoni, C.; Vera, L.; Devel, L.; Catalani, M.P.; Czarny, B.; Cassar-Lajeunesse, E.; Nuti, E.; Rossello, A.; Dive, V.; Stura, E.A. Crystallization of bi-functional ligand protein complexes. *J. Struct. Biol.* **2013**, *182*, 246–254. [CrossRef]

83. Thorens, B.; Mueckler, M. Glucose transporters in the 21st Century. *Am. J. Physiol. Endocrinol. Metab.* **2010**, *298*, E141–E145. [CrossRef]

84. Ponedel'kina, I.Y.; Gaskarova, A.R.; Khaybrakhmanova, E.A.; Lukina, E.S.; Odinokov, V.N. Hyaluronic acid based hydroxamate and conjugates with biologically active amines: In vitro effect on matrix metalloproteinase-2. *Carbohydr. Polym.* **2016**, *144*, 17–24. [CrossRef] [PubMed]

85. Compain, P.; Martin, O.R. *Iminosugars: From Synthesis to Therapeutic Applications*; Wiley-VCH: Weinheim, Germany, 2007.

86. Iftikhar, M.; Wang, L.; Fang, Z. Synthesis of 1-Deoxynojirimycin: Exploration of Optimised Conditions for Reductive Amidation and Separation of Epimers. *J. Chem. Res.* **2017**, *41*, 460–464. [CrossRef]

87. Wood, A.; Prichard, K.L.; Clarke, Z.; Houston, T.A.; Fleet, G.W.J.; Simone, M.I. Synthetic Pathways to 3,4,5-Trihydroxypiperidines from the Chiral Pool. *Eur. J. Org. Chem.* **2018**, *2018*, 6812–6829. [CrossRef]

88. Cuffaro, D.; Landi, M.; D'Andrea, F.; Guazzelli, L. Preparation of 1,6-di-deoxy-d-galacto and 1,6-di-deoxy-l-altro nojirimycin derivatives by aminocyclization of a 1,5-dicarbonyl derivative. *Carbohydr. Res.* **2019**, *482*, 107744. [CrossRef]

89. Dehoux-Baudoin, C.; Génisson, Y. C -Branched Imino Sugars: Synthesis and Biological Relevance. *Eur. J. Org. Chem.* **2019**, *2019*, 4765–4777. [CrossRef]

90. Afarinkia, K.; Bahar, A. Recent advances in the chemistry of azapyranose sugars. *Tetrahedron Asymmetry* **2005**, *16*, 1239–1287. [CrossRef]

91. Guazzelli, L.; Catelani, G.; D'Andrea, F.; Gragnani, T.; Griselli, A. Stereoselective Access to the β-D-N-Acetylhexosaminyl-(1→4)-1-deoxy-D-nojirimycin Disaccharide Series Avoiding the Glycosylation Reaction. *Eur. J. Org. Chem.* **2014**, *2014*, 6527–6537. [CrossRef]

92. Guazzelli, L.; D'Andrea, F.; Sartini, S.; Giorgelli, F.; Confini, G.; Quattrini, L.; Piano, I.; Nencetti, S.; Orlandini, E.; Gargini, C.; et al. Synthesis and investigation of polyhydroxylated pyrrolidine derivatives as novel chemotypes showing dual activity as glucosidase and aldose reductase inhibitors. *Bioorganic Chem.* **2019**, *92*, 103298. [CrossRef]

93. Bojarová, P.; Křen, V. Glycosidases: A key to tailored carbohydrates. *Trends Biotechnol.* **2009**, *27*, 199–209. [CrossRef]

94. Horne, G.; Wilson, F.X.; Tinsley, J.; Williams, D.H.; Storer, R. Iminosugars past, present and future: Medicines for tomorrow. *Drug Discov. Today* **2011**, *16*, 107–118. [CrossRef]

95. Gao, K.; Zheng, C.; Wang, T.; Zhao, H.; Wang, J.; Wang, Z.; Zhai, X.; Jia, Z.; Chen, J.; Zhou, Y.; et al. 1-Deoxynojirimycin: Occurrence, Extraction, Chemistry, Oral Pharmacokinetics, Biological Activities and In Silico Target Fishing. *Molecules* **2016**, *21*, 1600. [CrossRef] [PubMed]

96. Junge, B.; Matzke, M.; Stoltefuss, J. Chemistry and Structure—Activity Relationships of Glucosidase Inhibitors. In *Handbook of Experimental Pharmacology*; Springer: New York, NY, USA, 1996.

97. Campbell, L.K.; Baker, D.E.; Campbell, R.K. Miglitol: Assessment of its role in the treatment of patients with diabetes mellitus. *Ann. Pharmacother.* **2000**, *34*, 1291–1301. [CrossRef] [PubMed]

98. Lee, A.; Patrick, P.; Wishart, J.; Horo-witz, M.; Morley, J.E. The effects of miglitol on glucagon-like peptide-1 secretion and appetite sensations in obese type 2 diabetics. *Diabetes Obes. Metab.* **2002**, *4*, 329–335. [CrossRef] [PubMed]

99. Giraldo, P.; Andrade-Campos, M.; Alfonso, P.; Irún, P.; Atutxa, K.; Acedo, A.; Barez, A.; Blanes, M.; Diaz-Morant, V.; Fernández-Galán, M.A.; et al. Twelve years of experience with miglustat in the treatment of type 1 Gaucher disease: The Spanish ZAGAL project. *Blood Cells Mol. Dis.* **2018**, *68*, 173–179. [CrossRef]

100. Wang, R.-J.; Yang, C.-H.; Hu, M.-L. 1-Deoxynojirimycin Inhibits Metastasis of B16F10 Melanoma Cells by Attenuating the Activity and Expression of Matrix Metalloproteinases-2 and -9 and Altering Cell Surface Glycosylation. *J. Agric. Food Chem.* **2010**, *58*, 8988–8993. [CrossRef]

101. Chan, K.-C.; Lin, M.-C.; Huang, C.-N.; Chang, W.-C.; Wang, C.-J. Mulberry 1-Deoxynojirimycin Pleiotropically Inhibits Glucose-Stimulated Vascular Smooth Muscle Cell Migration by Activation of AMPK/RhoB and Down-regulation of FAK. *J. Agric. Food Chem.* **2013**, *61*, 9867–9875. [CrossRef]

102. Moriyama, H.; Tsukida, T.; Inoue, Y.; Kondo, H.; Yoshino, K.; Nishimura, S.-I. Structure–activity relationships of azasugar-based MMP/ADAM inhibitors. *Bioorganic Med. Chem. Lett.* **2003**, *13*, 2737–2740. [CrossRef]

103. Moriyama, H.; Tsukida, T.; Inoue, Y.; Kondo, H.; Yoshino, K.; Nishimura, S. Design, synthesis and evaluation of novel azasugar-based MMP/ADAM inhibitors. *Bioorganic Med. Chem. Lett.* **2003**, *13*, 2741–2744. [CrossRef]

104. Moriyama, H.; Tsukida, T.; Inoue, Y.; Yokota, K.; Yoshino, K.; Kondo, H.; Miura, A.N.; Nishimura, S.-I. Azasugar-Based MMP/ADAM Inhibitors as Antipsoriatic Agents. *J. Med. Chem.* **2004**, *47*, 1930–1938. [CrossRef]

105. Chikaraishi, Y.; Shimazawa, M.; Yokota, K.; Yoshino, K.; Hara, H. CB-12181, a New Azasugar-Based Matrix Metalloproteinase/Tumor Necrosis Factor-α Converting Enzyme Inhibitor, Inhibits Vascular Endothelial Growth Factor-Induced Angiogenesis in Vitro and Retinal Neovascularization in Vivo. *Curr. Neurovascular Res.* **2009**, *6*, 140–147. [CrossRef]

106. Sylte, I.; Dawadi, R.; Malla, N.; Von Hofsten, S.; Nguyen, T.-M.; Solli, A.I.; Berg, E.; Adekoya, O.A.; Svineng, G.; Winberg, J.-O. The selectivity of galardin and an azasugar-based hydroxamate compound for human matrix metalloproteases and bacterial metalloproteases. *PLoS ONE* **2018**, *13*, e0200237. [CrossRef] [PubMed]

107. Tsukida, T.; Moriyama, H.; Inoue, Y.; Kondo, H.; Yoshino, K.; Nishimura, S. Synthesis and biological activity of selective azasugar-based TACE inhibitors. *Bioorganic Med. Chem. Lett.* **2004**, *14*, 1569–1572. [CrossRef] [PubMed]

Publisher's Note: MDPI stays neutral with regard to jurisdictional claims in published maps and institutional affiliations.

pharmaceuticals

Review

The Pharmacological TAILS of Matrix Metalloproteinases and Their Inhibitors

Nabangshu Das [1,2,†], Colette Benko [2,3,†], Sean E. Gill [4,5] and Antoine Dufour [1,2,3,*]

1   Faculty of Kinesiology, University of Calgary, Calgary, AB T2N 4N1, Canada; nabangshu.das@ucalgary.ca
2   McCaig Institute for Bone and Join Healthy, 3280 Hospital Drive NW, Calgary, AB T2N 4Z6, Canada; cbenko3@uwo.ca
3   Department of Physiology and Pharmacology, Cumming School of Medicine, Foothills Hospital, 3330 Hospital Dr, Calgary, AB T2N 4N1, Canada
4   Centre for Critical Illness Research, Victoria Research Labs, Lawson Health Research Institute, A6-134, London, ON N6A 5W9, Canada; sgill8@uwo.ca
5   Division of Respirology, Department of Medicine, Western University, London, ON N6A 5W9, Canada
*   Correspondence: antoine.dufour@ucalgary.ca
†   These authors contributed equally.

Abstract: Matrix metalloproteinases (MMPs) have been demonstrated to have both detrimental and protective functions in inflammatory diseases. Several MMP inhibitors, with the exception of Periostat®, have failed in Phase III clinical trials. As an alternative strategy, recent efforts have been focussed on the development of more selective inhibitors or targeting other domains than their active sites through specific small molecule inhibitors or monoclonal antibodies. Here, we present some examples that aim to better understand the mechanisms of conformational changes/allosteric control of MMPs functions. In addition to MMP inhibitors, we discuss unbiased global approaches, such as proteomics and N-terminomics, to identify new MMP substrates. We present some examples of new MMP substrates and their implications in regulating biological functions. By characterizing the roles and substrates of individual MMP, MMP inhibitors could be utilized more effectively in the optimal disease context or in diseases never tested before where MMP activity is elevated and contributing to disease progression.

Keywords: matrix metalloproteinases (MMPs), protease; tissue inhibitors of metalloproteinases (TIMPs), exosite; small molecule inhibitors; monoclonal antibodies; proteomics; N-terminomics; terminal amine isotopic labeling of substrates (TAILS)

check for updates

Citation: Das, N.; Benko, C.; Gill, S.E.; Dufour, A. The Pharmacological TAILS of Matrix Metalloproteinases and Their Inhibitors. *Pharmaceuticals* **2021**, *14*, 31. https://doi.org/10.3390/ph14010031

Received: 5 December 2020
Accepted: 28 December 2020
Published: 31 December 2020

Publisher's Note: MDPI stays neutral with regard to jurisdictional claims in published maps and institutional affiliations.

1. Introduction

Matrix metalloproteinases (MMPs) are zinc-dependent proteases that have been extensively studied in the context of extracellular matrix (ECM) breakdown and remodelling [1]. Increasingly, non-ECM substrates are being investigated for MMPs as ECM substrates only account for approximately 30% of all known MMP substrates [2,3]. The dysregulation of MMPs, their substrates, and the tissue inhibitor of metalloproteinases (TIMPs) often results in the progression of numerous diseases [1,3,4]. Various MMPs have been implicated in multiple cancers including pancreas, brain, lung, prostate, breast, skin and gastrointestinal tract [3,5]. MMP12 has been studied in chronic obstructive pulmonary disease (COPD) and the minor allele of a single nucleotide polymorphism in MMP12 (rs2276109) was associated with a beneficial effect on lung function in smokers and children with asthma [6,7]. Multiple MMPs have been investigated in rheumatoid arthritis and osteoarthritis yet the precise functions of individual MMP remains to be better characterized (reviewed in [8]). MMPs have also been studied in context of periodontal diseases [9,10]. It is not surprising that MMP inhibitors were tested in clinical trials. However, to date, the only MMP inhibitor that is currently approved is Periostat® (doxycycline hyclate), which is used for treating periodontitis (Figure 1a). Despite their biological roles in multiple cancers, in addition to inflammatory and autoimmune diseases, most MMP inhibitors failed due to a combination of factors including poor study design, a lack of understanding of biological roles of MMPs and the substrates they cleave, and the lack of specific inhibitors [1,3,5,11]. The structures and amino acid sequence of the catalytic domain of the 23 MMPs are highly conserved, which initially resulted in the design of broad spectrum MMP inhibitors. MMPs, however, have both detrimental and protective functions, limiting the use of these broad-spectrum inhibitors and increasing the complexity of developing MMP inhibitors to treat human diseases.

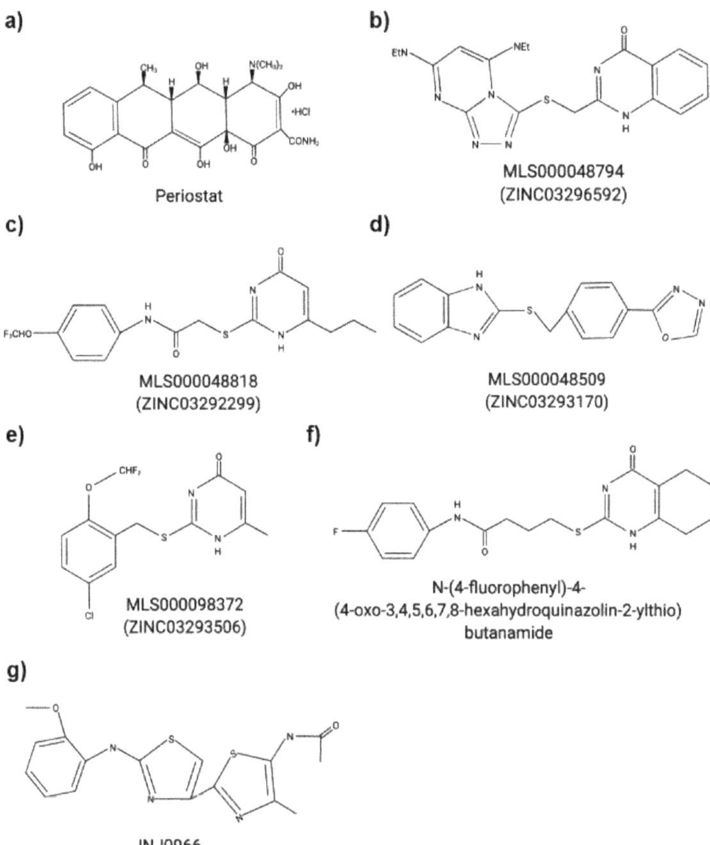

Figure 1. Chemical structures of small molecule MMP inhibitors. (**a**) Periostat®: doxycycline hyclate. (**b**) MLS000048794: 2-[[3,5bis(ethylamino)-2,4,6,8,9-pentazabicyclo[4 .3.0]nona-2,4,7,9-tetraen-7yl] sulfanyl-methyl]-1H-quinazolin4-one). (**c**) MLS000048818: *N*-[4-(difluoromethoxy)phenyl]-2-[(4-oxo- 6-propyl-1H-pyrimidin-2-yl)sulfanyl]-acetamide. (**d**) MLS000048509: 2-[[4-(1,3,4-oxadiazol-2-yl)phenyl] methylsulfanyl]-1Hbenzoimidazole). (**e**) MLS000098372: 2-[[5-chloro-2 (difluoromethoxy) phenyl]methylsulfanyl]-6-methyl-1H-pyrimidin-4-one). (**f**) N-(4-fluorophenyl)-4-(4-oxo-3,4,5,6,7,8-hexahydroquinazolin-2-ylthio)butanamide. (**g**) JNJ0966: *N*-{2-[(2-methoxyphenyl)amino]-4′-methyl-4,5′-bi-1,3-thiazol-2′-yl}acetamide.

2. Regulation of MMP Activity

The catalytic activity of MMPs is tightly regulated by endogenous TIMPs [12]. TIMPs are secreted proteins that inhibit metalloproteinases [13] through the formation of 1:1 stoichiometric complexes [12]. The C-terminus of TIMPs interacts with the hemopexin like domain, found in all MMPs except MMP7 and MMP26, whereas the N-terminus interacts with the zinc ion within the catalytic domains of MMPs [14]. When an imbalance between MMPs and TIMPs occurs, it often results in inflammation and immune responses, as seen in many inflammatory diseases and cancers [15]. Therefore, the reestablishment of MMP-TIMP homeostasis is of pharmacological value and supports the need for the development of effective MMP inhibitors. Moreover, a better understanding of the biological functions of TIMPs is also needed to clarify their roles in human pathologies.

3. Non-Proteolytic Functions of MMPs

As demonstrated in previous clinical trials, broad spectrum targeting of the catalytic domain of MMPs is challenging. Thus, alternative methods for the inhibition of MMP functions have been investigated such as targeting exosites and ectosites. Not only would this potentially enable greater specificity between MMPs, but some exosites may have unique functions distinct from proteolysis. One example is the hemopexin (PEX) domain that contributes to protein-protein interactions and can initiate cell signalling and increased cell migration [16–18]. Since the amino acid sequences of the PEX domain across MMPs is more divergent and less conserved than the catalytic domain, the PEX domain is a potential site to target with inhibitors to increase selectivity. Interestingly, MMP7 and MMP26 do not contain a PEX domain, therefore not all MMPs require this domain. The MMP1 PEX domain is essential for binding to collagen and in the modulation of the triple helical structure of the substrate to allow access to the catalytic cleft [19]. Both the catalytic and PEX domain of MMP1 are necessary for the cooperative binding of triple helix collagen, demonstrating the importance of the PEX domain in substrate binding. Additionally, the conserved collagen residue P10 interacts with MMP1 via a hydrophobic pocket or exosite composed of Phe301, Ile271, and Arg27 within the PEX domain [19]. Further, when double mutants of Ile271Ala/Arg272Ala were generated, the collagenolytic function was significantly reduced. Thus, the inhibition of this hydrophobic pocket could potentially be a therapeutic approach to regulate MMP1 activity as it is important in not only the binding of triple helix collagen but in the processing of collagen [19]. The PEX domain of MMP12 plays a critical role in clearance of various bacteria such as *Staphylococcus aureus*, *Klebsiella pneumoniae*, *Escherichia coli*, and *Salmonella enteriditis* in the phagolysosome [18]. In *Mmp12*$^{-/-}$ mice, there was an increase in mortality at lower titer concentration when infected with *S. aureus* as compared with wild type mice [18]. Anti-bacterial properties of MMP12 were determined to be the result of disruption of the bacterial outer membrane by amino acids 344-363 in blade II of the PEX domain [18]. Conversely, the catalytic domain of MMP12 may contribute to the cleavage of bacterial toxins but did not demonstrate antibacterial properties against *S. aureus* α-toxins [18]. Therefore, a better characterization of the PEX domains of MMP's may reveal new exciting functions in other MMP's.

The PEX domain of MMPs is also implicated in homo-/hetero-dimerization and can form multimers [20]. The propeller structure of the PEX domain includes 4 blades composed of two alpha-helices and four beta strands [21]. In MMP9, a mutation in blade IV of the PEX domain resulted in a loss of homodimer formation [16]. Mutations in blade I of the MMP9 PEX domain resulted in a loss of interactions with the cell surface CD44 [16]. This interaction between the outer blade I of the MMP9 PEX domain and CD44 was shown to increase cell migration via the activation of epidermal growth factor receptor (EGFR) and downstream kinase signaling [16]. Peptides generated to mimic the outer beta strand of blade I or IV resulted in decreased levels of MMP9 dimers and also a reduction cell migration [16]. MMP9 can also increase angiogenesis [22]. Using an allosteric inhibitor to the PEX domain, Hariono et al. [22] demonstrated that inhibition of ECM proteolysis, which decreases the release of vascular endothelial growth factor (VEGF) from within the ECM, significantly reduces the binding of VEGF to its membrane receptor, and subsequently decreases angiogenesis.

The catalytic domain of membrane type 1-matrix metalloproteinase/MT1-MMP (MMP14) has been implicated in pro-tumorigenic functions by processing type I collagen, in addition to increasing cell migration, angiogenesis, and cell invasion [23–25]. The PEX domain of MT1-MMP also forms hetero- (with CD44) and homo-dimers via blades I and IV of the PEX domain, respectively [25]. Synthetic peptides mimicking the outermost strand motifs within the PEX domain (blades I and IV) of MT1-MMP were shown to specifically inhibit MT1-MMP-enhanced cell migration, although the ability to directly prevent MT1-MMP proteolytic activity was not shown [25]. The PEX domain contributes to the tumor promoting nature of MT1-MMP as tumour volume was significantly larger in cancer cells containing the PEX domain compared to those without [24]. MT1-MMP also contains transmembrane and cytoplasmic tail domains that have been shown to have distinct functions from the catalytic domain and could be targeted with inhibitors to interfere with the biological functions of MT1-MMP. Targeting the PEX domain of MMPs could provide non-competitive inhibition as compared with active site inhibition with broad-spectrum compounds [26]. Each MMP is likely to have unique exosites or "hotspots" that may be targeted individually due to divergence of their amino acid sequences, chemical potential and geometry [27]. However, the binding affinity of most exosites for substrate is typically low (10^{-6}–10^{-7} M) making it potentially challenging to design an effective drug against that site [28,29].

4. Strategies for the Development of Protease Inhibitors

Multiple MMP inhibitors were originally designed with a substrate-based peptide, resembling the structure of type I collagen where MMPs cleave, aimed to interact with the necessary zinc ion in the MMPs' active site [30]. This active site zinc ion is a required component of their catalytic site activity [31], coordinated by three histidine residues, and calcium ions, which stabilize conformation of the active protease [32]. Examples of chemical groups with zinc chelating agents used in the development of MMP inhibitors include hydroxamates, carboxylates, aminocarboxylates, phosphonate, and sulfhydryl groups [28,33]. Tables 1–3 provides a summary of MMP inhibitors that were tested in clinical trials and pre-clinical studies. One example is Batimastat, a peptidomimetic composed of a hydroxamate group, that was investigated for the treatment of breast cancer and was terminated in clinical trials due to its poor solubility and low oral bioavailability [34,35]. Later, a chemical analogue, marimastat, with improved oral bioavailability, was taken further along in clinical trials and was terminated due to musculoskeletal pain and lack of efficacy [34]. Another hydroxamate derivative, Prinomastat, was also unsuccessful in phase III clinical trials due to lack of efficacy in patients with late-stage disease [36]. While the use of MMP inhibitors in combination with traditional chemotherapy drugs was reported to improve adverse side effects, the chemotherapeutics, in turn, surprisingly lowered the therapeutic effects of the MMP inhibitors [37–40]. Despite termination of clinical trials for MMP inhibitors for the treatment of cancer and arthritis, another MMP inhibitor, Periostat® (CollaGenex Pharmaceuticals Inc.), was successfully approved for the treatment of periodontitis [41–43]. Periostat® is a synthetic tetracycline, (doxycycline hyclate) but its precise mechanism of action on MMPs activity remains unclarified [42]. Periostat®'s ability to bind to the calcified surfaces of tooth roots may potentiate its efficacy in periodontal disease [44]. The gradual release of doxycycline from teeth in active form also may contribute to increased exposure, the maintained effectiveness during the post-treatment period [42]. Periostat® also reduces the level of localized and systemic inflammatory mediators in osteopenic patients in addition to improving on the clinical measurements of periodontitis [45]. Additionally, it has showed therapeutic effects in multiple sclerosis and type II diabetes. In multiple sclerosis (MS), in a combination therapy with intramuscular interferon-β (IFNβ), oral doxycycline was found to be effective, safe and well-tolerated [46]. In this study, outcome measures included number of lesion changes, relapse rates, safety and tolerability of the combination therapy in patients with MS. Multiple parameters were recorded including the Expanded Disability Status Scale scores, MMP9 levels in the serum, and the transendothelial migration of monocytes exposed to serum from patients with relapsing-remitting multiple sclerosis (RRMS). The inhibitory effect of doxycycline was associated with decreased serum level of MMP9 and was found to be corelated with reduction in brain lesion activity as measured by gadolinium-enhancing lesion number change [46]. When serum from RRMS patients was incubated with monocytes, their transendothelial migration was significantly diminished. Importantly, in this study, the adverse effects were mild, and one out of fifteen patients relapsed. In another clinical trial with obese people with type II diabetes, doxycycline was tested over a 12-week timepoint resulting in decreased inflammation and improved insulin sensitivity [47]. This effect was associated with a decrease in C-reactive protein and myeloperoxidase comparing to the placebo; it also increased 3′- phosphoinositide kinase-1, protein kinase B, and glycogen synthase kinase 3 ß [47]. However, these clinical trials were performed only on a small number of patients and further studies on a larger number of patients are needed to further test their efficacy. The failure of all MMP inhibitors in clinical trials, with the exception of Periostat®, led to the investigation of the roles of MMPs beyond their recognized roles in ECM remodeling [5]. Recent studies using animal model of disease coupled with high-throughput methods for substrate discovery (will be further discussed in Section 8) have revealed important roles of MMPs in inflammation and viral/bacterial infections [3,11,18,48,49]. In fact, multiple MMPs, such as MMP2, -3, -8, -9, and -12, play important roles in maintaining tissue homeostasis and have been demonstrated to have protective effects (full list is reviewed in [3]). Therefore, the ideal MMP inhibitor should be able to interfere with detrimental MMPs while sparing the beneficial MMPs. The catalytic domains of MMPs are highly conserved, therefore, targeting other MMP domains, which are unique to a single MMP, represents an alternative method of MMP inhibitor design.

5. Small Molecule MMP Inhibitors

Strategies to inhibit MMP functions with small molecules has been explored (Table 2) [16,22,24,25,50]. Italicize in silico analysis of MMP9 in which molecular docking programs were utilized to map potential ligand binding sites in the PEX domain at the dimerization interface were successful at identifying compound that interfered with MMP9 homodimerization and blocked a downstream signaling pathway critical for MMP9 mediated cell migration and invasion [50]. These compounds (Figure 1b–e) spared MMP9′s proteolytic activity, and 'compound 2′ (Figure 1c) from their study significantly diminished the phosphorylation of extracellular signal-regulated kinase 1/2 (ERK1/2). In a xenograft model of metastatic breast cancer cells (MDA-MB-435) stably transfected with a green fluorescent protein (GFP), a small molecule exosite MMP9 inhibitor, 'compound 2′, significantly decreased the tumor size and reduced the number of lung metastasis [50]. A follow-up study [17] demonstrated that treatment with their newly identified compound (Figure 1f) disrupted MMP9 homodimerization and prevented association with α4β1 integrin, CD44 and decreased phosphorylation of Src and its downstream target proteins focal adhesion kinase and paxillin. In the in vivo model of chick chorioallantoic membrane, treatment with a PEX domain small molecule inhibitor resulted in a reduction of cancer cell invasion and angiogenesis [17]. Using previously published active compounds (Figure 1c,f), Hariono et al. [22] evaluated ten additional arylamide compounds. Using molecular dynamic simulations, the mechanism of MMP inhibition via the hemopexin domain of MMP9 was investigated. Two compounds, (3-bromo-N-(4-nitrophenyl)propanamide) and (3-bromo-N-{4-[(pyrimidine-2-yl)sulfamoyl]phenyl}propanamide), demonstrated significant inhibition and cytotoxicity against 4T1 murine breast cancer cells. Using a biochemical and structural screening workflow, the compound JNJ0966 (Figure 1g) was found to selectively impede activation of proMMP9 into its active form via an interaction with a structural pocket in proximity to the MMP9 zymogen cleavage site near Arg106 [29]. JNJ0966 was unable to interact with the active forms of MMP1, -2, -3, -9, or -14. In a mouse experimental autoimmune encephalomyelitis model, JNJ0966 reduced disease severity at a dose of 10 and 30 mg/kg and to the same levels of dexamethasone [29]. Italicize in silico analysis of MT1-MMP's PEX domain identified a potentially targetable site that is distinct from the dimerization interface and located at the center of the PEX domain [24]. Subsequent docking studies of a small molecule inhibitor led to identification of a novel PEX inhibitor which is selective for MT1-MMP as compared to MMP2. It did not show any toxicity or interference to catalytic activities including MT1-MMP mediated activation of MMP2. This compound was effective in attenuating cancer cell migration and reduced tumor volume in vivo [24]. Despite promising results, there are still only a few studies that have tested small molecule MMP inhibitors in animal models and additional work in characterizing these inhibitors is needed to better understand whether this could be a viable approach to inhibit MMP function.

6. MMP Inhibition Using Selective Antibody

Monoclonal antibodies have emerged as potential enzyme inhibitors with numerous examples demonstrating them to be effective [51–54]. Initially, antibody generation was deemed challenging for the active site of MMPs due to the instability during presentation and lack of surface accessibility of the catalytic metal-protein cleft [55]. Sela-Passwell et al. [55] used a synthetic metal-baring mimicry complex and were able to generate a response not only to the metal-protein cleft, but also to the enzyme surface of MMP2 and MMP9 with high specificity which resulted in the generation of the monoclonal antibody SDS3. In a dextran sodium sulfate (DSS)-colitis mouse model, SDS3 was demonstrated to prevent colonic inflammation, release of proinflammatory cytokines, and tissue damage [55]. The applicability of SDS3 towards colitis and inflammatory bowel disorders was reinforced through clustering of the labelled SDS3 antibody in the intestine of the mice 24 h after injection. Another MMP9 monoclonal antibody inhibitor, REGA-3G12, was demonstrated to be selective towards MMP9 through recognition of the N-amino terminal catalytic domain subsequently inhibiting its catalysis [56,57]. REGA-3G12 recognizes the Trp116-Lys214 motif domain part of the Phe107-Gly223 outside of the Zn^{2+} binding site of MMP9, a site exploited by multiple MMP inhibitors. Although REGA-3G12 targets the catalytic domain, it was shown to be selective to MMP9 but not MMP2, therefore demonstrating the potential of the catalytic domain to be used to generate selective MMP inhibitors. The selectivity of REGA-3G12 is likely due to the vast number of interactions with the carboxy-terminal of the catalytic domain. REGA-3G12 did not show significant binding to synthetic linear fragments of the epitopes recognized by REGA-3G12, demonstrating the antibody may recognize a conformation instead of a linear residue. MT1-MMP inhibition is another attractive anti-cancer target as it is highly expressed in breast cancer and contributes to migration, invasion, and neovascularization [58]. Using a phage library display approach, an MT1-MMP active site inhibitor antibody, DX-2400, was developed to selectively inhibit active MT1-MMP. DX-2400 was also shown to inhibit pro-MMP2 activation due to the inhibition of MT1-MMP and reduced breast cancer cell invasion [58]. DX-2400 exhibited promising therapeutic results in an MDA-MB-231 breast cancer xenograft tumor mouse model associated with a decrease in cell growth and vascularization. Conversely, non-metastatic MCF-7 breast cancer cell that do not express MT1-MMP, did not show decreased growth in a xenograft tumor mouse model [58]. DX-2400 also delayed metastasis, and when in combination with bevacizumab delayed breast cancer cells' tumor growth [58]. Similarly, when applied to a BT-474 xenograft tumor mouse model, DX-2400 in combination with paclitaxel resulted in tumour growth delay. Surprisingly, while this study demonstrated promising results related to using selective MT1-MMP inhibitors to treat breast cancer, DX-2400 was unsuccessful in clinical trials.

Selective inhibition of MT1-MMP was also demonstrated in a model of influenza infections that resulted in ECM dysregulation and increased susceptibility for a bacterial co-infection [48]. The selective, potent and allosteric MT1-MMP inhibitor, LEM-2/15, was generated via mouse immunization with a cyclic peptide from the sequence of the V-B loop (residues 218–233), which displays a unique sequence divergence within the MMP family members [59]. This loop of MT1-MMP is flexible and likely undergoes conformational changes when binding to LEM-2/15 causing a narrower substrate binding cleft and constraining the flexibility of the loop. Therefore, the Fab fragment of LEM-2/15 interacted with the MT1-MMP expressed on the cell surface and inhibited its collagenase activity while not interfering significantly with the activation of proMMP2 and MT1-MMP homodimerization on the cell surface [59]. ECM remodeling is usually independent of the viral burden but rather linked to the proteolysis driven by the immune response. LEM-2/15 reduced inflammation and ECM remodeling during influenza infection [48]. When used prophylactically or therapeutically during a coinfection of influenza and *S. pneumoniae*, LEM-2/15 significantly improved survival in mice. Interestingly, when used in combination with Tamiflu® (Oseltamivir) both prophylactically or therapeutically, an approved anti-influenza inhibitor, 100% survival was achieved in a mouse model of viral infection [59]. Importantly, Tamiflu® alone resulted in increased survival only when used prophylactically. Thus, promoting ECM stability and homeostasis via MT1-MMP inhibition during influenza infection is an attractive target.

7. MMP Substrates Extend Beyond Matrix Proteins

Contrary to what their name suggests, MMPs have been shown to cleave substrates other than matrix (ECM) proteins [1,2,11]. MMPs cleave chemokines and cytokines to regulate their functions [49,60–63]. For example, the processing of monocyte chemoattractant proteins, CCL-7 and CCL-13, reduced the inflammatory response, as demonstrated in mouse model of inflammatory edema [64]. In a mouse model of asthma, MMP2 and MMP9 were found to be protective via disruption of transepithelial chemokine gradients regulated by CCL7, CCL11, and CCL17 [65]. In macrophages, MMP12 cleaved the C-terminus of IFNγ, removing the receptor binding site and thereby decreasing JAK–STAT1 signaling and IFNγ activation within the proinflammatory macrophage [63]. Genetic ablation of MMP12 or therapeutic inhibition of MMP12 using Rxp470.1 in murine models of autoimmune inflammatory diseases resulted in elevated IFNγ mediated inflammatory signatures compared to the control groups [63]. In virus-infected cells, MMP12 was shown to be localized in the nucleus and promoted *NFKBIA* transcriptional activity resulting in INFα secretion, a key mechanisms for antiviral immunity [49]. In parallel, extracellular MMP12 attenuated systemic IFNα, and use of an MMP12 inhibitor, Rxp470.1, which is unable to enter the cells, significantly reduced the viral load [49]. MMP processing of CCL15 and CCL23, implicated in inflammatory arthritis, resulted in an increase in monocyte recruitment during inflammation [62]. Collectively, these four lines of evidence support the critical role for MMPs in regulating the inflammatory response through direct cleavage of chemokines/cytokines. Further, only 30% of MMP substrates are linked with the ECM [2]. Therefore, additional roles of MMPs are likely to be identified when looking beyond the matrix.

8. Identification of Novel MMP Substrates Using N-Terminomics/TAILS

N-terminomics technologies have been used to profile and identify new MMP substrates in various cell systems and tissues [2,3,49,66,67]. One example is terminal amine isotopic labeling of substrates (TAILS), a high throughput quantitative proteomic platform that allows simultaneous quantitative analysis of the N-terminome and proteolysis on a proteome-wide scale, and hence allows for protease substrate discovery [68]. To study the substrate repertoire of a specific protease using TAILS, the protease of interest can be compared to an inactivated form of the protease or to a protease inhibitor-treated sample (Figure 2). Alternatively, tissue proteomes of protease knock-out mice can be compared to wild-type animals with or without induction of a specific infection, stress or disease. Once collected, proteomes are denatured, and it is important to avoid primary amine containing buffers. After the reduction and alkylation of cysteine residues, primary amines of both the N-termini and lysine residues are chemically labeled with formaldehyde. During this step, we incorporate stable isotope labeling in order to later compare the different conditions being tested. An example of isotopic labeling is light (+28 Da) and heavy (+34 Da) dimethylation used with the catalyst sodium cyanoborohydride (NaBH3CN) [69]. These isotopic modification can later be monitored using liquid chromatography and tandem mass spectrometry (LC-MS/MS) [68]. However, dimethylation reactions are limited to three distinct labels and other labelling such tandem mass tag (TMT) can label up to 11 different samples with a distinct isotope [67,70]. After isotopic labeling, the labeled proteomes are then mixed and digested with trypsin. During digestion, trypsin cleaves the peptide only after arginine (semi-ArgC specificity) as the blocked lysine residues are unreactive to trypsin. After trypsin cleavage, ~10% of the sample is collected and prepared for LC-MS/MS analysis; this is the pre-enrichment TAILS samples. Next, the TAILS aldehyde reactive polymer is used to remove the internal tryptic peptides that were generated during trypsin digestion. The unbound blocked and labeled peptides are recovered from the samples by size exclusion (10-kDa cut-off filters) filtration. The recovered peptides are then analyzed via LC-MS/MS analysis. The abundance ratio of blocked peptides from the TAILS samples can be compared to naturally blocked N-termini from the pre-enrichment TAILS samples. The neo-N-termini peptides specific to the protease of interest appear in higher ratios or only in the protease-treated sample and therefore show high protease/control abundance ratio. Therefore, the TAILS protocol is designed to identify new protease substrates and also identifies the precise cleavage site within the substrate sequence. TAILS has been used to profile the substrate repertoires of various MMPs and hundreds of new substrates have been identified in specific cell lines and tissues [66,71–75]. Using TAILS, the substrates of MMP2 and MMP9 were investigated in fibroblasts secretomes where 201 substrates were identified for MMP2 and only 19 for MMP9 [75]. Although, more MMP2 substrates were identified, most substrates can be cleaved by both MMP2 and MMP9 including thrombospondin-2, galectin-1, insulin-like growth factor-binding protein 4 (IGFBP4), dickkopf-related protein-3, and pyruvate kinase M1/M2 [75]. This study suggests that the regulation of MMP2 and MMP9's activity could be linked with differences in genetic expression, different rates of TIMP inhibition, alternate activation mechanisms and/or distinct kinetic activities that could explain the differences in the phenotypes of $Mmp2^{-/-}$ and $Mmp9^{-/-}$ mice. Additional italicize in vivo studies comparing various cells, tissues, and organs of WT counterparts, $Mmp2^{-/-}$ and $Mmp9^{-/-}$ mice using disease models will help better characterize the unique roles and substrates of these two MMPs.

The substrates of macrophage MMP12 were investigated using TAILS by incubating murine MMP12 with secretomes from $Mmp12^{-/-}$ murine embryonic fibroblasts (MEFs), murine macrophage cell line RAW264.7 secretomes and also by comparing WT and $Mmp12^{-/-}$ peritoneal macrophages from a peritonitis model using thioglycollate stimulation for 4 days [71]. Hundreds of MMP12 substrates were identified including pyruvate kinase, biglycan, vimentin, renin-receptor and alpha-2-HS-glycoprotein (see [71] for the full list of substrates). Using TAILS, new functions for MMP12 in coagulation, complement activation/deactivation and resolution of inflammation were identified. To further confirm these MMP12 substrates in human diseases, TAILS was used to investigate nine COPD patients at exacerbation and recovery [7]. When comparing MMP12 substrates from murine peritonitis and joint inflammation model to the sputum of COPD patients, multiple identical substrates were identified including alpha-2-HS-glycoprotein, complement C3 (C3), complement C4-B (C4b), hemopexin, antithrombin III (SERPINC1), but also new substrates were identified such as transmembrane protease serine 7 (TMPRSS7) and DEP domain-containing mTOR-interacting protein (DEPTOR) [7,71]. MMP12 can cleave hundreds of substrates, therefore, the regulation of its activity is likely driven and impacted by other proteases, TIMPs, tissue specific microenvironment and other immune cells present such as neutrophils, eosinophils, natural killer cells, mast cells, T and B cells. Interestingly, MMP12 has been predominantly identified as a beneficial/protective MMP in inflammatory diseases although it has been implicated as a potential drug target in certain cancers. However, its precise role in inflammation needs to be further characterized [3,49,63,71].

Using TAILS, numerous non-ECM substrates have been identified further demonstrating new roles for MMPs (reviewed in [1,2,76]). For example, 58 new substrates were identified when MT6-MMP (MMP25) was added to fibroblasts secretomes, including vimentin, cystatin C, galectin-1, secreted protein acidic and rich in cysteine (SPARC), and insulin-like growth factor-binding protein 7 (IGFBP7) [72]. These identified substrates indicated a novel role for MT6-MMP for the clearance of apoptotic neutrophils. Cleavage of vimentin by MT6-MMP resulted in a decrease in chemoattraction of THP-1 monocytic cells but an increase in phagocytosis activity in an assay where fluorescent microbeads were coated with vimentin or cleaved vimentin and added to THP-1 cells [72]. The identification of new MT6-MMP substrates using TAILS supported a key biological role for this MMP in innate immunity and resolution of inflammation. Most MMPs have yet to be investigated using unbiased N-terminomics approaches to identify the extent of their substrates. It is anticipated that the cell type producing MMPs and specific microenvironments where MMPs are present will greatly impact what substrates can be cleaved by MMPs.

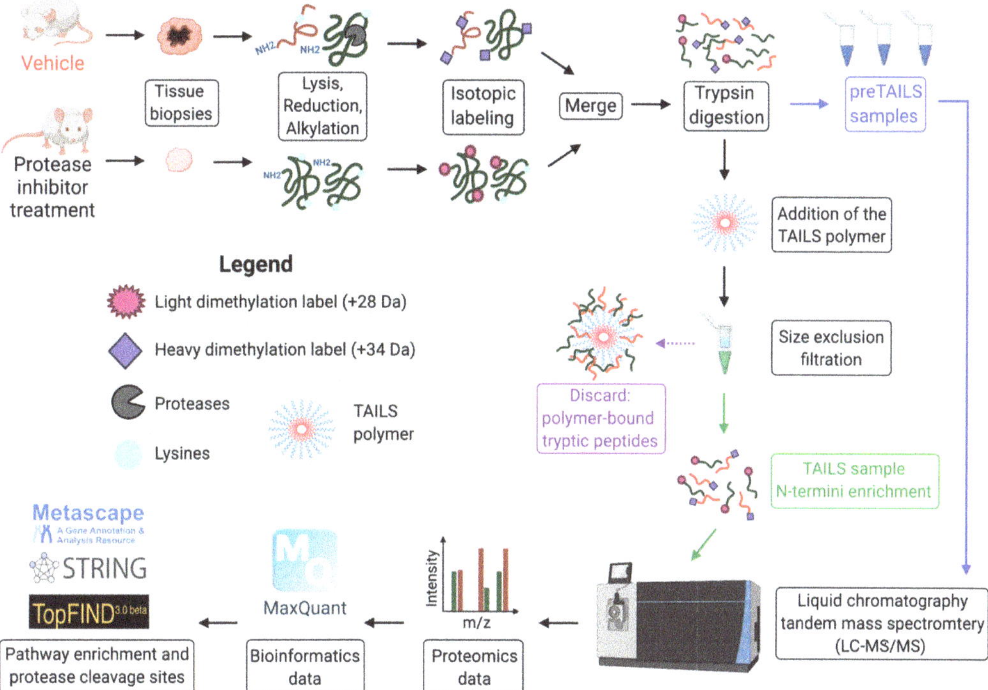

Figure 2. Experimental workflow of the N-terminomics/TAILS protocol using dimethylation isotopic labeling in a cancer mouse model treated with a protease inhibitor or a vehicle control. Tumors are lysed, reduced and alkylated before being isotopically labeled with light (+28 Da) or deuterated (+34 Da) formaldehyde. Proteins are then digested with trypsin and a pre-enrichment TAILS sample is collected. The remaining peptides are subjected to the TAILS polymer and N-termini are added to a size exclusion filter. Samples are subjected to liquid chromatography and tandem mass spectrometry (LC-MS/MS). The proteomics data is analyzed by bioinformatics software (e.g., MaxQuant). Data interpretation is analyzed with pathway enrichment tools (Metascape, STRING) or protease analysis software (TopFIND).

9. Conclusions and Perspective: Next Generation of MMP Inhibitors

MMPs have been shown to play a key biological role in numerous pathologies, however, the broad-spectrum inhibitors targeting multiple MMPs has potentially impeded their therapeutic applicability and use in the clinic to treat joint diseases or cancers. There are various pharmacological approaches that have been utilized to inhibit both the proteolytic and non-proteolytic functions of MMPs: peptides, monoclonal antibodies, and small molecule inhibitors (Figures 1 and 3, Tables 1–3). With the discovery and development of novel specific methods of targeting individual MMPs, there is renewed hope for other MMP inhibitors and a revival in better characterizing MMP functions. Moreover, additional opportunities for the use of MMP inhibitors may become apparent. For example, given the success of MMP inhibition in animal models, MMP inhibitors could play an emerging role in veterinary medicine. For example, MMP2 and MMP9 inhibitors may be beneficial for the treatment of canine chronic enteropathy [77]. Elevated levels of active MMP2 and MMP9 are found throughout the intestines of dogs with chronic enteropathy and are associated with increased inflammation and neutrophil infiltration. MMP2/9 inhibitors could be effective in lowering intestinal inflammation and decreased disease severity [77]. Another example is the use of MMP inhibitors to treat the elevated expression of MMP2 and MT1-MMP in the narrowing of myocardial vessels in canine myxomatous mitral valve disease [78]. When considering application to human health, one option could be to only inhibit MMPs in diseases requiring short-term treatments, therefore limiting the period over which the treatment is received effectively minimizing any detrimental consequences that could arise. For example, sepsis could be further investigated for the short-term implementation of MMP inhibitors. In sepsis, MMPs cleave and regulate cytokine storm and chemokine activity, thus, playing a role in many of the pathways resulting in complications [79]. Another example is MMP12 with its anti-bacterial functions and its involvement in the inflammatory pathway leading to reduced lethality of lipopolysaccharide (LPS) induced inflammation [18,80]. Selective MT1-MMP inhibitors alone or in combination with Tamiflu® also demonstrated efficacy in influenza infection models and could be further investigated for other viral infections. The utilization of shorter dosing and altering administration may reduce patient exposure to previously identified side effects of MMP inhibitors [28]. A more complete investigation into the protective and detrimental properties of MMPs and the subsequent development of therapeutics with high affinity for distinct MMPs may provide greater applications of MMP inhibition in the future.

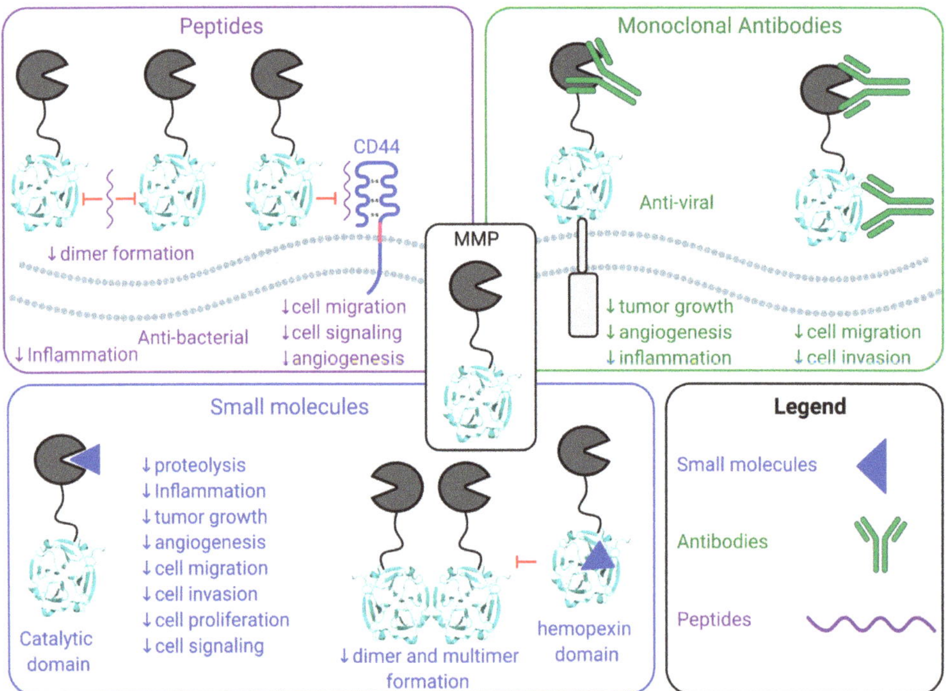

Figure 3. Schematic of three pharmacological approaches to inhibit the biological functions of MMPs: peptides (purple), monoclonal antibodies (green) and small molecules (blue).

Table 1. MMPs inhibitors in clinical trials.

Inhibitor Names	Class/Structure	Selectivity	Diseases	Clinical Trial	Outcomes/ Side Effects	References
Batimastat	Peptidomimetic/ Hydroxamate	Broad spectrum	Malignant tumor	Phase I	Local toxicities i.e., abdominal discomfort	[81]
Marimastat	Peptidomimetic/ Hydroxamate	Broad spectrum	Progressive ovarian, prostatic, pancreatic and colorectal cancer	Phase III	Adverse musculoskeletal (MS) syndrome	[82]
			Pancreatic cancer	Phase III	Musculoskeletal pain and inflammation	[83]
			Pancreatic cancer	Phase III, in combination with gemcitabine	Well tolerated but no therapeutic beneficial effects	[40]
			Gastric cancer	Phase III	Severe musculoskeletal (MS) syndrome	[84]
			Metastatic breast cancer	Phase I	Musculoskeletal pain associated with inferior survival	[85]
MMI-270	Hydroxamate/Small molecule	Broad spectrum	Advanced solid cancer	Phase I	Rash and musculoskeletal pain	[86]
Prinomastat	Hydroxamate/Small molecule	Broad spectrum	Advanced cancer	Phase I	No response in tumor growth	[36]
			Non-small cell lung cancer (stage IIIB or IV)	Phase III	Musculoskeletal syndrome	[87]
			Esophagus cancer	Phase II	Unexpected thromboembolic events	[88]
Tanomastat (Bay 12-9566)	Biphenyl, thioether zinc-binding group/small molecule	MMP2, -3 and -9	Solid tumor	Phase I	Mild toxicity, no musculoskeletal pain. No effect on tumor	[89]
			Pancreatic cancer without prior chemotherapy	Phase III	Poorer survival	[90]
			Ovarian cancer	Phase III	Well tolerated but did not impact patients' survival	[91]
Metastat	Tetracycline derivatives/small molecule	MMP2 and -9	Refractory solid tumors	Phase I	Subcutaneous phototoxicity	[92]
			AIDS related Kaposi's sarcoma	Phase I Applied with sun protection	Photosensitivity reaction	[93]
			Advanced soft tissue sarcoma	Phase II Applied with sun protection	Photosensitivity reaction	[94]
Periostat® / Doxycycline	Tetracycline derivatives/small molecule	Broad spectrum	Periodontitis	Phase III	Well tolerated; improved outcome	[41,43]
			Asymptomatic abdominal aortic aneurysms	Phase II	Well tolerated; but no significant therapeutic effects	[95]
			Multiple sclerosis	Phase II Along with IFNß-1a	Well tolerated; improved outcome	[46]
			Type II diabetes	Phase III	Reduced inflammation and better insulin sensitivity	[47]
Rebimastat (BMS- 275291)	Mercaptoacyl, thiol zinc-binding group/small molecule	MMP1, -2, -8, and MT1-MMP	Advanced cancer	Phase I	Well tolerated; no tumor response	[96]
			Early-stage breast cancer	Phase III	Study was terminated because of toxicity	[97]
			Non-small cell lung cancer	Phase II along with paclitaxel and carboplatin	Well tolerated but poor therapeutic response	[38]
			Non-small cell lung cancer	Phase II along with paclitaxel and carboplatin	Increased toxicity with no improved survival	[37]
S-3304	Sulfonamide derivatives/small molecule	MMP2 and -9	Advanced solid tumors	Phase I	Well tolerated	[98]
AZD1236		MMP9 and -12	Moderate to severe Chronic obstructive pulmonary disease (COPD)	Phase II	Well tolerated but no therapeutic efficacy	[99,100]
Neovastat (AE-941)	Mixed extract from shark cartilage	Broad spectrum	Non-small cell lung cancer (stage III)	Phase III along with chemotherapy	Well tolerated but no therapeutic effect	[39]

Table 2. Small molecule exosite MMP inhibitors.

Compounds	Target	Binding Site	Mechanism of Action	Assays and Models Tested on	References
N-[4 (difluoromethoxy) phenyl] 2-[(4-oxo-6-propyl 1Hpyrimidin-2yl) sulfanyl]-acetamide	MMP9	Hemopexin (PEX)	Interfered with homodimerization; inhibition of cell migration and proliferation	Tumor growth and metastasis (xenograft mouse model)	[16]
NSC405020	MT1-MMP	Hemopexin (PEX)	Interfered with homodimerization and interaction with catalytic domain	Tumor growth (xenograft mouse model)	[24]
JNJ0966: (*N*-{2-[(2-methoxyphenyl) amino]-4-methyl-4,5-bi-1,3-thiazol-2-yl} acetamide)	MMP9	Pro-peptide domain	Inhibit activation of MMP9 without affecting MMP1, -2, -3, -9 and MT1-MMP	Autoimmune encephalomyelitis (mouse model)	[29]
N-(4-fluorophenyl)-4-(4-oxo-3,4,5,6,7,8-hexahydroquinazolin-2-ylthio) butanamide	MMP9	Hemopexin (PEX)	Inhibition of homodimerization; decreased cancer cell migration-blocks cancer cell invasion of basement membrane and angiogenesis	in vitro migration assay, Tumor growth (xenograft mouse model), angiogenesis (chicken chorioallantoic membrane)	[17]
Synthesized acrylamides (1) 3-bromo-*N*-(4-nitrophenyl) propenamide (2) 3-bromo-*N*-{4-[(pyrimidine-2-yl) sulfamoyl] phenyl} propenamide (3) 3-bromo-*N*-{4-[(4,6 dimethylpyrimidin-2-yl) sulfamoyl]-phenyl} propanamide	MMP9	Hemopexin (PEX)	Inhibition of 4T1 breast cancer cell growth; inhibition of MMP9 gelatinolytic activity	in vitro migration assay and (xenograft mouse model)	[22]

Table 3. Regulation of MMP inhibition by antibodies.

Name	Antibody Type	Target	Epitope/Domains	Assays and Models Tested on	References
LEM-2/5	Monoclonal	MT1-MMP	Surface epitope; V-P loop	Migrating cancer cells; lung pathology; influenza	[59] [48]
SD3	Monoclonal	MMP2 and -9	Catalytic domain	Inflammatory bowel disease (mouse model); colitis	[55]
REGA-3G12	Monoclonal	MMP-9	Catalytic domain other than Zn^{2+} binding	Inhibited MMP9 proteolytic activity	[56,57]
DX-2400	Fab fragment	MT1-MMP	Catalytic domain	Breast cancer	[58]
Multiple (A4-7 Fc-ScFv, E2_C6 Fc-ScFv)	Antibody fragments	MT1-MMP	Catalytic domain outside the active site cleft, inhibiting binding to triple helical collagen	Tumor growth and proliferation (xenograft mouse model)	[101]

Funding: Antoine Dufour and Sean Gill received funding from NSERC Discovery grants for this work.

Institutional Review Board Statement: Not applicable.

Informed Consent Statement: Not applicable.

Conflicts of Interest: The authors declare no conflict of interest.

References

1. Young, D.; Das, N.; Anowai, A.; Dufor, A. Matrix Metalloproteases as Influencers of the Cells' Social Media. *Int. J. Mol. Sci.* **2019**, *20*, 3847. [CrossRef]
2. Dufour, A.; Overall, C.M. Subtracting Matrix Out of the Equation: New Key Roles of Matrix Metalloproteinases in Innate Immunity and Disease. In *Matrix Metalloproteinase Biology*, 1st ed.; Sagi, I., Gaffney, J., Eds.; John Wiley & Sons, Inc.: Franklin Township, NJ, USA, 2015; pp. 131–152. [CrossRef]
3. Dufour, A.; Overall, C.M. Missing the target: Matrix metalloproteinase antitargets in inflammation and cancer. *Trends Pharmacol. Sci.* **2013**, *34*, 233–242. [CrossRef]
4. Mainoli, B.; Hirota, S.; Edgington-Mitchell, L.E. Proteomics and Imaging in Crohn's Disease: TAILS of Unlikely Allies. *Trends Pharmacol. Sci.* **2020**, *41*, 74–84. [CrossRef]
5. Overall, C.M.; Kleifeld, O. Validating matrix metalloproteinases as drug targets and anti-targets for cancer therapy. *Nat. Rev. Cancer* **2006**, *6*, 227–239. [CrossRef]

6.	Hunninghake, G.M.; Cho, M.H.; Tesfaigzi, Y.; Soto-Quiros, M.E.; Avila, L.; Lasky-Su, J.; Stidley, C.; Melén, E.; Söderhäll, C.; Hallberg, J.; et al. MMP12, lung function, and COPD in high-risk populations. *N. Engl. J. Med.* **2009**, *361*, 2599–2608. [CrossRef]

7.	Mallia-Milanes, B.; Dufour, A.; Philp, C.; Solis, N.; Klein, T.; Fischer, M.; Bolton, C.E.; Shapiro, S.; Overall, C.M.; Johnson, S.R. TAILS proteomics reveals dynamic changes in airway proteolysis controlling protease activity and innate immunity during COPD exacerbations. *Am. J. Physiol. Cell. Mol. Physiol.* **2018**, *315*, L1003–L1014. [CrossRef]

8.	Murphy, G.; Nagase, H. Reappraising metalloproteinases in rheumatoid arthritis and osteoarthritis: Destruction or repair? *Nat. Clin. Pract. Rheumatol.* **2008**, *4*, 128–135. [CrossRef]

9.	Checchi, V.; Maravic, T.; Bellini, P.; Generali, L.; Consolo, U.; Breschi, L.; Mazzoni, A. The role of matrix metalloproteinases in periodontal disease. *Int. J. Environ. Res. Public Health* **2020**, *17*, 4923. [CrossRef]

10.	Franco, C.; Patricia, H.-R.; Timo, S.; Claudia, B.; Marcela, H. Matrix metalloproteinases as regulators of periodontal inflammation. *Int. J. Mol. Sci.* **2017**, *18*, 440. [CrossRef]

11.	Chopra, S.; Overall, C.M.; Dufour, A. Matrix metalloproteinases in the CNS: Interferons get nervous. *Cell. Mol. Life Sci.* **2019**, *76*, 3083–3095. [CrossRef]

12.	Moore, C.S.; Crocker, S.J. An alternate perspective on the roles of TIMPs and MMPs in pathology. *Am. J. Pathol.* **2012**, *180*, 12–16. [CrossRef] [PubMed]

13.	Crocker, S.J.; Pagenstecher, A.; Campbell, I.L. The TIMPs tango with MMPs and more in the central nervous system. *J. Neurosci. Res.* **2004**, *75*, 1–11. [CrossRef] [PubMed]

14.	Fischer, T.; Riedl, R. Challenges with matrix metalloproteinase inhibition and future drug discovery avenues. *Expert Opin. Drug Discov.* **2020**, 1–14. [CrossRef] [PubMed]

15.	de Bruyn, M.; Vandooren, J.; Ugarte-Berzal, E.; Arijs, I.; Vermeire, S.; Opdenakker, G. The molecular biology of matrix metalloproteinases and tissue inhibitors of metalloproteinases in inflammatory bowel diseases. *Crit. Rev. Biochem. Mol. Biol.* **2016**, *51*, 295–358. [CrossRef]

16.	Dufour, A.; Zucker, S.; Sampson, N.S.; Kuscu, C.; Cao, J. Role of matrix metalloproteinase-9 dimers in cell migration design of inhibitory peptides. *J. Biol. Chem.* **2010**, *285*, 35944–35956. [CrossRef]

17.	Alford, V.M.; Kamat, A.; Ren, X.; Kumar, K.; Gan, Q.; Awwa, M.; Tong, M.; Seeliger, M.A.; Cao, J.; Ojima, I.; et al. Targeting the hemopexin-like domain of latent matrix metalloproteinase-9 (proMMP-9) with a small molecule inhibitor prevents the formation of focal adhesion junctions. *ACS Chem. Biol.* **2017**, *12*, 2788–2803. [CrossRef]

18.	Houghton, A.M.; Hartzell, W.O.; Robbins, C.S.; Gomis-Rüth, F.X.; Shapiro, S.D. Macrophage elastase kills bacteria within murine macrophages. *Nature* **2009**, *460*, 637–641. [CrossRef]

19.	Manka, S.W.; Carafoli, R.; Visse, R.; Bihan, D.; Raynal, N.; Farndale, R.W.; Murphy, G.; Enghild, J.J.; Hohenester, E.; Nagase, H. Structural insights into triple-helical collagen cleavage by matrix metalloproteinase 1. *Proc. Natl. Acad. Sci. USA* **2012**, *109*, 12461–12466. [CrossRef]

20.	Vandenbroucke, R.E.; Libert, C. Is there new hope for therapeutic matrix metalloproteinase inhibition? *Nat. Rev. Drug Discov.* **2014**, *13*, 904–927. [CrossRef]

21.	Cha, H.; Kopetzki, E.; Huber, R.; Lanzendörfer, M.; Brandstetter, H. Structural basis of the adaptive molecular recognition by MMP9. *J. Mol. Biol.* **2002**, *320*, 1065–1079. [CrossRef]

22.	Hariono, M.; Nuwarda, R.F.; Yusuf, M.; Rollando, R.; Jenie, R.I.; Al-Najjar, B.; Julianus, J.; Putra, K.C.; Nugroho, E.S.; Wisnumurti, Y.K.; et al. Arylamide as Potential Selective Inhibitor for Matrix Metalloproteinase 9 (MMP9): Design, Synthesis, Biological Evaluation, and Molecular Modeling. *J. Chem. Inf. Model.* **2019**, *60*, 349–359. [CrossRef] [PubMed]

23.	Feinberg, T.Y.; Zheng, H.; Lui, R.; Wicha, M.S.; Yu, S.M.; Weiss, S.J. Divergent matrix-remodeling strategies distinguish developmental from neoplastic mammary epithelial cell invasion programs. *Dev. Cell* **2018**, *47*, 145–160. [CrossRef] [PubMed]

24.	Remacle, A.G.; Golubkov, V.S.; Shiryaev, S.A.; Dahl, R.; Stebbins, J.L.; Chernov, A.V.; Cheltsov, A.V.; Pellecchia, M.; Strongin, A.Y. Novel MT1-MMP small-molecule inhibitors based on insights into hemopexin domain function in tumor growth. *Cancer Res.* **2012**, *72*, 2339–2349. [CrossRef] [PubMed]

25.	Zarrabi, K.; Dufour, A.; Li, J.; Kuscu, C.; Pulkoski-Gross, A.; Zhi, J.; Hu, Y.; Sampson, N.S.; Zucker, S.; Cao, J. Inhibition of matrix metalloproteinase 14 (MMP-14)-mediated cancer cell migration. *J. Biol. Chem.* **2011**, *286*, 33167–33177. [CrossRef] [PubMed]

26.	Lu, S.; Li, S.; Zhang, J. Harnessing allostery: A novel approach to drug discovery. *Med. Res. Rev.* **2014**, *34*, 1242–1285. [CrossRef]

27.	Levin, M.; Udi, Y.; Solomonov, I.; Siga, I. Next generation matrix metalloproteinase inhibitors—Novel strategies bring new prospects. *Biochim. Biophys. Acta (BBA) Mol. Cell Res.* **2017**, *1864*, 1927–1939. [CrossRef]

28.	Overall, C.M.; Kleifeld, O. Towards third generation matrix metalloproteinase inhibitors for cancer therapy. *Br. J. Cancer* **2006**, *94*, 941–946. [CrossRef]

29.	Scannevin, R.H.; Alexander, R.; Haarlander, T.M.; Burke, S.L.; Singer, M.; Huo, C.; Zhang, Y.M.; Maguire, D.; Spurlino, J.; Deckman, I.; et al. Discovery of a highly selective chemical inhibitor of matrix metalloproteinase-9 (MMP-9) that allosterically inhibits zymogen activation. *J. Biol. Chem.* **2017**, *292*, 17963–17974. [CrossRef]

30.	Fields, G.B. The rebirth of matrix metalloproteinase inhibitors: Moving beyond the dogma. *Cells* **2019**, *8*, 984. [CrossRef]

31.	Van Wart, H.E.; Birkedal-Hansen, H. The cysteine switch: A principle of regulation of metalloproteinase activity with potential applicability to the entire matrix metalloproteinase gene family. *Proc. Natl. Acad. Sci. USA* **1990**, *87*, 5578–5582. [CrossRef]

32.	Wetmore, D.R.; Hardman, K.D. Roles of the propeptide and metal ions in the folding and stability of the catalytic domain of stromelysin (matrix metalloproteinase 3). *Biochemistry* **1996**, *35*, 6549–6558. [CrossRef] [PubMed]

33. Pirard, B. Insight into the structural determinants for selective inhibition of matrix metalloproteinases. *Drug Discov. Today* **2007**, *12*, 640–646. [CrossRef] [PubMed]

34. Coussens, L.M.; Fingleton, B.; Matrisian, L.M. Matrix metalloproteinase inhibitors and cancer—Trials and tribulations. *Science* **2002**, *295*, 2387–2392. [CrossRef] [PubMed]

35. Cathcart, J.; Pulkoski-Gross, A.; Cao, J. Targeting matrix metalloproteinases in cancer: Bringing new life to old ideas. *Genes Dis.* **2015**, *2*, 26–34. [CrossRef]

36. Hande, K.R.; Collier, M.; Paradiso, L.; Stuart-Smith, J.; Dixon, M.; Clendeninn, N.; Yeun, G.; Alberti, D.; Binger, K.; Wilding, G. Phase I and pharmacokinetic study of prinomastat, a matrix metalloprotease inhibitor. *Clin. Cancer Res.* **2004**, *10*, 909–915. [CrossRef]

37. Leighl, N.B.; Paz-Ares, L.; Douillard, J.-Y.; Peschel, C.; Arnold, A.; Depierre, A.; Santoro, A.; Betticher, D.C.; Gatzemeier, U.; Jassem, J.; et al. Randomized phase III study of matrix metalloproteinase inhibitor BMS-275291 in combination with paclitaxel and carboplatin in advanced non-small-cell lung cancer: National Cancer Institute of Canada-Clinical Trials Group Study BR. 18. *J. Clin. Oncol.* **2005**, *23*, 2831–2839. [CrossRef]

38. Douillard, J.-Y.; Peschel, C.; Shepherd, F.; Paz-Ares, L.; Arnold, A.; Davis, M.; Tonato, M.; Smylie, M.; Tu, D.; Voi, M.; et al. Randomized phase II feasibility study of combining the matrix metalloproteinase inhibitor BMS-275291 with paclitaxel plus carboplatin in advanced non-small cell lung cancer. *Lung Cancer* **2004**, *46*, 361–368. [CrossRef]

39. Lu, C.; Lee, J.; Komaki, R.; Herbst, R.S.; Feng, L.; Evans, W.K.; Choy, H.; Desjardins, P.; Esparaz, B.T.; Truong, M.T.; et al. Chemoradiotherapy with or without Ae-941 in Stage III Non–Small cell lung cancer: A randomized Phase III trial. *JNCI J. Natl. Cancer Inst. C* **2020**, *102*, 859–865. [CrossRef]

40. Bramhall, S.R.; Schulz, J.; Nemunaitis, J.; Brown, P.D.; Baillet, M.; Buckels, J.A. A double-blind placebo-controlled, randomised study comparing gemcitabine and marimastat with gemcitabine and placebo as first line therapy in patients with advanced pancreatic cancer. *Br. J. Cancer* **2002**, *87*, 161–167. [CrossRef]

41. Preshaw, P.M.; Hefti, A.F.; Novak, M.J.; Michalowicz, B.S.; Pihlstrom, B.L.; Schoor, R.; Trummel, C.L.; Dean, J.; Van Dyke, T.E.; Walker, C.B.; et al. Subantimicrobial dose doxycycline enhances the efficacy of scaling and root planing in chronic periodontitis: A multicenter trial. *J. Periodontol.* **2004**, *75*, 1068–1076. [CrossRef]

42. Peterson, J.T. Matrix metalloproteinase inhibitor development and the remodeling of drug discovery. *Heart Fail. Rev.* **2004**, *9*, 63–79. [CrossRef] [PubMed]

43. Golub, L.M.; McNamara, T.F.; Ryan, M.E.; Kohut, B.; Blieden, T.; Payonk, G.; Sipos, T.; Baron, H.J. Adjunctive treatment with subantimicrobial doses of doxycycline: Effects on gingival fluid collagenase activity and attachment loss in adult periodontitis. *J. Clin. Periodontol.* **2001**, *28*, 146–156. [CrossRef] [PubMed]

44. Baker, P.J.; Evans, R.T.; Coburn, R.A.; Genco, R.J. Tetracycline and its derivatives strongly bind to and are released from the tooth surface in active form. *J. Periodontol.* **1983**, *54*, 580–585. [CrossRef] [PubMed]

45. Payne, J.B.; Stoner, J.A.; Nummikoski, P.V.; Reinhardt, R.A.; Goren, A.D.; Wolff, M.S.; Lee, H.M.; Lynch, J.C.; Valente, R.; Golub, L.M. Subantimicrobial dose doxycycline effects on alveolar bone loss in post-menopausal women. *J. Clin. Periodontol.* **2007**, *34*, 776–787. [CrossRef] [PubMed]

46. Minagar, A.; Alexander, J.S.; Schwendimann, R.N.; Kelley, R.E.; Gonzalez-Toledo, E.; Jimenez, J.J.; Mauro, L.; Jy, W.; Smith, S.J. Combination therapy with interferon beta-1a and doxycycline in multiple sclerosis: An open-label trial. *Arch. Neurol.* **2008**, *65*, 199–204. [CrossRef]

47. Frankwich, K.; Tibble, C.; Torres-Gonzalez, M.; Bonner, M.; Lefkowitz, R.; Tyndall, M.; Schmid-Schönbein, G.W.; Villarreal, F.; Heller, M.; Herbst, K. Proof of Concept: Matrix metalloproteinase inhibitor decreases inflammation and improves muscle insulin sensitivity in people with type 2 diabetes. *J. Inflamm.* **2012**, *9*, 35. [CrossRef]

48. Talmi-Frank, D.; Altboum, Z.; Solomonov, I.; Udi, Y.; Jaitin, D.A.; Klepfish, M.; David, E.; Zhuravlev, A.; Keren-Shaul, H.; Winter, D.R.; et al. Extracellular matrix proteolysis by MT1-MMP contributes to influenza-related tissue damage and mortality. *Cell Host Microbe* **2016**, *20*, 458–470. [CrossRef]

49. Marchant, D.J.; Bellac, C.L.; Moraes, T.J.; Wadsworth, S.J.; Dufour, A.; Butler, G.S.; Bilawchuk, L.M.; Hendry, R.G.; Robertson, A.G.; Cheung, C.T.; et al. A new transcriptional role for matrix metalloproteinase-12 in antiviral immunity. *Nat. Med.* **2014**, *20*, 493–502. [CrossRef]

50. Dufour, A.; Sampson, N.S.; Li, J.; Kuscu, C.; Rizzo, R.C.; Deleon, J.L.; Zhi, J.; Jaber, N.; Liu, E.; Zucker, S.; et al. Small-molecule anticancer compounds selectively target the hemopexin domain of matrix metalloproteinase-9. *Cancer Res.* **2011**, *71*, 4877–4888. [CrossRef]

51. Lipsky, P.E.; van der Heijde, D.M.; Clair, E.W.S.; Furst, D.E.; Breedveld, F.C.; Kalden, J.R.; Smolen, J.S.; Weisman, M.; Emery, P.; Feldmann, M.; et al. Anti-Tumor Necrosis Factor Trial in Rheumatoid Arthritis with Concomitant Therapy Study Group. Infliximab and methotrexate in the treatment of rheumatoid arthritis. *N. Engl. J. Med.* **2000**, *343*, 1594–1602. [CrossRef]

52. Starnes, H.; Pearce, M.K.; Tewari, A.; Yim, J.H.; Zou, J.C.; Abrams, J.S. Anti-IL-6 monoclonal antibodies protect against lethal Escherichia coli infection and lethal tumor necrosis factor-alpha challenge in mice. *J. Immunol.* **1990**, *145*, 4185–4191. [PubMed]

53. Hudziak, R.M.; Lewis, G.D.; Winget, M.; Fendly, B.M.; Shepard, H.M.; Ullrich, A. p185HER2 monoclonal antibody has antiproliferative effects in vitro and sensitizes human breast tumor cells to tumor necrosis factor. *Mol. Cell. Biol.* **1989**, *9*, 1165–1172. [CrossRef] [PubMed]

54. Milgrom, H.; Fick, R.B.; Su, J.Q.; Reimann, J.D.; Bush, R.K.; Watrous, M.L.; Metzger, W.J. Treatment of allergic asthma with monoclonal anti-IgE antibody. *N. Engl. J. Med.* **1999**, *341*, 1966–1973. [CrossRef] [PubMed]

55. Sela-Passwell, N.; Kikkeri, R.; Dym, O.; Rozenberg, H.; Margalit, R.; Arad-Yellin, R.; Eisenstein, M.; Brenner, O.; Shoham, T.; Danon, T.; et al. Antibodies targeting the catalytic zinc complex of activated matrix metalloproteinases show therapeutic potential. *Nat. Med.* **2012**, *18*, 143–147. [CrossRef]

56. Martens, E.; Leyssen, A.; Aelst, I.V.; Fiten, P.; Piccard, H.; Hu, J.; Descamps, F.J.; van den Steen, P.E.; Proost, P.; van Damme, J.; et al. A monoclonal antibody inhibits gelatinase B/MMP-9 by selective binding to part of the catalytic domain and not to the fibronectin or zinc binding domains. *Biochim. Biophys. Acta (BBA) Gen. Subj.* **2007**, *1770*, 178–186. [CrossRef]

57. Paemen, L.; Martens, E.; Masure, S.; Opdenakker, G. Monoclonal antibodies specific for natural human neutrophil gelatinase B used for affinity purification, quantitation by two-site ELISA and inhibition of enzymatic activity. *Eur. J. Biochem.* **1995**, *234*, 759–765. [CrossRef]

58. Devy, L.; Huang, L.; Naa, L.; Yanamandra, N.; Pieters, H.; Frans, N.; Chang, E.; Tao, Q.; Vanhove, M.; Lejeune, A.; et al. Selective inhibition of matrix metalloproteinase-14 blocks tumor growth, invasion, and angiogenesis. *Cancer Res.* **2009**, *69*, 1517–1526. [CrossRef]

59. Udi, Y.; Grossman, M.; Solomonov, I.; Dym, O.; Rozenberg, H.; Moreno, V.; Cuniasse, P.; Dive, V.; Arroyo, A.G.; Sagi, I. Inhibition mechanism of membrane metalloprotease by an exosite-swiveling conformational antibody. *Structure* **2015**, *23*, 104–115. [CrossRef]

60. Dean, R.A.; Cox, J.H.; Bellac, C.L.; Doucet, A.; Starr, A.E.; Overal, C.M. Macrophage-specific metalloelastase (MMP-12) truncates and inactivates ELR + CXC chemokines and generates CCL2, -7, -8, and -13 antagonists: Potential role of the macrophage in terminating polymorphonuclear leukocyte influx. *Blood* **2008**, *112*, 3455–3464. [CrossRef]

61. McQuibban, G.A.; Gong, J.-H.; Tam, E.M.; McCulloch, C.A.; Clark-Lewis, I.; Overall, C.M. Inflammation dampened by gelatinase a cleavage of monocyte chemoattractant protein-3. *Science* **2000**, *289*, 1202–1206. [CrossRef]

62. Starr, A.E.; Dufour, A.; Maier, J.; Overall, C.M. Biochemical analysis of matrix metalloproteinase activation of chemokines CCL15 and CCL23 and increased glycosaminoglycan binding of CCL16. *J. Biol. Chem.* **2012**, *287*, 5848–5860. [CrossRef] [PubMed]

63. Dufour, A.; Bellac, C.L.; Eckhard, U.; Solis, N.; Klein, T.; Kappelhoff, R.; Fortelny, N.; Jobin, P.; Rozmus, J.; Mark, J.; et al. C-terminal truncation of IFN-γ inhibits proinflammatory macrophage responses and is deficient in autoimmune disease. *Nat. Commun.* **2018**, *9*, 2416. [CrossRef] [PubMed]

64. McQuibban, G.A.; Gong, J.-H.; Wong, J.P.; Wallace, J.L.; Clark-Lewis, I.; Overall, C.M. Matrix metalloproteinase processing of monocyte chemoattractant proteins generates CC chemokine receptor antagonists with anti-inflammatory properties in vivo. *Blood J. Am. Soc. Hematol.* **2002**, *100*, 1160–1167. [CrossRef]

65. Corry, D.B.; Kiss, A.; Song, L.-Z.; Song, L.; Xu, J.; Lee, S.; Werb, Z.; Kheradmand, F. Overlapping and independent contributions of MMP2 and MMP9 to lung allergic inflammation cell egression through decreased CC chemokines. *FASEB J.* **2004**, *18*, 995–997. [CrossRef]

66. Kleifeld, O.; Doucet, A.; auf dem Keller, U.; Prudova, A.; Schilling, O.; Kainthan, R.K.; Starr, A.E.; Foster, L.J.; Kizhakkedathu, J.N.; Overall, C.M. Isotopic labeling of terminal amines in complex samples identifies protein N-termini and protease cleavage products. *Nat. Biotechnol.* **2010**, *28*, 281–288. [CrossRef]

67. Prudova, A.; Gocheva, V.; auf dem Keller, U.; Eckhard, U.; Olson, O.C.; Akkari, L.; Butler, G.S.; Fortelny, N.; Lange, P.F.; Mark, J.C.; et al. TAILS N-terminomics and proteomics show protein degradation dominates over proteolytic processing by cathepsins in pancreatic tumors. *Cell Rep.* **2016**, *16*, 1762–1773. [CrossRef]

68. Kleifeld, O.; Doucet, A.; Prudova, A.; Keller, U.a.d.; Gioia, M.; Kizhakkedathu, J.N.; Overall, C.M. Identifying and quantifying proteolytic events and the natural N terminome by terminal amine isotopic labeling of substrates. *Nat. Protoc.* **2011**, *6*, 1578–1611. [CrossRef]

69. Wold, F. In vivo chemical modification of proteins (post-translational modification). *Annu. Rev. Biochem.* **1981**, *50*, 783–814. [CrossRef]

70. Biancur, D.E.; Paulo, J.A.; Matachowska, B.; del Rey, M.Q.; Sousa, C.M.; Wang, X.; Sohn, A.S.W.; Chu, G.C.; Gygi, S.P.; Harper, J.W.; et al. Compensatory metabolic networks in pancreatic cancers upon perturbation of glutamine metabolism. *Nat. Commun.* **2017**, *8*, 1–15. [CrossRef]

71. Bellac, C.L.; Dufour, A.; Krisinger, J.M.; Loonchanta, A.; Starr, A.E.; Keller, U.A.d.; Lange, P.F.; Goebeler, V.; Kappelhoff, R.; Butler, G.S.; et al. Macrophage matrix metalloproteinase-12 dampens inflammation and neutrophil influx in arthritis. *Cell Rep.* **2014**, *9*, 618–663. [CrossRef]

72. Starr, A.E.; Bellac, C.L.; Dufour, A.; Goebeler, V.; Overall, C.M. Biochemical characterization and N-terminomics analysis of leukolysin, the membrane-type 6 matrix metalloprotease (MMP25) chemokine and vimentin cleavages enhance cell migration and macrophage phagocytic activities. *J. Biol. Chem.* **2012**, *287*, 13382–13395. [CrossRef] [PubMed]

73. Schlage, P.; Egli, F.E.; Nanni, P.; Wang, L.W.; Kizhakkedathu, J.N.; Apte, S.S.; Keller, U.a.d. Time-resolved analysis of the matrix metalloproteinase 10 substrate degradome. *Mol. Cell. Proteom.* **2014**, *13*, 580–593. [CrossRef] [PubMed]

74. auf dem Keller, U.; Bellac, C.L.; Li, Y.; Lou, Y.; Lange, P.F.; Ting, R.; Harwig, C.; Kappelhoff, R.; Dedhar, S.; Adam, M.J.; et al. Novel matrix metalloproteinase inhibitor [18F] marimastat-aryltrifluoroborate as a probe for in vivo positron emission tomography imaging in cancer. *Cancer Res.* **2010**, *70*, 7562–7569. [CrossRef] [PubMed]

75. Prudova, A.; auf dem Keller, U.; Butler, G.S.; Overall, C.M. Multiplex N-terminome analysis of MMP-2 and MMP-9 substrate degradomes by iTRAQ-TAILS quantitative proteomics. *Mol. Cell. Proteom.* **2010**, *9*, 894–911. [CrossRef]
76. Butler, G.S.; Overall, C.M. Proteomic identification of multitasking proteins in unexpected locations complicates drug targeting. *Nat. Rev. Drug Discov.* **2009**, *8*, 935–948. [CrossRef]
77. Hanifeh, M.; Rajamaki, M.M.; Makitalo, L.; Syrjä, P.; Sankari, S.; Kilpinen, S.; Spillmann, T. Identification of matrix metalloproteinase-2 and-9 activities within the intestinal mucosa of dogs with chronic enteropathies. *Acta Vet. Scand.* **2018**, *60*, 16. [CrossRef]
78. Moesgaard, S.G.; Aupperle, H.; Rajamaki, M.M.; Falk, T.; Rasmussen, C.E.; Zois, N.E.; Olsen, L.H. Matrix metalloproteinases (MMPs), tissue inhibitors of metalloproteinases (TIMPs) and transforming growth factor-β (TGF-β) in advanced canine myxomatous mitral valve disease. *Res. Vet. Sci.* **2014**, *97*, 560–567. [CrossRef]
79. Qiu, Z.; Hu, J.; Van den Steen, P.E.; Opdenakker, G. Targeting matrix metalloproteinases in acute inflammatory shock syndromes. *Comb. Chem. High Throughput Screen.* **2012**, *15*, 555–570. [CrossRef]
80. Xia, L.; Zhu, Z.; Zhang, L.; Xu, Y.; Chen, G.; Luo, J. EZH2 enhances expression of CCL5 to promote recruitment of macrophages and invasion in lung cancer. *Biotechnol. Appl. Biochem.* **2019**. [CrossRef]
81. Wojtowicz-Praga, S.; Low, J.; Marshall, J.; Ness, E.; Dickson, R.; Barter, J.; Sale, M.; McCann, P.; Moore, J.; Cole, A.; et al. Phase I trial of a novel matrix metalloproteinase inhibitor batimastat (BB-94) in patients with advanced cancer. *Invest. New Drugs* **1996**, *14*, 193–202. [CrossRef]
82. Nemunaitis, J.; Poole, C.; Primrose, J.; Rosemurgy, A.; Malfetano, J.; Brown, P.; Berrington, A.; Cornish, A.; Lynch, K.; Rasmussen, H.; et al. Combined analysis of studies of the effects of the matrix metalloproteinase inhibitor marimastat on serum tumor markers in advanced cancer: Selection of a biologically active and tolerable dose for longer-term studies. *Clin. Cancer Res.* **1998**, *4*, 1101–1109. [PubMed]
83. Bramhall, S.R.; Rosemurgy, A.; Brwon, P.D.; Bowry, C.; Buckels, J.A.; Marimastat Pancreatic Cancer Study Group. Marimastat as first-line therapy for patients with unresectable pancreatic cancer: A randomized trial. *J. Clin. Oncol.* **2001**, *19*, 3447–3455. [CrossRef]
84. Bramhall, S.R.; Hallissey, M.T.; Whiting, J.; Scholefield, J.; Tierney, G.; Stuart, R.C.; Hawkins, R.E.; McCulloch, P.; Maughan, T.; Brown, P.D.; et al. Marimastat as maintenance therapy for patients with advanced gastric cancer: A randomised trial. *Br. J. Cancer* **2002**, *86*, 1864–1870. [CrossRef] [PubMed]
85. Sparano, J.A.; Bernardo, P.; Stephenson, P.; Gradishar, W.J.; Ingle, J.N.; Zucker, S.; Davidson, N.E. Randomized phase III trial of marimastat versus placebo in patients with metastatic breast cancer who have responding or stable disease after first-line chemotherapy: Eastern Cooperative Oncology Group trial E2196. *J. Clin. Oncol.* **2004**, *22*, 4683–4690. [CrossRef]
86. Levitt, N.C.; Eskens, F.A.L.M.; O'Byrne, K.J.; Propper, D.J.; Denis, L.J.; Owen, S.J.; Choi, L.; Foekens, J.A.; Wilner, S.; Wood, J.M.; et al. Phase I and pharmacological study of the oral matrix metalloproteinase inhibitor, MMI270 (CGS27023A), in patients with advanced solid cancer. *Clin. Cancer Res.* **2001**, *7*, 1912–1922. [PubMed]
87. Bissett, D.; O'Byrne, K.J.; von Pawel, J.; Gatzemeier, U.; Price, A.; Nicolson, M.; Mercier, R.; Mazabel, E.; Penning, C.; Zhang, M.H.; et al. Phase III study of matrix metalloproteinase inhibitor prinomastat in non-small-cell lung cancer. *J. Clin. Oncol.* **2004**, *23*, 842–849. [CrossRef]
88. Heath, E.I.; Burtness, B.A.; Kleinberg, L.; Salem, R.R.; Yang, S.C.; Heitmiller, R.F.; Canto, M.I.; Knisely, J.P.S.; Topazian, M.; Montgomery, E.; et al. Phase II, parallel-design study of preoperative combined modality therapy and the matrix metalloprotease (mmp) inhibitor prinomastat in patients with esophageal adenocarcinoma. *Invest. New Drugs* **2006**, *24*, 135–140. [CrossRef]
89. Erlichman, C.; Adjei, A.A.; Alberts, S.R.; Sloan, J.A.; Goldberg, R.M.; Pitot, H.C.; Rubin, J.; Atherton, P.J.; Klee, G.G.; Humphrey, R. Phase I study of the matrix metal loproteinase inhibitor, BAY 12-9566. *Ann. Oncol.* **2001**, *12*, 389–395. [CrossRef]
90. Moore, M.J.; Hamm, J.; Dancey, J.; Eisenberg, P.D.; Dagenais, M.; Fields, A.; Hagan, K.; Greenberg, B.; Colwell, B.; Zee, B.; et al. Comparison of gemcitabine versus the matrix metalloproteinase inhibitor BAY 12-9566 in patients with advanced or metastatic adenocarcinoma of the pancreas: A phase III trial of the National Cancer Institute of Canada Clinical Trials Group. *J. Clin. Oncol.* **2003**, *21*, 3296–3302. [CrossRef]
91. Hirte, H.; Vergote, I.B.; Jeffrey, J.R.; Grimshaw, R.N.; Coppieters, S.; Schwartz, B.; Tu, D.; Sadura, A.; Brundage, M.; Seymour, L. A phase III randomized trial of BAY 12-9566 (tanomastat) as maintenance therapy in patients with advanced ovarian cancer responsive to primary surgery and paclitaxel/platinum containing chemotherapy: A National Cancer Institute of Canada Clinical Trials Group Study. *Gynecol. Oncol.* **2006**, *102*, 300–308. [CrossRef]
92. Rudek, M.A.; Figg, W.D.; Dyer, V.; Dahut, W.; Turner, M.L.; Steinberg, S.M.; Liewehr, D.J.; Kohler, D.R.; Pluda, J.M.; Reed, E. Phase I clinical trial of oral COL-3, a matrix metalloproteinase inhibitor, in patients with refractory metastatic cancer. *J. Clin. Oncol.* **2001**, *19*, 584–592. [CrossRef] [PubMed]
93. Cianfrocca, M.; Cooley, T.P.; Lee, J.Y.; Rudek, M.A.; Scadden, D.T.; Ratner, L.; Pluda, J.M.; Figg, W.D.; Krown, S.E.; Dezube, B.J. Matrix metalloproteinase inhibitor COL-3 in the treatment of AIDS-related Kaposi's sarcoma: A phase I AIDS malignancy consortium study. *J. Clin. Oncol.* **2001**, *20*, 153–159.
94. Chu, Q.S.C.; Forouzesh, B.; Syed, S.; Mita, M.; Schwartz, G.; Cooper, J.; Curtright, J.; Rowinsky, E.K. A phase II and pharmacological study of the matrix metalloproteinase inhibitor (MMPI) COL-3 in patients with advanced soft tissue sarcomas. *Invest. New Drugs* **2007**, *25*, 359. [CrossRef] [PubMed]

95. Baxter, B.T.; Pearce, W.H.; Waltke, E.A.; Littooy, F.N.; Hallett, J.W., Jr.; Kent, K.C.; Upchurch, G.R., Jr.; Chaikof, E.L.; Mills, J.L.; Fleckten, B.; et al. Prolonged administration of doxycycline in patients with small asymptomatic abdominal aortic aneurysms: Report of a prospective (Phase II) multicenter study. *J. Vasc. Surg.* **2002**, *36*, 1–12. [CrossRef]

96. Rizvi, N.A.; Humphrey, J.S.; Ness, E.A.; Johnson, M.D.; Gupta, E.; Williams, K.; Daly, D.J.; Sonnichsen, D.; Conway, D.; Marshall, J.; et al. A phase I study of oral BMS-275291, a novel nonhydroxamate sheddase-sparing matrix metalloproteinase inhibitor, in patients with advanced or metastatic cancer. *Clin. Cancer Res.* **2004**, *10*, 1963–1970. [CrossRef]

97. Miller, K.D.; Saphner, R.J.; Waterhouse, D.M.; Chen, T.-T.; Rush-Taylor, A.; Sparano, J.A.; Wolff, A.C.; Cobleigh, M.A.; Galbraith, S.; Sledge, G.W. A randomized phase II feasibility trial of BMS-275291 in patients with early stage breast cancer. *Clin. Cancer Res.* **2004**, *10*, 1971–1975. [CrossRef]

98. Chiappori, A.A.; Eckhardt, S.G.; Bukowski, R.; Sullivan, D.M.; Ikeda, M.; Yano, Y.; Yamada-Sawada, T.; Kambayashi, Y.; Tanaka, K.; Javle, M.M.; et al. A phase I pharmacokinetic and pharmacodynamic study of s-3304, a novel matrix metalloproteinase inhibitor, in patients with advanced and refractory solid tumors. *Clin. Cancer Res.* **2007**, *13*, 2091–2099. [CrossRef]

99. Magnussen, H.; Watz, H.; Kirsten, A.; Wang, M.; Wray, H.; Samuelsson, V.; Mo, J.; Kay, R. Safety and tolerability of an oral MMP-9 and-12 inhibitor, AZD1236, in patients with moderate-to-severe COPD: A randomised controlled 6-week trial. *Pulm. Pharmacol. Ther.* **2011**, *24*, 563–570. [CrossRef]

100. Dahl, R.; Titlestad, I.; Lindqvist, A.; Wielders, P.; Wray, H.; Wang, M.; Samuelsson, V.; Mo, J.; Holt, A. Effects of an oral MMP-9 and-12 inhibitor, AZD1236, on biomarkers in moderate/severe COPD: A randomised controlled trial. *Pulm. Pharmacol. Ther.* **2012**, *25*, 169–177. [CrossRef]

101. Botkjaer, K.A.; Kwok, H.F.; Terp, M.G.; Karatt-Vellatt, A.; Santamaria, S.; McCafferty, J.; Andreasen, P.A.; Itoh, Y.; Ditzel, H.J.; Murphy, G. Development of a specific affinity-matured exosite inhibitor to MT1-MMP that efficiently inhibits tumor cell invasion in vitro and metastasis in vivo. *Oncotarget* **2016**, *7*, 16773. [CrossRef]

pharmaceuticals

MDPI

Review

Matrix Metalloproteinase 2 as a Pharmacological Target in Heart Failure

Pricila Rodrigues Gonçalves [1], Lisandra Duarte Nascimento [1], Raquel Fernanda Gerlach [2], Keuri Eleutério Rodrigues [1] and Alejandro Ferraz Prado [1,*]

[1] Cardiovascular System Pharmacology and Toxicology Laboratory, Institute of Biological Sciences, Federal University of Pará, Belém 66075-110, PA, Brazil; pri.pharma.12@gmail.com (P.R.G.); duartelis18@gmail.com (L.D.N.); keuri13@yahoo.com.br (K.E.R.)

[2] Department of Basic and Oral Biology, Faculty of Dentistry of Ribeirao Preto, University of Sao Paulo (FORP/USP), Ribeirao Preto 14040-904, SP, Brazil; rfgerlach@forp.usp.br

* Correspondence: alejandrofp@ufpa.br

Abstract: Heart failure (HF) is an acute or chronic clinical syndrome that results in a decrease in cardiac output and an increase in intracardiac pressure at rest or upon exertion. The pathophysiology of HF is heterogeneous and results from an initial harmful event in the heart that promotes neurohormonal changes such as autonomic dysfunction and activation of the renin-angiotensin-aldosterone system, endothelial dysfunction, and inflammation. Cardiac remodeling occurs, which is associated with degradation and disorganized synthesis of extracellular matrix (ECM) components that are controlled by ECM metalloproteinases (MMPs). MMP-2 is part of this group of proteases, which are classified as gelatinases and are constituents of the heart. MMP-2 is considered a biomarker of patients with HF with reduced ejection fraction (HFrEF) or preserved ejection fraction (HFpEF). The role of MMP-2 in the development of cardiac injury and dysfunction has clearly been demonstrated in animal models of cardiac ischemia, transgenic models that overexpress MMP-2, and knockout models for this protease. New research to minimize cardiac structural and functional alterations using non-selective and selective inhibitors for MMP-2 demonstrates that this protease could be used as a possible pharmacological target in the treatment of HF.

Keywords: MMP-2 inhibitor; cardiac dysfunction; ischemia

check for updates

Citation: Gonçalves, P.R.; Nascimento, L.D.; Gerlach, R.F.; Rodrigues, K.E.; Prado, A.F. Matrix Metalloproteinase 2 as a Pharmacological Target in Heart Failure. *Pharmaceuticals* **2022**, *15*, 920. https://doi.org/10.3390/ph15080920

Academic Editor: Salvatore Santamaria

Received: 3 June 2022
Accepted: 11 July 2022
Published: 25 July 2022

Publisher's Note: MDPI stays neutral with regard to jurisdictional claims in published maps and institutional affiliations.

1. Introduction

HF is highly prevalent, affecting approximately 26 million people worldwide every year, with high rates of hospitalization and death [1]. According to the annual report on cardiovascular disease of the American Heart Association (AHA), the lifetime risk of developing HF is high, ranging from 20 to 45% after the age of 45 years. Estimates indicate that six million American adults aged 20 years or older have HF [2,3]. In Europe, the prevalence of HF is around 17.2 cases per 1000 individuals. It is an important public health problem with an average number of hospitalizations of 2671 per million people [4]. Furthermore, it is one of the most expensive syndromes in the US and Europe, consuming around 1–2% of the overall healthcare budget [5]. Global spending on CI is around US $108 billion per year [6].

HF is an acute or chronic clinical syndrome that results in a decrease in cardiac output and an increase in intracardiac pressure at rest or upon exertion. In this condition, the heart is unable to pump enough blood to meet the metabolic needs of tissue [7]. HF can be determined according to the left ventricle (LV) ejection fraction (EF), characterized as the percentage of blood ejected from the LV with each systole. HF can be classified as reduced ejection fraction (HFrEF), HF with intermediate ejection fraction (HFiEF) or HF with preserved ejection fraction (HFpEF). Patients with HFrEF have a left ventricular ejection fraction <40%, with inadequate stroke volume and cardiac output as the primary manifestation. Patients with HFpEF have a left ventricular ejection fraction ≥50%, with impaired left ventricular relaxation. Patients with EF ranging from 41% to 49% were classified as HFiEF, presenting clinical characteristics similar to the population with HFpEF (Table 1). The New York Heart Association (NYHA) HF classification system, which stratifies the patient into classes I–IV, is based on the symptoms presented by the patient and the level of tolerated physical activity (Table 2) [8,9]. The etiology of HF stems from several conditions, and the leading causes are hypertension, valvular diseases, genetic cardiomyopathies, myocarditis, extracardiac diseases, and ischemia [10,11] (Figure 1).

The pathophysiology of HF is complex and results from an initial harmful event in the heart. The event can occur acutely (such as ischemic events, valvular diseases, and viral and bacterial myocarditis) or chronically (in arterial hypertension, genetic cardiomyopathies, and extracardiac diseases) and promotes functional and structural changes that compromise both systolic and diastolic blood pumping [7,9,12]. Diastolic dysfunction occurs due to structural changes resulting from fibrosis, promoting increased stiffness, decreased cardiac compliance, and hypertrophic cardiac remodeling, which causes an increase in LV filling pressure [12]. Systolic and diastolic electrical and mechanical asynchronies are related to the extent of diastolic dysfunction and exercise tolerance. Neurohormonal changes such as autonomic dysfunction and activation of the renin-angiotensin-aldosterone system are also implicated, as are endothelial dysfunction and inflammation. This makes HF heterogeneous and creates difficulties in choosing the therapeutic approach [13] (Figure 1).

In HF, cardiac remodeling occurs with degradation and disorganized synthesis of extracellular matrix (ECM) components. The ECM content is divided into fibrillar components (collagen, elastin, and reticular) and non-fibrillar components (glycoproteins and proteoglycans), which are responsible for tissue resistance and elasticity. Furthermore, their breakdown promotes functional changes. ECM metalloproteinases (MMPs) are proteases specialized in controlling the content of ECM [14–16] (Figure 1).

Table 1. Definition of HF, according to left ventricular ejection fraction.

Classification	Left Ventricle Ejection Fraction (LVEF)	Main Cardiac Alterations (Ecodoppler)
HFrEF	<40%	Structural change and systolic dysfunction
HFpEF	≥50%	Structural change and diastolic dysfunction
HFiEF	41% to 49%	Structural change and diastolic dysfunction

HFrEF: heart failure with reduced ejection fraction; HFpEF: heart failure with preserved ejection fraction; HFiEF: heart failure with intermediate ejection fraction.

Table 2. New York Heart Association (NYHA) classification of heart failure based on symptoms and level of tolerated physical activity.

Class	General Description	Patient Symptoms
I	Asymptomatic	No limitation of physical activity; regular physical activity does not cause undue fatigue, palpitation and dyspnea.
II	Mild symptoms	Slight limitation of physical activity; comfortable at rest; activity results in fatigue, palpitation and dyspnea.
III	Moderate symptoms	Marked limitation of physical activity; comfortable at rest; regular exercise causes fatigue, palpitation and dyspnea.
IV	Severe symptoms	Unable to perform any physical activity without discomfort; HF symptoms at rest; if any physical activity is performed, the pain increases.

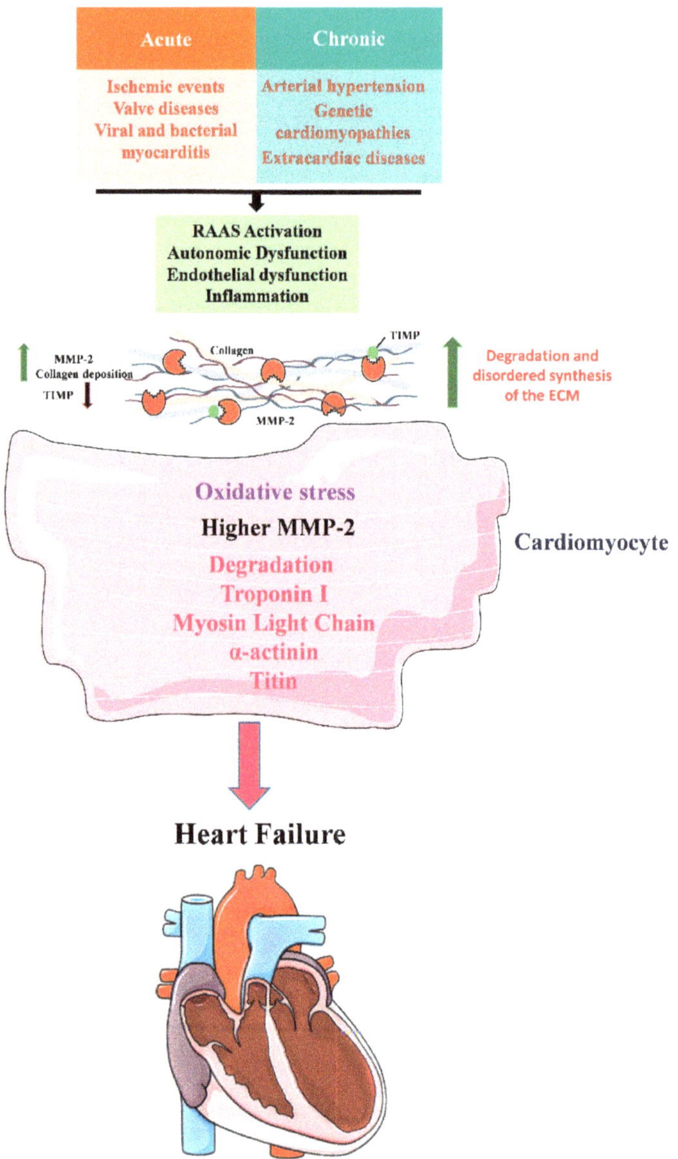

Figure 1. Cardiac remodeling in HF. HF occurs after an acute or chronic harmful event. The conditions that trigger this disease are hypertension, valvular diseases, genetic cardiomyopathies, myocarditis, extracardiac diseases and ischemia, generating autonomic dysfunction and activation of the renin-angiotensin-aldosterone system (RAAS), endothelial dysfunction and inflammation. In addition, the degradation and disordered synthesis of the extracellular matrix (ECM) occurs due to increased activity of MMP-2 and decreased activity of endogenous tissue inhibitors (TIMP), leading to collagen deposition and oxidative stress, causing degradation of components of the contractile apparatus, troponin I, light chain myosin, alpha-actinin and titin, promoting structural and functional changes; image elements from smart.server.com.

MMP-2 is a gelatinase constitutive of the heart that is considered a biomarker of patients with HFrEF and HFpEF. MMP-2 can digest components of the contractile apparatus, such as troponin I and light chain myosin 1, which contributes to the reduction in cardiac contractility [17]. At the transcriptional level, MMP-2 expression is controlled by transcription factors [18]. At the post-transcriptional level, MMP-2 activity and expression are regulated by inflammatory stimuli, oxidative stress, and alteration of the renin-angiotensin-aldosterone (RAAS) axis [19–22]. There is also a class of endogenous tissue inhibitors of metalloproteinase (TIMPs) that participate in the control of MMPs, and the balance between MMPs and TIMPs plays an essential role in the pathophysiology of heart disease [23,24] (Figure 1).

MMP-2 can be produced and secreted in the heart by cardiomyocytes, fibroblasts, endothelial cells and inflammatory cells present during the progression of HF. Although present in greater volume, cardiomyocytes are in smaller numbers than non-myocyte cells, which comprise 70% of the cells in cardiac tissue, most of which include fibroblasts. Cardiac fibroblasts maintain cardiac structural integrity by controlling cardiac extracellular matrix content. In addition, fibroblasts surround cardiomyocytes, and the myocyte function depends on the fibroblast. In pathological processes, cytokines and growth factors alter the fibroblast phenotype by increasing the secretion of ECM proteins that lead to fibrosis [25–27]. Fibroblasts and myofibroblasts are the central MMP-secreting cells in the heart and are indicated as a therapeutic target for the treatment of myocardial infarction, hypertension and HF [26].

Before introducing neurohormonal therapies in treating patients with HF, more than a third of deaths were attributed to sudden cardiac death. However, evidence-based clinical trials using neurohormonal treatments in patients not using a defibrillator showed a reduction in the rate of premature death. Furthermore, in addition to current pharmacological therapies, resynchronization devices, cardioverter-defibrillators and ventricular assist devices drastically reduced the risk of death. Indeed, HFrEF shows a different trajectory in response to drug therapy and cardiac resynchronization devices compared to HFpEF. However, both patient groups have benefited from available treatments, showing improvements in myocyte function, normalization of action potential duration, and improvement in mitochondrial energy metabolism. However, an unacceptable number of patients suffer impairment of functional capacity, low quality of life and early death due to HF. Thus, therapies that can stop or minimize the progression of HF continue to be challenging [8,9,28].

In this mini-review, we will highlight the participation of MMP-2 in cardiac alterations related to HF. Next, we will present some MMP inhibitors used in pre-clinical and clinical trials as a pharmacological tool for treating various diseases. Then we will address the use of MMP-2 inhibitors as an alternative treatment for HF in animal and human models. Finally, we will briefly explain the possible use of MMP-2 inhibitors and new technologies as an adjuvant treatment associated with standard therapy and their impact on the progression of HF.

2. Matrix Metalloproteinase 2 (MMP-2)

MMPs are a family of proteases specialized in degrading ECM components, which have a highly homologous protein structure. Most have four basic domains: signal peptide, pro-peptide, catalytic, and hemopexin-like domains [29]. Based on their substrate affinity and structural organization, MMPs are commonly classified as collagenases, gelatinases, stromelysins, matrilysins, membrane MMPs, and others [30]. MMP-2 belongs to a group of gelatinases that have a unique catalytic domain among MMPs composed of a triple repeat of type II fibronectin, which forms a collagen affinity domain. This allows the binding and degradation of type IV collagen and denatured collagen (gelatin) [29] (Figure 2).

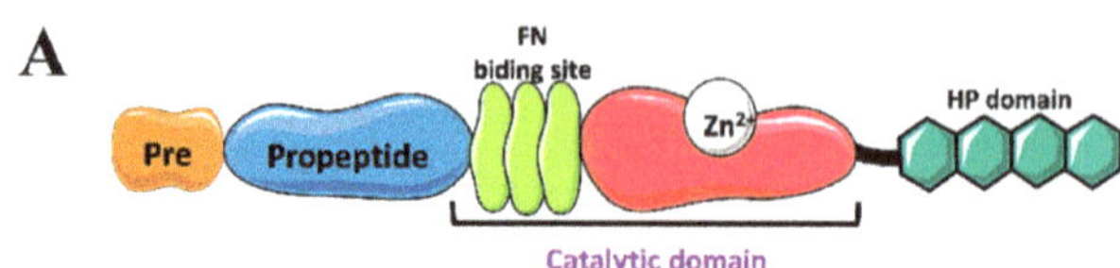

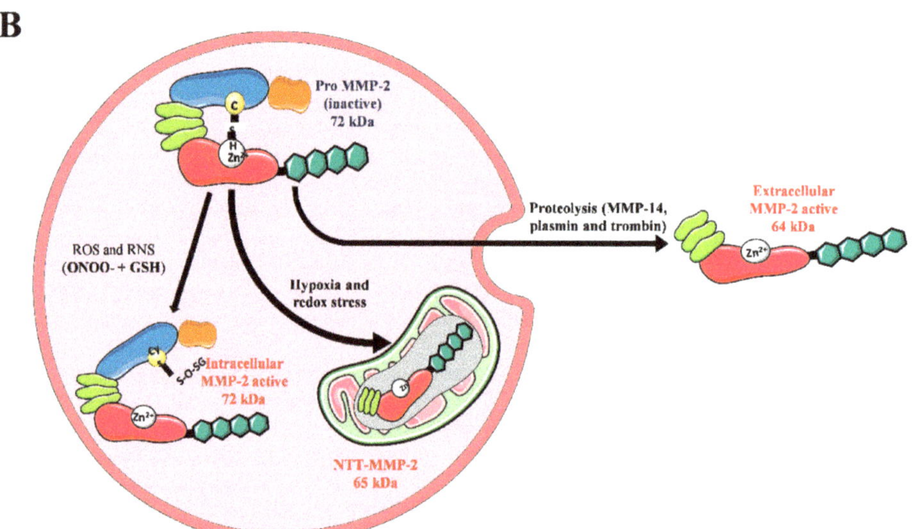

Figure 2. Structure, activation and isoforms of MMP-2. (**A**) MMP-2 has in its structure a signal peptide (Pre), a propeptide (Pro), a catalytic domain (having a zinc ion and three fibronectin repeats that confer affinity to collagen) and hemopexin (HP). (**B**) The inactive isoform of MMP-2 has a molecular size of 72 kDa (pro-MMP-2). Inactivity is guaranteed by a cysteine residue in the propeptide domain, which binds to Zn^{2+} in the catalytic domain, preventing the binding and proteolysis of substrates. The active intracellular isoform of MMP-2 has a molecular size of 72 kDa and occurs when $ONOO^-$ or GSH reacts with $ONOO^-$. The reaction product binds to the cysteine residue of the propeptide, which prevents it from complexing with the Zn^{2+} atom in the catalytic domain, allowing the catalytic domain to interact with substrates. The 65 kDa NTT-MMP-2 is constitutively active, formed under conditions of hypoxia and oxidative stress, and leads to activation of alternative MMP-2 promoters that do not translate the first 77 amino acids. This isoform is not secreted into the extracellular environment, found in mitochondria and cytosol. The extracellular isoform of MMP-2 with a molecular size of 72 kDa, activated by the proteolytic removal of the propeptide domain by MMP-14, thrombin and plasmin, produces an active isoform of 64 kDa. FN: fibronectin GSH: reduced glutathione; $ONOO^-$: peroxynitrite; ROS: reactive oxygen species; RNS: reactive nitrogen species; NTT-MMP-2: truncated N-terminal isoform of MMP-2; image elements from smart.server.com.

MMP-2 is encoded by a 27-kb gene that has 13 exons and 12 introns located on chromosome 16. The gene has consensus sequences for the transcription factors AP-2 and SpI [31]. The gene is transcribed into a 3.1-kb mRNA [32], which is translated into a 660-residue protein that contains a 29-residue signal peptide. This peptide is responsible for translocating MMP-2 to the endoplasmic reticulum, followed by secretion into the extracellular medium, giving rise to a latent enzyme of about 72 kDa [29,33] (Figure 2).

The absence of catalytic activity of the enzyme is maintained by the interaction between a sulfhydryl bond between a cysteine residue present in the pro-peptide and zinc at the catalytic site [34]. In the enzymatic activation, the catalytic site is exposed, which can occur through proteolysis of the propeptide by other proteases (MMP-2, plasmin, and thrombin) or through the interruption of the sulfhydryl bond by reactive species [35,36]. In proteolytic activation, an MMP-2 with a molecular size of 64 kDa is formed [29]. In the process of activation by reactive species, the molecular size of 72 Kda is maintained due to the permanence of the pro-peptide [36]. MMP-2 is expressed in most body tissues and modulates several physiological processes, such as cell migration, angiogenesis, and wound healing [29] (Figure 2). However, increased expression and activity of MMP-2 is involved in cardiovascular diseases such as atherosclerosis, aneurysm, hypertension, and HF [37,38].

3. Role of MMP-2 in HF

The knowledge of specific biomarkers used in the clinic to determine pathological states, define diagnoses or even prognoses are of great value since the symptoms of HF are often not pathognomonic, making diagnosis difficult. The main biomarkers to diagnose HF are the natriuretic peptides BNP and NT-pro-BNP [9,39]. However, BNP can be altered due to several factors such as kidney disorders, advanced age, obesity, diabetes, sepsis, Cushing's syndrome and hyperthyroidism. Thus, the search for biomarkers that can help in the diagnosis and prognosis of HF becomes imperative.

MMP-2 can be considered a biomarker of HF, as higher plasma MMP-2 levels were found in patients with congestive HF, resulting from different etiologies (acute myocardial infarction, dilated cardiomyopathy and valvular disease). Higher levels of MMP-2 are correlated with patients with a worse prognosis for HF (NYHA class II–IV) [40], as well as an increased risk of death or hospitalizations for HF [41]. MMP-2 and TIMP-1 are higher in the plasma of patients with acute HF [42]. MMP-2 is considered the best biomarker among the other proteases in the ECM, as its levels varied only slightly during a temporal evaluation [43].

MMP-2 has also been considered a biomarker of LV remodeling in patients with HF who have suffered an acute myocardial infarction, with extensive areas of injury and decreased ejection fraction [44]. A study that evaluated control patients (who did not have cardiovascular disease), patients with LV hypertrophy without HF, and patients with diastolic HF and LV hypertrophy, showed that the dosage of MMP-2 and procollagen III N-terminal propeptide (PIIINP) together, were shown to be biomarkers that predict HFpEF better than NT-proBNP dosing alone [45]. MMP-2, MMP-9 and TIMP-1 have also been recognized as highly valuable biomarkers for predicting the risk of death in patients with HF [46].

In addition to being a biomarker, the participation of MMP-2 in the development of HF was suggested due to its role in the degradation of components of the myocardial matrix and the regulation of the fibrotic process, contributing to progressive dilation of the cardiac chambers, reduction in heart compliance and driving problems. In this context, MMP-2 is a key protease in the maladaptive remodeling process of the heart [20,40,41]

The role of MMP-2 in developing cardiac injury and dysfunction was demonstrated in a transgenic model of MMP-2 overexpression. Interestingly, this model showed that an increase in MMP-2 levels in the heart is sufficient to induce ventricular dysfunction, with myocyte hypertrophy, contractile protein lysis and cardiac fibrosis, even without a pathological process [47]. Furthermore, when subjected to ischemia and reperfusion injury, those animals that overexpress MMP-2 have higher infarction areas, lipid peroxidation and cardiac dysfunction compared to normal animals [48].

For a better understanding of the involvement of MMP-2 in pathological processes that lead to HF, the hearts of rats that suffered ischemia and reperfusion were evaluated, and there was an increase in the production of reactive species, including peroxynitrite ($ONOO^-$) associated with MMP activation, before being secreted into the extracellular environment [36,49]. As a result, MMP-2 promoted proteolysis of contractile machinery proteins, including titin [50], troponin I [51], myosin light chain [52,53] and alpha-actinin [54]. In this experimental model, the disruption of the cell cytoskeleton by MMP-2 is involved in the decrease in myocardial contractility, with oxidative stress being the main factor related to the increase in MMP-2 activity. Thus, decreasing the oxidative stress or inhibiting the catalytic activity of MMP-2 may be a therapeutic target.

The redox imbalance in cardiomyocytes during ischemia also leads to the activation of alternative MMP-2 promoters, producing an N-Terminal Truncated isoform called NTT-MMP-2, constitutively active and present in mitochondria, altering energy metabolism and mitochondrial function and activation of the innate immune response [33]. In addition, overexpression of NTT-MMP-2 in mouse hearts results in LV hypertrophy, intense inflammatory cell infiltration, cardiomyocyte apoptosis and cardiac dysfunction [55].

Studies with knockout mice for MMP-2 were performed to confirm the participation of MMP-2 in HF after acute myocardial infarction. The absence of MMP-2 did not change the infarct area. However, the animals showed less LV dilation and increased survival than wild animals [56]. MMP-2 deletion was also beneficial in mice subjected to increased cardiac preload, in which they showed decreased myocyte hypertrophy and improved fibrosis and cardiac dysfunction [57].

On the other hand, inhibition of MMP-2 at below baseline levels can become an issue [58]. This was demonstrated in a preclinical study, using MMP-2$^{-/-}$ mice with cardiac overexpression of TNF-α showing decreased survival, LV contractile dysfunction, and increased infiltration of inflammatory cells of the myocardium [59]. Another study with *MMP-2$^{-/-}$* mice infused with angiotensin II showed that MMP-2 deletion did not affect the severity of hypertension but caused cardiac hypertrophy to develop earlier and to a greater extent than in wild-type animals [60]. Furthermore, clinical studies have shown that patients with loss of MMP-2 function due to mutations in the *MMP-2* gene are predisposed to a complex multisystem syndrome involving abnormalities of cardiac development [61,62].

The manipulation of MMP-2 genes helped us to confirm its participation in the pathophysiology of HF. However, when we analyzed the studies mentioned above, we realized that an exacerbated increase in activity or expression and the complete deletion of this protease is harmful during HF's evolution. In this way, the ideal would be to modulate the levels and the activity of MMP-2 to prevent functional dysfunction caused by the remodeling of the heart. Therefore, the use of molecules that can inhibit MMP-2 may work as an effective treatment for the progression of HF.

4. The Development of Inhibitors for MMPs and Their Use as a Pharmacological Tool in Disease

The understanding of the role of MMPs in the pathophysiology of cardiovascular diseases, neurodegenerative disorders such as Parkinson's and Alzheimer's and cancer raised the hypothesis of the importance of regulating these prostheses as a way to stop changes in physiological processes such as angiogenesis, tissue remodeling, healing, migration cell, activation of signaling molecules and immunity [63]. Table 3 summarizes the different MMP inhibitors and their characteristics.

MMPs have endogenous inhibitors, including α2-macroglobulin, a protease secreted by the liver, that binds to MMPs in plasma, preventing them from degrading their substrates. Tissue MMPs are inactivated by TIMPs, with four members that make up this family: TIMP-1 to 4. TIMPs can inhibit all MMPs, but with different specificities for each one of them [64]. TIMPs have been used as a therapeutic tool in several diseases to modulate MMPs. However, without favorable results, possibly because they share similar pathways without directly interfering with each other's role [63].

Table 3. Development of MMP inhibitors and their characteristics.

Class of MMP Inhibitors	Inhibitor (Alternative Names)	Characteristics
Endogenous inhibitors	α2-macroglobulin and TIMPs	It traps MMPs in the plasma, preventing them from degrading their substrates. It inhibits tissue MMPs and has four members: TIMP-1 to 4. TIMPs can inhibit all MMPs, but with different specificities.
Hydroxamate-based inhibitors	Batimastat and Marimastat	They are designed to mimic the natural peptide substrate (collagen) of MMPs. It targets the catalytic site of MMPs.
The new generation of hydroxamate-based inhibitors	Cipemastat and MMI-166	They were developed with a sulfonamide and a zinc-binding hydroxamate group, in addition to the substitution of an aryl group, generating a compound with more specificity. It targets the catalytic site of MMPs.
Non-hydroxamate inhibitors	Rebimastat and Tanomastat	They were designed with various peptidomimetics and non-mimetics, not limited to mimicking the substrate of MMPs. It targets the catalytic site of MMPs.
Inhibitors targeting alternative binding sites	BMS-275291 and specific MMP-13 inhibitor (provided by Pfizer, Ann Arbor, MI, USA)	Highly selective, unlike previous MMP inhibitors, because it does not bind to catalytic zinc ion and is not competitive for substrate binding. They target alternative, less conserved binding sites.

The first class of synthetic molecules used as inhibitors of MMPs was based on hydroxamate. These compounds were designed to mimic the natural peptide substrate (collagen) of MMPs associated with a group that can chelate the Zn^{2+} ion of the catalytic site, thus favoring broad-spectrum inhibition [65,66]. An example of this type of drug was batimastat, which could inhibit MMP-9 and slow the growth of solid tumors in preclinical trials [67]. In a double-blind prospective clinical trial, this drug was used to inhibit MMPs, as an adjuvant treatment in patients with small-cell lung cancer. However, it did not improve survival, decreasing the quality of life of patients who used it [68]. Another example of a hydroxamate-based drug was marimastat, a structural analog of batimastat, which showed favorable results in most preclinical studies in tumor models, thus serving as a basis for clinical trials [69]. A series of phase I, phase II and phase III tests were performed against solid metastatic tumors, with significant depletions in the increase in tumor markers being observed. However, in phase III studies, patients showed musculoskeletal toxicity, which may be associated with inhibition of ADAM and ADAMTS family members [66,70–72]. This class of hydroxamate-based inhibitors was able to inhibit several MMPs, including MMP-1, MMP-2, MMP-7 and MMP-9, showing benefits in preclinical trials but failed in clinical trials mainly due to their nonspecificity promoting combined inhibition of several MMPs resulting in musculoskeletal syndrome (arthralgia, myalgia and tendinitis) [63,66]. In addition, these drugs may have been introduced too late to modify a pathological condition, which could explain their failure in clinical trials. Therefore, first-generation hydroxamate-based molecules were discontinued after clinical trials failed.

However, a new generation of hydroxamate-based molecules was developed, presenting a sulfonamide and a zinc hydroxamate linking group and substituting an aryl group, generating a compound with more specificity to minimize the adverse effects of the previous generation triggered. Its development uses structure-activity relationship analysis (SAR), which helps identify molecular substructures related to the presence or absence of biological activity [63]. Cipemastat, used in treating patients with rheumatoid arthritis and osteoarthritis, is an example of this class. It was able to inhibit MMP-1, MMP-3 and MMP-9 more selectively. However, it did not prevent the progression of joint damage [66]. MMI-166 is selective for MMP-2, MMP-9 and MMP-14 and decreases the cellular invasion of cervical carcinoma in vitro. However, there was no suppression of the proliferation of tumor cells [66,73]. A recurring limitation in the use of these inhibitors was the premature metabolism suffered by the drug, leading to the loss of the hydroxamate group that binds with zinc. Despite the difficulties encountered in the therapeutic use of this new generation of inhibitors, there is still significant interest in developing drugs derived from hydroxamic acid, as these compounds are the most potent inhibitors of MMP available to date [63].

The search for molecules with lower metabolic lability and more stable bonds to the Zn^{2+} ion of the catalytic site led to the development of compounds derived from phosphoric acid, hydantoins and carboxylates, usually called non-hydroxamate MMP inhibitors. Next-generation MMP inhibitors have been designed with a variety of peptidomimetics and non-mimetics, not limited to mimicking their substrates [63]. One of the first non-hydroxamate MMP inhibitors developed was Rebimastat, a broad-spectrum inhibitor containing a thiol group that binds to zinc. Phase II clinical trials in early-stage breast cancer and a phase III study in lung carcinoma using this compound as adjuvant therapy discontinued treatment because patients experienced arthralgia consistent with MMP inhibitor-induced toxicity [74,75]. Another tanomastat inhibitor showed good tolerability and variable efficacy that depended on the timing of administration concerning disease progression [76,77]. Several biphenyl sulfonamide carboxylate-based MMP inhibitors have been designed to treat osteoarthritis by inhibiting MMP-13.

Substances such as polyphenols, flavonoids and carotenoids, obtained from natural products, can inhibit MMPs, with photoprotective and antioxidant properties. For example, *P. leucotomos* inhibits the expression of MMPs in epidermal keratinocytes and fibroblasts and stimulates TGF-β in skin fibroblasts by decreasing lipid peroxidation and oxidative stress [78,79]. Xanthohumol directly inhibits MMP-1, MMP-3 and MMP-9 while increasing the expression of collagen types I, III and V, fibrillin-1 and 2 in dermal fibroblasts [80]. Lutein prevents photoaging by inhibiting MMPs and oxidative stress, reducing epidermal hyperproliferation, expanding mutant keratinocytes, and mast cell infiltration in response to solar radiation [81,82]. The photoprotective activity presented by these compounds is probably associated with a decrease in the degradation of collagen fibers by MMP-1 and MMP-2 and of elastin fibers by MMP-2 and MMP-9 as well as by stimulating TGF-β in fibroblasts, which inhibits MMP-1 and stimulates collagen production [78]. Tetracycline antibiotics, such as doxycycline and minocycline, have an innate inhibitory capacity for MMPs [63,66].

To reduce off-target effects and to avoid broad inhibition of MMPs, due to their high structural homology, current inhibitors are being designed to target alternative, less conserved binding sites. Using crystallography and X-rays combined with computational methods allows the modeling of drug-protein interactions with inhibitors that bind to other sites. In addition, combining techniques with computational prediction revealed hidden sites in the structure of MMPs that can be explored for the rational design of new molecular effectors and therapeutic agents [63,66]

5. Use of Nonspecific and Specific Inhibitors for MMP-2 in HF

Preclinical studies demonstrate that a therapeutic approach to MMP-2 inhibition may be a promising strategy for treating patients with HF. ONO-4817 a selective inhibitor, has shown beneficial results in an ischemia and reperfusion model, improving contractile dysfunction, associated with decreased MMP-2 activity and titin proteolysis [50]. In addition, it showed promising results in attenuating LV remodeling and myocardial fibrosis in mice treated with doxorubicin, a drug used in cancer patients that is cardiotoxic and leads to HF, by increasing oxidative stress and MMP-2 activity [83]. Collectively these studies support the hypothesis that inhibiting MMPs by selectively using ONO-4817 has therapeutic potential, as this compound selectively inhibits MMP2, significantly decreasing the extent of lesions and disease severity. Furthermore, it cannot inhibit MMP-1, which has been associated with adverse effects triggered by hydroxamate-based inhibitors.

Antibiotics of the tetracycline class, such as doxycycline, which has an innate inhibitory capacity for MMPs, with greater specificity for MMP-2, MMP-9 and MMP-8, have also been used in models of cardiac injury. In preclinical trials, doxycycline prevented the conversion of concentric hypertrophy to eccentric hypertrophy of the LV during hypertension. This effect was associated with decreased MMP-2 activity and reduced troponin I and dystrophin proteolysis, thus improving the mechanical stability of cardiomyocytes and the contractile function [84]. On the other hand, doxycycline could not reduce scar thinning and compensatory LV hypertrophy, despite having decreased MMP-2 and MMP-9 activity in a model of acute myocardial infarction with LV dysfunction. These findings draw attention to the non-selective inhibition of MMPs in the initial healing phase after IM [85]. Doxycycline is an antibiotic capable of inhibiting MMPs at subantimicrobial doses and is currently the only FDA-approved MMP inhibitor for the treatment of periodontal disease [86]. In addition, it shows benefits in other conditions such as abdominal aortic aneurysm [87], arterial hypertension [88–92] and acute myocardial infarction [93].

Clinical studies that evaluated the effects of doxycycline on HF showed results dependent on the dose used and the etiology. For example, patients with acute myocardial infarction (40% of patients with HFrEF) were treated with doxycycline 100 mg as adjunctive therapy. They improved diastolic function and decreased the infarct area [94]. On the other hand, two randomized clinical trials that evaluated the effects of adjuvant treatment with doxycycline at a dose of 20 mg in patients with coronary artery disease and atherosclerosis showed no improvement in cardiac dysfunction parameters and sudden death outcomes [95,96]. The pathophysiological processes that trigger the damage and cardiac remodeling are closely correlated with the therapeutic response and the dose used since the preferential inhibition of specific MMPs are associated with the dose of doxycycline used as well time of therapy instituted.

The adverse effects of non-selective MMP inhibition are reinforced by the clinical study in patients with acute myocardial infarction and HFpEF, which evaluated the impact of the inhibitor PG-116800 (oral MMP inhibitor with an affinity for MMP-2, MMP-3, MMP-8, MMP-9, MMP-13 and MMP-14 and low affinity for MMP-1 and MMP-7). The PG-116800-treated group showed no improvement in heart function and death rates compared to the placebo. Unfortunately, this study was discontinued due to the development of musculoskeletal toxicity with no apparent benefit following administration of PG-116800 [97]. Therefore, we emphasize the importance of using inhibitors as an adjuvant therapy with greater specificity and fewer off-target effects.

In this way, using more selective inhibitors aimed at binding to alternative, less conserved sites can be a therapeutic strategy in HF, minimizing the adverse effects of nonspecific inhibitors. For example, TISAM, an N-sulfonylamino acid derivative, which selectively inhibits MMP-2, was used in a model of acute myocardial infarction, improving survival rate, preventing cardiac rupture and delaying post-infarction remodeling. These benefits were associated with decreased MMP-2 activity and macrophage infiltration into cardiac tissue [56].

Using chemical modeling to inhibit MMP-2 selectively, the inhibitors MMPI-1154, MMPI-1260 and MMPI-1248 were developed. MMPI-1154 and MMPI-1260 showed efficacy in reducing infarct size associated with decreased MMP-2 activity [98,99]. However, despite the potential of the TISAM, MMPI-1154 and MMPI-1260, these compounds have not been tested in other preclinical, experimental models of HF and clinical studies have not been conducted.

The selective inhibition of MMP-2 was also evaluated with siRNA technology. MMP-2 deletion in cardiomyocytes isolated from adult rats undergoing ischemia and reperfusion injury prevented contractile dysfunction associated with decreased MLC1/2 degradation [100]. The encapsulation of siRNA for MMP-2 in a hydrogel to improve cell penetration was also investigated in an in vivo acute infarction model. Positive effects on heart hemodynamics were observed, where the reduction in MMP-2 in cardiomyocytes led to the maintenance of cardiac output and ejection fraction [101]. Together, the studies that used compounds or technology of selective inhibition of MMP-2 reinforce that this protease can be a therapeutic target in the treatment of HF.

Clinical studies evaluating the effects of statins on MMP-2 inhibition (atorvastatin, rosuvastatin and pravastatin) in patients with HFrEF after acute myocardial infarction showed decreased serum MMP-2 levels associated with a reduction in the number of deaths and hospital readmission [102–104]. These studies did not assess whether these drugs directly inhibit MMP-2, but statins are known to decrease inflammation and oxidative stress [105,106], which may be related to the results found in reduced serum levels of MMP-2.

Drugs used in the treatment of hypertension, such as verapamil, carvedilol and trimetazidine, have shown positive effects on cardiac function and remodeling associated with decreased activity and expression of MMP-2 [107–109]. Low-dose carvedilol is cardio-protective and inhibits MMP-2 activity in an ischemia/reperfusion model [109]. Verapamil has been shown to reduce MMP-2 activity by decreasing oxidative stress and calpain-1 that regulates MMP-2 activity in a model of HF induced by hypertension [108]. At the same time, trimetazidine has reduced MMP-2 expression by decreasing oxidative stress in an animal model of myocardial infarction [107]. Table 4 summarizes the non-selective and selective MMP-2 inhibitors evaluated in preclinical and clinical studies of HF.

It is evident that MMP-2 plays an essential role in cardiac injury and HF development. Preclinical studies better explain the pathophysiological mechanisms involving extracellular matrix proteins. Clinical evidence indicates that selective inhibition of MMPs is optimal, as non-selective inhibition with MMP-inhibiting compounds is associated with loss of response and adverse reactions, possibly through inhibition of MMPs essential for body homeostasis. It is noteworthy that more clinical studies involving selective and non-selective MMP-2 compounds are needed, as the use of inhibitors for MMP-2 has shown promise as adjuvant therapy for HF in preclinical models. In addition, the use of drugs already used in the clinic can be an alternative for inhibiting MMP-2, having the advantage of safety in its use.

Table 4. Non-selective and selective MMP-2 inhibitors were evaluated in preclinical and clinical studies of HF.

Non-Selective Inhibitor	Species	Disease	Comments	References
Doxycycline	Rats	Renovascular hypertension with HF	Prevented the conversion of concentric hypertrophy to eccentric hypertrophy in the LV, associated with decreased MMP-2 activity and reduced troponin I and dystrophin proteolysis	[84]
Doxycycline	Mice	Model of acute myocardial infarction with HF	It has not reduced scar thinning and compensatory LV hypertrophy, despite having decreased MMP-2 and MMP-9 activity	[85]
Doxycycline (Adjuvant therapy)	Humans	Acute myocardial infarction (40% of patients with HFrEF)	Improved diastolic function and reduced infarct area	[94]
Doxycycline (Adjuvant therapy)	Humans	Coronary artery disease and atherosclerosis	There was no improvement in cardiac dysfunction parameters and sudden death outcomes	[95,96]
PG-116800 (Adjuvant therapy)	Humans	Acute myocardial infarction (HFpEF) with HF	No improvement in heart function and death rates Development of musculoskeletal toxicity	[97]

Table 4. *Cont.*

Non-Selective Inhibitor	Species	Disease	Comments	References
MMP-2 selective inhibitor				
ONO-4817	Mice	Ischemia and reperfusion model with HF	Shown to improve contractile dysfunction associated with decreased MMP-2 activity and titin proteolysis	[50]
ONO-4817	Mice	Model of doxorubicin-induced cardiotoxicity	Attenuated LV remodeling and myocardial fibrosis	[83]
TISAM	Mice	Model of acute myocardial infarction with HF	It improved survival rate by preventing cardiac rupture and delaying post-infarction remodeling	[56]
MMPI-1154, MMPI-1260 and MMPI-1248 (Chemical Modeling)	Mice	Model of acute myocardial infarction with HF	They showed inhibitory activity on MMP-2, associated with a reduction in the infarct area	[98,99]
siRNA for MMP-2	Mice	Ischemia and reperfusion model with HF	It prevented contractile dysfunction associated with decreased degradation of MLC1/2	[100]
Hydrogel encapsulated siRNA for MMP-2	Mice	Model of acute myocardial infarction	Improved cardiac output and ejection fraction	[101]
Statins (Atorvastatin, Rosuvastatin and Pravastatin)	Humans	Acute myocardial infarction (HFrEF)	Decreased serum MMP-2 levels are associated with a reduced number of deaths and hospital readmission	[102–104]
Antihypertensive drugs (Verapamil, Carvedilol and Trimethazine)	Mice and rats	Ischemia/reperfusion model; Model of HF induced by hypertension and Myocardial Infarction Model	Positive effects on cardiac function and remodeling associated with decreased activity and expression of MMP-2	[107–109]

Selective inhibitors for MMP-2 tested in a multitude of diseases may be a plausible alternative in the treatment of HF, but the safety profile, possible adverse effects, and clinical results must be taken into account [110–115]. In addition, as a future perspective, the use of computational modeling as a tool can help predict the behavior of a molecule in non-living systems. Besides that, structure-activity relationship (SAR) analysis helps identify molecular substructures related to the presence or absence of biological activity. Finally, genomic engineering, such as clustered regularly interspaced short palindromic repeat (CRISPR), has a potential future in treating cardiovascular diseases, including HF.

6. Conclusions

Modulation of MMP-2 by non-selective or specific inhibitors has the potential to provide new directions for studying the mechanisms underlying various heart diseases, including HF. In addition, it has potential as a therapeutic tool for clinical practice and could have a significant impact on the development of new approaches to protect against cardiac remodeling and dysfunction in HF.

Author Contributions: Writing—original draft preparation, P.R.G., L.D.N., R.F.G., K.E.R. and A.F.P.; writing—review and editing, R.F.G., K.E.R. and A.F.P.; supervision, K.E.R. and A.F.P.; project administration, A.F.P.; funding acquisition, R.F.G. and A.F.P. All authors have read and agreed to the published version of the manuscript.

Funding: Fundação de Amparo a Pesquisa do Estado de São Paulo (FAPESP Grant number 2014-23888-0), Conselho Nacional de Desenvolvimento científico e Tecnológico, Coordenação de Aperfeiçoamento de Pessoal de Nível Superior—Brasil (CAPES)—Finance Code 001. The APC payment was supported by Pró-reitoria de Pesquisa e Pós-Graduação (PROPESP) from Federal University of Pará (UFPA).

Institutional Review Board Statement: Not applicable.

Informed Consent Statement: Not applicable.

Data Availability Statement: Not applicable.

Conflicts of Interest: The authors declare no conflict of interest.

References

1. Savarese, G.; Lund, L.H. Global Public Health Burden of Heart Failure. *Card. Fail Rev.* **2017**, *3*, 7–11. [CrossRef]
2. Tsao, C.W.; Aday, A.W.; Almarzooq, Z.I.; Alonso, A.; Beaton, A.Z.; Bittencourt, M.S.; Boehme, A.K.; Buxton, A.E.; Carson, A.P.; Commodore-Mensah, Y.; et al. Heart Disease and Stroke Statistics-2022 Update: A Report From the American Heart Association. *Circulation* **2022**, *145*, e153–e639. [CrossRef] [PubMed]
3. Virani, S.S.; Alonso, A.; Aparicio, H.J.; Benjamin, E.J.; Bittencourt, M.S.; Callaway, C.W.; Carson, A.P.; Chamberlain, A.M.; Cheng, S.; Delling, F.N.; et al. Heart Disease and Stroke Statistics-2021 Update: A Report From the American Heart Association. *Circulation* **2021**, *143*, e254–e743. [CrossRef] [PubMed]
4. Seferovic, P.M.; Vardas, P.; Jankowska, E.A.; Maggioni, A.P.; Timmis, A.; Milinkovic, I.; Polovina, M.; Gale, C.P.; Lund, L.H.; Lopatin, Y.; et al. The Heart Failure Association Atlas: Heart Failure Epidemiology and Management Statistics 2019. *Eur. J. Heart Fail.* **2021**, *23*, 906–914. [CrossRef] [PubMed]
5. Liao, L.; Allen, L.A.; Whellan, D.J. Economic burden of heart failure in the elderly. *PharmacoEconomics* **2008**, *26*, 447–462. [CrossRef] [PubMed]
6. Cook, C.; Cole, G.; Asaria, P.; Jabbour, R.; Francis, D.P. The annual global economic burden of heart failure. *Int. J. Cardiol.* **2014**, *171*, 368–376. [CrossRef]
7. Malik, A.; Brito, D.; Vaqar, S.; Chhabra, L. *Congestive Heart Failure*; StatPearls: Treasure Island, FL, USA, 2022.
8. Mazurek, J.A.; Jessup, M. Understanding Heart Failure. *Heart Fail Clin.* **2017**, *13*, 1–19. [CrossRef]
9. Snipelisky, D.; Chaudhry, S.P.; Stewart, G.C. The Many Faces of Heart Failure. *Card Electrophysiol. Clin.* **2019**, *11*, 11–20. [CrossRef]
10. Pagliaro, B.R.; Cannata, F.; Stefanini, G.G.; Bolognese, L. Myocardial ischemia and coronary disease in heart failure. *Heart Fail. Rev.* **2020**, *25*, 53–65. [CrossRef]
11. Yancy, C.W.; Jessup, M.; Bozkurt, B.; Butler, J.; Casey, D.E., Jr.; Drazner, M.H.; Fonarow, G.C.; Geraci, S.A.; Horwich, T.; Januzzi, J.L.; et al. 2013 ACCF/AHA guideline for the management of heart failure: Executive summary: A report of the American College of Cardiology Foundation/American Heart Association Task Force on practice guidelines. *Circulation* **2013**, *128*, 1810–1852. [CrossRef]
12. Lalande, S.; Johnson, B.D. Diastolic dysfunction: A link between hypertension and heart failure. *Drugs Today* **2008**, *44*, 503–513. [CrossRef] [PubMed]
13. Sanchez-Marteles, M.; Rubio Gracia, J.; Gimenez Lopez, I. Pathophysiology of acute heart failure: A world to know. *Rev. Clin. Esp* **2016**, *216*, 38–46. [CrossRef] [PubMed]
14. Frantz, C.; Stewart, K.M.; Weaver, V.M. The extracellular matrix at a glance. *J. Cell Sci.* **2010**, *123*, 4195–4200. [CrossRef] [PubMed]
15. Hutchinson, K.R.; Stewart, J.A., Jr.; Lucchesi, P.A. Extracellular matrix remodeling during the progression of volume overload-induced heart failure. *J. Mol. Cell. Cardiol.* **2010**, *48*, 564–569. [CrossRef]
16. Segura, A.M.; Frazier, O.H.; Buja, L.M. Fibrosis and heart failure. *Heart Fail. Rev.* **2014**, *19*, 173–185. [CrossRef]
17. Kobusiak-Prokopowicz, M.; Krzysztofik, J.; Kaaz, K.; Jolda-Mydlowska, B.; Mysiak, A. MMP-2 and TIMP-2 in Patients with Heart Failure and Chronic Kidney Disease. *Open Med.* **2018**, *13*, 237–246. [CrossRef]
18. Piccoli, M.T.; Gupta, S.K.; Viereck, J.; Foinquinos, A.; Samolovac, S.; Kramer, F.L.; Garg, A.; Remke, J.; Zimmer, K.; Batkai, S.; et al. Inhibition of the Cardiac Fibroblast-Enriched lncRNA Meg3 Prevents Cardiac Fibrosis and Diastolic Dysfunction. *Circ. Res.* **2017**, *121*, 575–583. [CrossRef]
19. Hughes, B.G.; Schulz, R. Targeting MMP-2 to treat ischemic heart injury. *Basic Res. Cardiol.* **2014**, *109*, 424. [CrossRef]
20. Massaro, M.; Scoditti, E.; Carluccio, M.A.; De Caterina, R. Oxidative stress and vascular stiffness in hypertension: A renewed interest for antioxidant therapies? *Vasc. Pharmacol.* **2019**, *116*, 45–50. [CrossRef]
21. Wang, C.; Qian, X.; Sun, X.; Chang, Q. Angiotensin II increases matrix metalloproteinase 2 expression in human aortic smooth muscle cells via AT1R and ERK1/2. *Exp. Biol. Med.* **2015**, *240*, 1564–1571. [CrossRef]
22. White, D.A.; Su, Y.; Kanellakis, P.; Kiriazis, H.; Morand, E.F.; Bucala, R.; Dart, A.M.; Gao, X.M.; Du, X.J. Differential roles of cardiac and leukocyte derived macrophage migration inhibitory factor in inflammatory responses and cardiac remodelling post myocardial infarction. *J. Mol. Cell. Cardiol.* **2014**, *69*, 32–42. [CrossRef] [PubMed]
23. Hendry, R.G.; Bilawchuk, L.M.; Marchant, D.J. Targeting matrix metalloproteinase activity and expression for the treatment of viral myocarditis. *J. Cardiovasc. Transl. Res.* **2014**, *7*, 212–225. [CrossRef] [PubMed]
24. Squire, I.B.; Evans, J.; Ng, L.L.; Loftus, I.M.; Thompson, M.M. Plasma MMP-9 and MMP-2 following acute myocardial infarction in man: Correlation with echocardiographic and neurohumoral parameters of left ventricular dysfunction. *J. Card. Fail.* **2004**, *10*, 328–333. [CrossRef]
25. Dostal, D.; Glaser, S.; Baudino, T.A. Cardiac fibroblast physiology and pathology. *Compr. Physiol.* **2015**, *5*, 887–909. [CrossRef]
26. Turner, N.A.; Porter, K.E. Regulation of myocardial matrix metalloproteinase expression and activity by cardiac fibroblasts. *IUBMB Life* **2012**, *64*, 143–150. [CrossRef]

27. Pytliak, M.; Vaník, V.; Bojčík, P. Heart Remodelation: Role of MMPs. In *The Role of Matrix Metalloproteinase in Human Body Pathologies*; Travascio, F., Ed.; IntechOpen: London, UK, 2017.
28. Gouveia, M.R.A.; Ascencao, R.; Fiorentino, F.; Costa, J.; Broeiro-Goncalves, P.M.; Fonseca, M.; Borges, M. Current costs of heart failure in Portugal and expected increases due to population aging. *Rev. Port. Cardiol. Engl. Ed.* **2020**, *39*, 3–11. [CrossRef]
29. Nagase, H.; Woessner, J.F., Jr. Matrix metalloproteinases. *J. Biol. Chem.* **1999**, *274*, 21491–21494. [CrossRef]
30. Liu, J.; Khalil, R.A. Matrix Metalloproteinase Inhibitors as Investigational and Therapeutic Tools in Unrestrained Tissue Remodeling and Pathological Disorders. *Prog. Mol. Biol. Transl. Sci.* **2017**, *148*, 355–420. [CrossRef]
31. Huhtala, P.; Chow, L.T.; Tryggvason, K. Structure of the human type IV collagenase gene. *J. Biol. Chem.* **1990**, *265*, 11077–11082. [CrossRef]
32. Collier, I.E.; Wilhelm, S.M.; Eisen, A.Z.; Marmer, B.L.; Grant, G.A.; Seltzer, J.L.; Kronberger, A.; He, C.S.; Bauer, E.A.; Goldberg, G.I. H-ras oncogene-transformed human bronchial epithelial cells (TBE-1) secrete a single metalloprotease capable of degrading basement membrane collagen. *J. Biol. Chem.* **1988**, *263*, 6579–6587. [CrossRef]
33. Lovett, D.H.; Mahimkar, R.; Raffai, R.L.; Cape, L.; Maklashina, E.; Cecchini, G.; Karliner, J.S. A novel intracellular isoform of matrix metalloproteinase-2 induced by oxidative stress activates innate immunity. *PLoS ONE* **2012**, *7*, e34177. [CrossRef] [PubMed]
34. Van Wart, H.E.; Birkedal-Hansen, H. The cysteine switch: A principle of regulation of metalloproteinase activity with potential applicability to the entire matrix metalloproteinase gene family. *Proc. Natl. Acad. Sci. USA* **1990**, *87*, 5578–5582. [CrossRef] [PubMed]
35. Page-McCaw, A.; Ewald, A.J.; Werb, Z. Matrix metalloproteinases and the regulation of tissue remodelling. *Nat. Rev. Mol. Cell Biol.* **2007**, *8*, 221–233. [CrossRef] [PubMed]
36. Viappiani, S.; Nicolescu, A.C.; Holt, A.; Sawicki, G.; Crawford, B.D.; Leon, H.; van Mulligen, T.; Schulz, R. Activation and modulation of 72kDa matrix metalloproteinase-2 by peroxynitrite and glutathione. *Biochem. Pharmacol.* **2009**, *77*, 826–834. [CrossRef] [PubMed]
37. Azevedo, A.; Prado, A.F.; Antonio, R.C.; Issa, J.P.; Gerlach, R.F. Matrix metalloproteinases are involved in cardiovascular diseases. *Basic Clin. Pharmacol. Toxicol.* **2014**, *115*, 301–314. [CrossRef]
38. Prado, A.F.; Batista, R.I.M.; Tanus-Santos, J.E.; Gerlach, R.F. Matrix Metalloproteinases and Arterial Hypertension: Role of Oxidative Stress and Nitric Oxide in Vascular Functional and Structural Alterations. *Biomolecules* **2021**, *11*, 585. [CrossRef]
39. Willerson, J.T. The Medical and Device-Related Treatment of Heart Failure. *Circ. Res.* **2019**, *124*, 1519. [CrossRef]
40. Yamazaki, T.; Lee, J.D.; Shimizu, H.; Uzui, H.; Ueda, T. Circulating matrix metalloproteinase-2 is elevated in patients with congestive heart failure. *Eur. J. Heart Fail.* **2004**, *6*, 41–45. [CrossRef]
41. George, J.; Patal, S.; Wexler, D.; Roth, A.; Sheps, D.; Keren, G. Circulating matrix metalloproteinase-2 but not matrix metalloproteinase-3, matrix metalloproteinase-9, or tissue inhibitor of metalloproteinase-1 predicts outcome in patients with congestive heart failure. *Am. Heart J.* **2005**, *150*, 484–487. [CrossRef]
42. Biolo, A.; Fisch, M.; Balog, J.; Chao, T.; Schulze, P.C.; Ooi, H.; Siwik, D.; Colucci, W.S. Episodes of acute heart failure syndrome are associated with increased levels of troponin and extracellular matrix markers. *Circulation. Heart Fail.* **2010**, *3*, 44–50. [CrossRef]
43. Tager, T.; Wiebalck, C.; Frohlich, H.; Corletto, A.; Katus, H.A.; Frankenstein, L. Biological variation of extracellular matrix biomarkers in patients with stable chronic heart failure. *Clin. Res. Cardiol. Off. J. Ger. Card. Soc.* **2017**, *106*, 974–985. [CrossRef] [PubMed]
44. Cogni, A.L.; Farah, E.; Minicucci, M.F.; Azevedo, P.S.; Okoshi, K.; Matsubara, B.B.; Zanati, S.; Haggeman, R.; Paiva, S.A.; Zornoff, L.A. Metalloproteinases-2 and -9 predict left ventricular remodeling after myocardial infarction. *Arq. Bras. De Cardiol.* **2013**, *100*, 315–321. [CrossRef] [PubMed]
45. Zile, M.R.; Desantis, S.M.; Baicu, C.F.; Stroud, R.E.; Thompson, S.B.; McClure, C.D.; Mehurg, S.M.; Spinale, F.G. Plasma biomarkers that reflect determinants of matrix composition identify the presence of left ventricular hypertrophy and diastolic heart failure. *Circulation. Heart Fail.* **2011**, *4*, 246–256. [CrossRef]
46. Gao, Y.; Bai, X.; Lu, J.; Zhang, L.; Yan, X.; Huang, X.; Dai, H.; Wang, Y.; Hou, L.; Wang, S.; et al. Prognostic Value of Multiple Circulating Biomarkers for 2-Year Death in Acute Heart Failure With Preserved Ejection Fraction. *Front. Cardiovasc. Med.* **2021**, *8*, 779282. [CrossRef] [PubMed]
47. Bergman, M.R.; Teerlink, J.R.; Mahimkar, R.; Li, L.; Zhu, B.Q.; Nguyen, A.; Dahi, S.; Karliner, J.S.; Lovett, D.H. Cardiac matrix metalloproteinase-2 expression independently induces marked ventricular remodeling and systolic dysfunction. *Am. J. Physiol. Heart Circ. Physiol.* **2007**, *292*, H1847–H1860. [CrossRef]
48. Zhou, H.Z.; Ma, X.; Gray, M.O.; Zhu, B.Q.; Nguyen, A.P.; Baker, A.J.; Simonis, U.; Cecchini, G.; Lovett, D.H.; Karliner, J.S. Transgenic MMP-2 expression induces latent cardiac mitochondrial dysfunction. *Biochem. Biophys. Res. Commun.* **2007**, *358*, 189–195. [CrossRef]
49. Jacob-Ferreira, A.L.; Schulz, R. Activation of intracellular matrix metalloproteinase-2 by reactive oxygen-nitrogen species: Consequences and therapeutic strategies in the heart. *Arch. Biochem. Biophys.* **2013**, *540*, 82–93. [CrossRef]
50. Ali, M.A.; Cho, W.J.; Hudson, B.; Kassiri, Z.; Granzier, H.; Schulz, R. Titin is a target of matrix metalloproteinase-2: Implications in myocardial ischemia/reperfusion injury. *Circulation* **2010**, *122*, 2039–2047. [CrossRef]
51. Wang, W.; Schulze, C.J.; Suarez-Pinzon, W.L.; Dyck, J.R.; Sawicki, G.; Schulz, R. Intracellular action of matrix metalloproteinase-2 accounts for acute myocardial ischemia and reperfusion injury. *Circulation* **2002**, *106*, 1543–1549. [CrossRef]

52. Cadete, V.J.; Sawicka, J.; Jaswal, J.S.; Lopaschuk, G.D.; Schulz, R.; Szczesna-Cordary, D.; Sawicki, G. Ischemia/reperfusion-induced myosin light chain 1 phosphorylation increases its degradation by matrix metalloproteinase 2. *FEBS J.* **2012**, *279*, 2444–2454. [CrossRef]

53. Sawicki, G.; Leon, H.; Sawicka, J.; Sariahmetoglu, M.; Schulze, C.J.; Scott, P.G.; Szczesna-Cordary, D.; Schulz, R. Degradation of myosin light chain in isolated rat hearts subjected to ischemia-reperfusion injury: A new intracellular target for matrix metalloproteinase-2. *Circulation* **2005**, *112*, 544–552. [CrossRef] [PubMed]

54. Sung, M.M.; Schulz, C.G.; Wang, W.; Sawicki, G.; Bautista-Lopez, N.L.; Schulz, R. Matrix metalloproteinase-2 degrades the cytoskeletal protein alpha-actinin in peroxynitrite mediated myocardial injury. *J. Mol. Cell. Cardiol.* **2007**, *43*, 429–436. [CrossRef]

55. Lovett, D.H.; Mahimkar, R.; Raffai, R.L.; Cape, L.; Zhu, B.Q.; Jin, Z.Q.; Baker, A.J.; Karliner, J.S. N-terminal truncated intracellular matrix metalloproteinase-2 induces cardiomyocyte hypertrophy, inflammation and systolic heart failure. *PLoS ONE* **2013**, *8*, e68154. [CrossRef] [PubMed]

56. Matsumura, S.; Iwanaga, S.; Mochizuki, S.; Okamoto, H.; Ogawa, S.; Okada, Y. Targeted deletion or pharmacological inhibition of MMP-2 prevents cardiac rupture after myocardial infarction in mice. *J. Clin. Investig.* **2005**, *115*, 599–609. [CrossRef]

57. Matsusaka, H.; Ide, T.; Matsushima, S.; Ikeuchi, M.; Kubota, T.; Sunagawa, K.; Kinugawa, S.; Tsutsui, H. Targeted deletion of matrix metalloproteinase 2 ameliorates myocardial remodeling in mice with chronic pressure overload. *Hypertension* **2006**, *47*, 711–717. [CrossRef] [PubMed]

58. Hardy, E.; Hardy-Sosa, A.; Fernandez-Patron, C. MMP-2: Is too low as bad as too high in the cardiovascular system? *Am. J. Physiol. Heart Circ. Physiol.* **2018**, *315*, H1332–H1340. [CrossRef]

59. Matsusaka, H.; Ikeuchi, M.; Matsushima, S.; Ide, T.; Kubota, T.; Feldman, A.M.; Takeshita, A.; Sunagawa, K.; Tsutsui, H. Selective disruption of MMP-2 gene exacerbates myocardial inflammation and dysfunction in mice with cytokine-induced cardiomyopathy. *Am. J. Physiol. Heart Circ. Physiol.* **2005**, *289*, H1858–H1864. [CrossRef]

60. Wang, X.; Berry, E.; Hernandez-Anzaldo, S.; Takawale, A.; Kassiri, Z.; Fernandez-Patron, C. Matrix metalloproteinase-2 mediates a mechanism of metabolic cardioprotection consisting of negative regulation of the sterol regulatory element-binding protein-2/3-hydroxy-3-methylglutaryl-CoA reductase pathway in the heart. *Hypertension* **2015**, *65*, 882–888. [CrossRef]

61. Martignetti, J.A.; Aqeel, A.A.; Sewairi, W.A.; Boumah, C.E.; Kambouris, M.; Mayouf, S.A.; Sheth, K.V.; Eid, W.A.; Dowling, O.; Harris, J.; et al. Mutation of the matrix metalloproteinase 2 gene (MMP2) causes a multicentric osteolysis and arthritis syndrome. *Nat. Genet.* **2001**, *28*, 261–265. [CrossRef]

62. Tuysuz, B.; Mosig, R.; Altun, G.; Sancak, S.; Glucksman, M.J.; Martignetti, J.A. A novel matrix metalloproteinase 2 (MMP2) terminal hemopexin domain mutation in a family with multicentric osteolysis with nodulosis and arthritis with cardiac defects. *Eur. J. Hum. Genet. EJHG* **2009**, *17*, 565–572. [CrossRef]

63. Cabral-Pacheco, G.A.; Garza-Veloz, I.; Castruita-De la Rosa, C.; Ramirez-Acuna, J.M.; Perez-Romero, B.A.; Guerrero-Rodriguez, J.F.; Martinez-Avila, N.; Martinez-Fierro, M.L. The Roles of Matrix Metalloproteinases and Their Inhibitors in Human Diseases. *Int. J. Mol. Sci.* **2020**, *21*, 9739. [CrossRef] [PubMed]

64. Nagase, H.; Visse, R.; Murphy, G. Structure and function of matrix metalloproteinases and TIMPs. *Cardiovasc. Res.* **2006**, *69*, 562–573. [CrossRef] [PubMed]

65. Davies, B.; Brown, P.D.; East, N.; Crimmin, M.J.; Balkwill, F.R. A synthetic matrix metalloproteinase inhibitor decreases tumor burden and prolongs survival of mice bearing human ovarian carcinoma xenografts. *Cancer Res.* **1993**, *53*, 2087–2091. [PubMed]

66. Vandenbroucke, R.E.; Libert, C. Is there new hope for therapeutic matrix metalloproteinase inhibition? *Nat. Rev. Drug Discov.* **2014**, *13*, 904–927. [CrossRef]

67. Shepherd, F.A.; Giaccone, G.; Seymour, L.; Debruyne, C.; Bezjak, A.; Hirsh, V.; Smylie, M.; Rubin, S.; Martins, H.; Lamont, A.; et al. Prospective, randomized, double-blind, placebo-controlled trial of marimastat after response to first-line chemotherapy in patients with small-cell lung cancer: A trial of the National Cancer Institute of Canada-Clinical Trials Group and the European Organization for Research and Treatment of Cancer. *J. Clin. Oncol. Off. J. Am. Soc. Clin. Oncol.* **2002**, *20*, 4434–4439. [CrossRef]

68. Wada, C.K.; Holms, J.H.; Curtin, M.L.; Dai, Y.; Florjancic, A.S.; Garland, R.B.; Guo, Y.; Heyman, H.R.; Stacey, J.R.; Steinman, D.H.; et al. Phenoxyphenyl sulfone N-formylhydroxylamines (retrohydroxamates) as potent, selective, orally bioavailable matrix metalloproteinase inhibitors. *J. Med. Chem.* **2002**, *45*, 219–232. [CrossRef]

69. Bramhall, S.R.; Hallissey, M.T.; Whiting, J.; Scholefield, J.; Tierney, G.; Stuart, R.C.; Hawkins, R.E.; McCulloch, P.; Maughan, T.; Brown, P.D.; et al. Marimastat as maintenance therapy for patients with advanced gastric cancer: A randomised trial. *Br. J. Cancer* **2002**, *86*, 1864–1870. [CrossRef]

70. Das, S.; Amin, S.A.; Jha, T. Inhibitors of gelatinases (MMP-2 and MMP-9) for the management of hematological malignancies. *Eur. J. Med. Chem.* **2021**, *223*, 113623. [CrossRef]

71. Edwards, D.R.; Handsley, M.M.; Pennington, C.J. The ADAM metalloproteinases. *Mol. Asp. Med.* **2008**, *29*, 258–289. [CrossRef]

72. Tan Ide, A.; Ricciardelli, C.; Russell, D.L. The metalloproteinase ADAMTS1: A comprehensive review of its role in tumorigenic and metastatic pathways. *Int. J. Cancer. J. Int. Du Cancer* **2013**, *133*, 2263–2276. [CrossRef]

73. Iwasaki, M.; Nishikawa, A.; Fujimoto, T.; Akutagawa, N.; Manase, K.; Endo, T.; Yoshida, K.; Maekawa, R.; Yoshioka, T.; Kudo, R. Anti-invasive effect of MMI-166, a new selective matrix metalloproteinase inhibitor, in cervical carcinoma cell lines. *Gynecol. Oncol.* **2002**, *85*, 103–107. [CrossRef] [PubMed]

74. Leighl, N.B.; Paz-Ares, L.; Douillard, J.Y.; Peschel, C.; Arnold, A.; Depierre, A.; Santoro, A.; Betticher, D.C.; Gatzemeier, U.; Jassem, J.; et al. Randomized phase III study of matrix metalloproteinase inhibitor BMS-275291 in combination with paclitaxel and carboplatin in advanced non-small-cell lung cancer: National Cancer Institute of Canada-Clinical Trials Group Study BR.18. *J. Clin. Oncol. Off. J. Am. Soc. Clin. Oncol.* **2005**, *23*, 2831–2839. [CrossRef] [PubMed]

75. Miller, K.D.; Saphner, T.J.; Waterhouse, D.M.; Chen, T.T.; Rush-Taylor, A.; Sparano, J.A.; Wolff, A.C.; Cobleigh, M.A.; Galbraith, S.; Sledge, G.W. A randomized phase II feasibility trial of BMS-275291 in patients with early stage breast cancer. *Clin. Cancer Res. Off. J. Am. Assoc. Cancer Res.* **2004**, *10*, 1971–1975. [CrossRef] [PubMed]

76. Hirte, H.; Vergote, I.B.; Jeffrey, J.R.; Grimshaw, R.N.; Coppieters, S.; Schwartz, B.; Tu, D.; Sadura, A.; Brundage, M.; Seymour, L. A phase III randomized trial of BAY 12-9566 (tanomastat) as maintenance therapy in patients with advanced ovarian cancer responsive to primary surgery and paclitaxel/platinum containing chemotherapy: A National Cancer Institute of Canada Clinical Trials Group Study. *Gynecol. Oncol.* **2006**, *102*, 300–308. [CrossRef]

77. Wong, M.S.; Sidik, S.M.; Mahmud, R.; Stanslas, J. Molecular targets in the discovery and development of novel antimetastatic agents: Current progress and future prospects. *Clin. Exp. Pharmacol. Physiol.* **2013**, *40*, 307–319. [CrossRef]

78. Philips, N.; Conte, J.; Chen, Y.J.; Natrajan, P.; Taw, M.; Keller, T.; Givant, J.; Tuason, M.; Dulaj, L.; Leonardi, D.; et al. Beneficial regulation of matrixmetalloproteinases and their inhibitors, fibrillar collagens and transforming growth factor-beta by Polypodium leucotomos, directly or in dermal fibroblasts, ultraviolet radiated fibroblasts, and melanoma cells. *Arch. Dermatol. Res.* **2009**, *301*, 487–495. [CrossRef]

79. Philips, N.; Smith, J.; Keller, T.; Gonzalez, S. Predominant effects of Polypodium leucotomos on membrane integrity, lipid peroxidation, and expression of elastin and matrixmetalloproteinase-1 in ultraviolet radiation exposed fibroblasts, and keratinocytes. *J. Dermatol. Sci.* **2003**, *32*, 1–9. [CrossRef]

80. Philips, N.; Samuel, M.; Arena, R.; Chen, Y.J.; Conte, J.; Natarajan, P.; Haas, G.; Gonzalez, S. Direct inhibition of elastase and matrixmetalloproteinases and stimulation of biosynthesis of fibrillar collagens, elastin, and fibrillins by xanthohumol. *J. Cosmet. Sci.* **2010**, *61*, 125–132. [CrossRef]

81. Astner, S.; Wu, A.; Chen, J.; Philips, N.; Rius-Diaz, F.; Parrado, C.; Mihm, M.C.; Goukassian, D.A.; Pathak, M.A.; Gonzalez, S. Dietary lutein/zeaxanthin partially reduces photoaging and photocarcinogenesis in chronically UVB-irradiated Skh-1 hairless mice. *Ski. Pharm. Physiol.* **2007**, *20*, 283–291. [CrossRef]

82. Philips, N.; Keller, T.; Hendrix, C.; Hamilton, S.; Arena, R.; Tuason, M.; Gonzalez, S. Regulation of the extracellular matrix remodeling by lutein in dermal fibroblasts, melanoma cells, and ultraviolet radiation exposed fibroblasts. *Arch. Dermatol. Res.* **2007**, *299*, 373–379. [CrossRef]

83. Chan, B.Y.H.; Roczkowsky, A.; Cho, W.J.; Poirier, M.; Sergi, C.; Keschrumrus, V.; Churko, J.M.; Granzier, H.; Schulz, R. MMP inhibitors attenuate doxorubicin cardiotoxicity by preventing intracellular and extracellular matrix remodelling. *Cardiovasc. Res.* **2021**, *117*, 188–200. [CrossRef] [PubMed]

84. Parente, J.M.; Blascke de Mello, M.M.; Silva, P.; Omoto, A.C.M.; Pernomian, L.; Oliveira, I.S.; Mahmud, Z.; Fazan, R., Jr.; Arantes, E.C.; Schulz, R.; et al. MMP inhibition attenuates hypertensive eccentric cardiac hypertrophy and dysfunction by preserving troponin I and dystrophin. *Biochem. Pharmacol.* **2021**, *193*, 114744. [CrossRef] [PubMed]

85. Tessone, A.; Feinberg, M.S.; Barbash, I.M.; Reich, R.; Holbova, R.; Richmann, M.; Mardor, Y.; Leor, J. Effect of matrix metalloproteinase inhibition by doxycycline on myocardial healing and remodeling after myocardial infarction. *Cardiovasc. Drugs Ther.* **2005**, *19*, 383–390. [CrossRef]

86. Novak, M.J.; Johns, L.P.; Miller, R.C.; Bradshaw, M.H. Adjunctive benefits of subantimicrobial dose doxycycline in the management of severe, generalized, chronic periodontitis. *J. Periodontol.* **2002**, *73*, 762–769. [CrossRef]

87. Mata, K.M.; Tefe-Silva, C.; Floriano, E.M.; Fernandes, C.R.; Rizzi, E.; Gerlach, R.F.; Mazzuca, M.Q.; Ramos, S.G. Interference of doxycycline pretreatment in a model of abdominal aortic aneurysms. *Cardiovasc. Pathol. Off. J. Soc. Cardiovasc. Pathol.* **2015**, *24*, 110–120. [CrossRef] [PubMed]

88. Antonio, R.C.; Ceron, C.S.; Rizzi, E.; Coelho, E.B.; Tanus-Santos, J.E.; Gerlach, R.F. Antioxidant effect of doxycycline decreases MMP activity and blood pressure in SHR. *Mol. Cell. Biochem.* **2014**, *386*, 99–105. [CrossRef]

89. Castro, M.M.; Rizzi, E.; Ceron, C.S.; Guimaraes, D.A.; Rodrigues, G.J.; Bendhack, L.M.; Gerlach, R.F.; Tanus-Santos, J.E. Doxycycline ameliorates 2K-1C hypertension-induced vascular dysfunction in rats by attenuating oxidative stress and improving nitric oxide bioavailability. *Nitric Oxide Biol. Chem. Off. J. Nitric Oxide Soc.* **2012**, *26*, 162–168. [CrossRef]

90. Castro, M.M.; Rizzi, E.; Figueiredo-Lopes, L.; Fernandes, K.; Bendhack, L.M.; Pitol, D.L.; Gerlach, R.F.; Tanus-Santos, J.E. Metalloproteinase inhibition ameliorates hypertension and prevents vascular dysfunction and remodeling in renovascular hypertensive rats. *Atherosclerosis* **2008**, *198*, 320–331. [CrossRef]

91. Guimaraes, D.A.; Rizzi, E.; Ceron, C.S.; Oliveira, A.M.; Oliveira, D.M.; Castro, M.M.; Tirapelli, C.R.; Gerlach, R.F.; Tanus-Santos, J.E. Doxycycline dose-dependently inhibits MMP-2-mediated vascular changes in 2K1C hypertension. *Basic Clin. Pharmacol. Toxicol.* **2011**, *108*, 318–325. [CrossRef]

92. Rizzi, E.; Castro, M.M.; Prado, C.M.; Silva, C.A.; Fazan, R., Jr.; Rossi, M.A.; Tanus-Santos, J.E.; Gerlach, R.F. Matrix metalloproteinase inhibition improves cardiac dysfunction and remodeling in 2-kidney, 1-clip hypertension. *J. Card. Fail.* **2010**, *16*, 599–608. [CrossRef]

93. Villarreal, F.J.; Griffin, M.; Omens, J.; Dillmann, W.; Nguyen, J.; Covell, J. Early short-term treatment with doxycycline modulates postinfarction left ventricular remodeling. *Circulation* **2003**, *108*, 1487–1492. [CrossRef] [PubMed]

94. Cerisano, G.; Buonamici, P.; Valenti, R.; Sciagra, R.; Raspanti, S.; Santini, A.; Carrabba, N.; Dovellini, E.V.; Romito, R.; Pupi, A.; et al. Early short-term doxycycline therapy in patients with acute myocardial infarction and left ventricular dysfunction to prevent the ominous progression to adverse remodelling: The TIPTOP trial. *Eur. Heart J.* **2014**, *35*, 184–191. [CrossRef]

95. Brown, D.L.; Desai, K.K.; Vakili, B.A.; Nouneh, C.; Lee, H.M.; Golub, L.M. Clinical and biochemical results of the metalloproteinase inhibition with subantimicrobial doses of doxycycline to prevent acute coronary syndromes (MIDAS) pilot trial. *Arterioscler. Thromb. Vasc. Biol.* **2004**, *24*, 733–738. [CrossRef]

96. Schulze, C.J.; Castro, M.M.; Kandasamy, A.D.; Cena, J.; Bryden, C.; Wang, S.H.; Koshal, A.; Tsuyuki, R.T.; Finegan, B.A.; Schulz, R. Doxycycline reduces cardiac matrix metalloproteinase-2 activity but does not ameliorate myocardial dysfunction during reperfusion in coronary artery bypass patients undergoing cardiopulmonary bypass. *Crit. Care Med.* **2013**, *41*, 2512–2520. [CrossRef]

97. Hudson, M.P.; Armstrong, P.W.; Ruzyllo, W.; Brum, J.; Cusmano, L.; Krzeski, P.; Lyon, R.; Quinones, M.; Theroux, P.; Sydlowski, D.; et al. Effects of selective matrix metalloproteinase inhibitor (PG-116800) to prevent ventricular remodeling after myocardial infarction: Results of the PREMIER (Prevention of Myocardial Infarction Early Remodeling) trial. *J. Am. Coll. Cardiol.* **2006**, *48*, 15–20. [CrossRef]

98. Bencsik, P.; Kupai, K.; Gorbe, A.; Kenyeres, E.; Varga, Z.V.; Paloczi, J.; Gaspar, R.; Kovacs, L.; Weber, L.; Takacs, F.; et al. Development of Matrix Metalloproteinase-2 Inhibitors for Cardioprotection. *Front. Pharmacol.* **2018**, *9*, 296. [CrossRef] [PubMed]

99. Gomori, K.; Szabados, T.; Kenyeres, E.; Pipis, J.; Foldesi, I.; Siska, A.; Dorman, G.; Ferdinandy, P.; Gorbe, A.; Bencsik, P. Cardioprotective Effect of Novel Matrix Metalloproteinase Inhibitors. *Int. J. Mol. Sci.* **2020**, *21*, 6990. [CrossRef] [PubMed]

100. Lin, H.B.; Cadete, V.J.; Sra, B.; Sawicka, J.; Chen, Z.; Bekar, L.K.; Cayabyab, F.; Sawicki, G. Inhibition of MMP-2 expression with siRNA increases baseline cardiomyocyte contractility and protects against simulated ischemic reperfusion injury. *BioMed Res. Int.* **2014**, *2014*, 810371. [CrossRef] [PubMed]

101. Wang, L.L.; Chung, J.J.; Li, E.C.; Uman, S.; Atluri, P.; Burdick, J.A. Injectable and protease-degradable hydrogel for siRNA sequestration and triggered delivery to the heart. *J. Control. Release Off. J. Control. Release Soc.* **2018**, *285*, 152–161. [CrossRef]

102. Nakaya, R.; Uzui, H.; Shimizu, H.; Nakano, A.; Mitsuke, Y.; Yamazaki, T.; Ueda, T.; Lee, J.D. Pravastatin suppresses the increase in matrix metalloproteinase-2 levels after acute myocardial infarction. *Int. J. Cardiol.* **2005**, *105*, 67–73. [CrossRef]

103. Shirakabe, A.; Asai, K.; Hata, N.; Yokoyama, S.; Shinada, T.; Kobayashi, N.; Tomita, K.; Tsurumi, M.; Matsushita, M.; Mizuno, K. Immediate administration of atorvastatin decreased the serum MMP-2 level and improved the prognosis for acute heart failure. *J. Cardiol.* **2012**, *59*, 374–382. [CrossRef]

104. Tousoulis, D.; Andreou, I.; Tentolouris, C.; Antoniades, C.; Papageorgiou, N.; Gounari, P.; Kotrogiannis, I.; Miliou, A.; Charakida, M.; Trikas, A.; et al. Comparative effects of rosuvastatin and allopurinol on circulating levels of matrix metalloproteinases and tissue inhibitors of metalloproteinases in patients with chronic heart failure. *Int. J. Cardiol.* **2010**, *145*, 438–443. [CrossRef]

105. Cortese, F.; Gesualdo, M.; Cortese, A.; Carbonara, S.; Devito, F.; Zito, A.; Ricci, G.; Scicchitano, P.; Ciccone, M.M. Rosuvastatin: Beyond the cholesterol-lowering effect. *Pharmacol. Res.* **2016**, *107*, 1–18. [CrossRef]

106. Mason, R.P. Molecular basis of differences among statins and a comparison with antioxidant vitamins. *Am. J. Cardiol.* **2006**, *98*, 34P–41P. [CrossRef]

107. Gong, W.; Ma, Y.; Li, A.; Shi, H.; Nie, S. Trimetazidine suppresses oxidative stress, inhibits MMP-2 and MMP-9 expression, and prevents cardiac rupture in mice with myocardial infarction. *Cardiovasc. Ther.* **2018**, *36*, e12460. [CrossRef]

108. Mendes, A.S.; Blascke de Mello, M.M.; Parente, J.M.; Omoto, A.C.M.; Neto-Neves, E.M.; Fazan, R., Jr.; Tanus-Santos, J.E.; Castro, M.M. Verapamil decreases calpain-1 and matrix metalloproteinase-2 activities and improves hypertension-induced hypertrophic cardiac remodeling in rats. *Life Sci.* **2020**, *244*, 117153. [CrossRef]

109. Skrzypiec-Spring, M.; Urbaniak, J.; Sapa-Wojciechowska, A.; Pietkiewicz, J.; Orda, A.; Karolko, B.; Danielewicz, R.; Bil-Lula, I.; Wozniak, M.; Schulz, R.; et al. Matrix Metalloproteinase-2 Inhibition in Acute Ischemia-Reperfusion Heart Injury-Cardioprotective Properties of Carvedilol. *Pharmaceuticals* **2021**, *14*, 1276. [CrossRef]

110. Dang, Y.; Gao, N.; Niu, H.; Guan, Y.; Fan, Z.; Guan, J. Targeted Delivery of a Matrix Metalloproteinases-2 Specific Inhibitor Using Multifunctional Nanogels to Attenuate Ischemic Skeletal Muscle Degeneration and Promote Revascularization. *ACS Appl. Mater. Interfaces* **2021**, *13*, 5907–5918. [CrossRef]

111. Harwood, S.L.; Nielsen, N.S.; Diep, K.; Jensen, K.T.; Nielsen, P.K.; Yamamoto, K.; Enghild, J.J. Development of selective protease inhibitors via engineering of the bait region of human alpha2-macroglobulin. *J. Biol. Chem.* **2021**, *297*, 100879. [CrossRef]

112. Higashi, S.; Hirose, T.; Takeuchi, T.; Miyazaki, K. Molecular design of a highly selective and strong protein inhibitor against matrix metalloproteinase-2 (MMP-2). *J. Biol. Chem.* **2013**, *288*, 9066–9076. [CrossRef]

113. Huang, Y.H.; Chu, P.Y.; Chen, J.L.; Huang, C.T.; Huang, C.C.; Tsai, Y.F.; Wang, Y.L.; Lien, P.J.; Tseng, L.M.; Liu, C.Y. Expression pattern and prognostic impact of glycoprotein non-metastatic B (GPNMB) in triple-negative breast cancer. *Sci. Rep.* **2021**, *11*, 12171. [CrossRef] [PubMed]

114. Sarkar, P.; Li, Z.; Ren, W.; Wang, S.; Shao, S.; Sun, J.; Ren, X.; Perkins, N.G.; Guo, Z.; Chang, C.A.; et al. Inhibiting Matrix Metalloproteinase-2 Activation by Perturbing Protein-Protein Interactions Using a Cyclic Peptide. *J. Med. Chem.* **2020**, *63*, 6979–6990. [CrossRef] [PubMed]

115. Webb, A.H.; Gao, B.T.; Goldsmith, Z.K.; Irvine, A.S.; Saleh, N.; Lee, R.P.; Lendermon, J.B.; Bheemreddy, R.; Zhang, Q.; Brennan, R.C.; et al. Inhibition of MMP-2 and MMP-9 decreases cellular migration, and angiogenesis in in vitro models of retinoblastoma. *BMC Cancer* **2017**, *17*, 434. [CrossRef] [PubMed]

 *pharmaceuticals*

Article

Matrix Metalloproteinase-2 Inhibition in Acute Ischemia-Reperfusion Heart Injury—Cardioprotective Properties of Carvedilol

Monika Skrzypiec-Spring [1,*], Joanna Urbaniak [2], Agnieszka Sapa-Wojciechowska [3], Jadwiga Pietkiewicz [4], Alina Orda [5], Bożena Karolko [5], Regina Danielewicz [4], Iwona Bil-Lula [3], Mieczysław Woźniak [3], Richard Schulz [6] and Adam Szeląg [1]

1. Department of Pharmacology, Wrocław Medical University, 50-345 Wrocław, Poland; adam.szelag@umed.wroc.pl
2. Lower Silesian Oncology Centre, 53-413 Wrocław, Poland; urbaniak.joanna@dco.com.pl
3. Department of Clinical Chemistry, Wrocław Medical University, 50-556 Wrocław, Poland; agnieszka.sapa-wojciechowska@umw.edu.pl (A.S.-W.); iwona.bil-lula@umw.edu.pl (I.B.-L.); mieczyslaw.wozniak@umed.wroc.pl (M.W.)
4. Department of Biochemistry, Wrocław Medical University, 50-368 Wrocław, Poland; jadwiga.pietkiewicz@umed.wroc.pl (J.P.); regina.danielewicz@umed.wroc.pl (R.D.)
5. Department of Cardiology, Wrocław Medical University, 50-556 Wrocław, Poland; alina.orda@op.pl (A.O.); bozena.karolko@umed.wroc.pl (B.K.)
6. Departments of Pediatrics and Pharmacology, University of Alberta, Edmonton, AB T6G 2S2, Canada; richard.schulz@ualberta.ca
* Correspondence: monika.skrzypiec-spring@umed.wroc.pl; Tel.: +48-717841450

Citation: Skrzypiec-Spring, M.; Urbaniak, J.; Sapa-Wojciechowska, A.; Pietkiewicz, J.; Orda, A.; Karolko, B.; Danielewicz, R.; Bil-Lula, I.; Woźniak, M.; Schulz, R.; et al. Matrix Metalloproteinase-2 Inhibition in Acute Ischemia-Reperfusion Heart Injury—Cardioprotective Properties of Carvedilol. *Pharmaceuticals* **2021**, *14*, 1276. https://doi.org/10.3390/ph14121276

Academic Editor: Salvatore Santamaria

Received: 11 October 2021
Accepted: 2 December 2021
Published: 7 December 2021

Publisher's Note: MDPI stays neutral with regard to jurisdictional claims in published maps and institutional affiliations.

Abstract: Matrix metalloproteinase 2 (MMP-2) is activated in hearts upon ischemia-reperfusion (IR) injury and cleaves sarcomeric proteins. It was shown that carvedilol and nebivolol reduced the activity of different MMPs. Hence, we hypothesized that they could reduce MMPs activation in myocytes, and therefore, protect against cardiac contractile dysfunction related with IR injury. Isolated rat hearts were subjected to either control aerobic perfusion or IR injury: 25 min of aerobic perfusion, followed by 20 min global, no-flow ischemia, and reperfusion for 30 min. The effects of carvedilol, nebivolol, or metoprolol were evaluated in hearts subjected to IR injury. Cardiac mechanical function and MMP-2 activity in the heart homogenates and coronary effluent were assessed along with troponin I content in the former. Only carvedilol improved the recovery of mechanical function at the end of reperfusion compared to IR injury hearts. IR injury induced the activation and release of MMP-2 into the coronary effluent during reperfusion. MMP-2 activity in the coronary effluent increased in the IR injury group and this was prevented by carvedilol. Troponin I levels decreased by 73% in IR hearts and this was abolished by carvedilol. These data suggest that the cardioprotective effect of carvedilol in myocardial IR injury may be mediated by inhibiting MMP-2 activation.

Keywords: β-blockers; carvedilol; matrix metalloproteinase-2; ischemia-reperfusion injury; isolated heart perfusion

---

1. Introduction

β-blockers improve the oxygen supply/demand ratio by reducing myocardial oxygen consumption via a decrease in heart rate and myocardial contractility [1]. Several clinical studies showed the ability of β-blockers to decrease the incidence of coronary heart disease and mortality in patients with coronary heart disease and acute coronary syndromes [2,3]. The first generation β-blockers are represented by propranolol, sotalol, oxprenolol and nadolol, the second by metoprolol, atenolol, and bisoprolol, and the third by carvedilol, nebivolol, and labetalol [3]. The cardioprotective effects of different β-blockers were tested on animal models. It was shown that carvedilol and nebivolol possess superior cardioprotective efficacy as compared to other β-blockers which failed to protect the myocardium against IR injury [4–9]. These observations suggest that the protective activity of some β-blockers is most probably unrelated to β-adrenoceptor inhibition but to other'-ancillary'- properties.

Carvedilol (CAR) is a third generation, non-selective β-blocker which also possesses α-blocker and anti-oxidant properties. It may prevent oxidative stress-induced activation of transcription factors associated with inflammatory and remodeling processes and inhibit the direct cytotoxicity of reactive oxygen and nitrogen species [10]. It was shown to inhibit MMP-8 in coxackie virus-induced myocarditis and MMP-2 and MMP-9 expression in experimental atherosclerosis [11,12]. As we previously described, CAR reduces the contractile dysfunction of heart muscle by attenuation of MMP-2 activity and decreased degradation of troponin and myofilaments in hearts subjected to experimental autoimmune myocarditis [13]. Nebivolol (NEB) is a third generation, highly selective β1-adrenoceptor antagonist endowed with the ability to induce nitric oxide release from the endothelium [14]. Moreover, it was shown to attenuate MMP-2 and MMP-9 activities in experimental renovascular hypertension and renal IR injury [15,16]. In contrast, metoprolol (MET) represents a second generation, selective β1-adrenoceptor antagonist with no pleiotropic actions disclosed.

IR injury occurs as a result of acute oxidative stress following reperfusion after ischemia. It can be described as a cascade of pathophysiological events including the biosynthesis of reactive oxygen and nitrogen species and cytokines and activation of MMPs, in particular MMP-2 [17–19]. Although MMPs are primarily known for their ability to cleave substrates in the extracellular matrix, it was also shown that the acute contractile dysfunction in IR injury is also caused by the activation of intracellularly localized MMP-2 which results in troponin I, myosin light chain-1, α-actinin and titin degradation [20,21].

The properties of CAR and NEB to inhibit MMPs, and thus, prevent the degradation of sarcomeric proteins, such as troponin I may provide the underlying pharmacologic rationale for the use of these drugs as a first choice for acute coronary syndromes. The present study was, therefore, performed to test the hypothesis that the cardioprotective action of CAR or NEB in the setting of IR injury is related to their ability to inhibit MMPs activity and troponin I degradation in heart tissue.

2. Results

2.1. Carvedilol Protects Hearts from Ischemia-Reperfusion Injury

Stable mechanical function of hearts perfused for 75 min in aerobic conditions was observed (data not shown). IR injury caused a significant reduction in the heart rate and mechanical function recovery expressed as LVDP and RPP in the reperfusion period (Figure 1a–c). CAR 0.1 μM significantly improved the recovery of mechanical function at the end of the reperfusion period in comparison to the control IR injury group ($p < 0.01$; Figure 1a). The higher CAR concentrations, as well NEB and MET in all tested concentrations, did not significantly improve the recovery after IR injury (Figure 1a–c).

2.2. Carvedilol Influences MMP-2 Activity

The differences when comparing MMP-2 activities in the coronary effluent between carvedilol treated groups were not significant (Figure 2a). Additionally, at higher CAR concentrations, as well as with NEB and MET at all tested concentrations, there were no significant differences between groups (data not shown).

Nevertheless, a significant increase of MMP-2 activity in the coronary effluent, calculated as the ratio of 45 and 25 min perfusion time values was seen in IR injury group, as well as CAR 1 and CAE 10 groups in comparison with the C group ($p < 0.0001$; Figure 2c–e). This increase in MMP-2 activity was significantly lower in the IR injury with CAR 0.1 μM group (Figure 2c). NEB and MET in all tested concentrations had no significant influence on the increase of MMP-2 activity evoked by IR which was significantly higher than in the C group ($p < 0.0001$, Figure 2d,e).

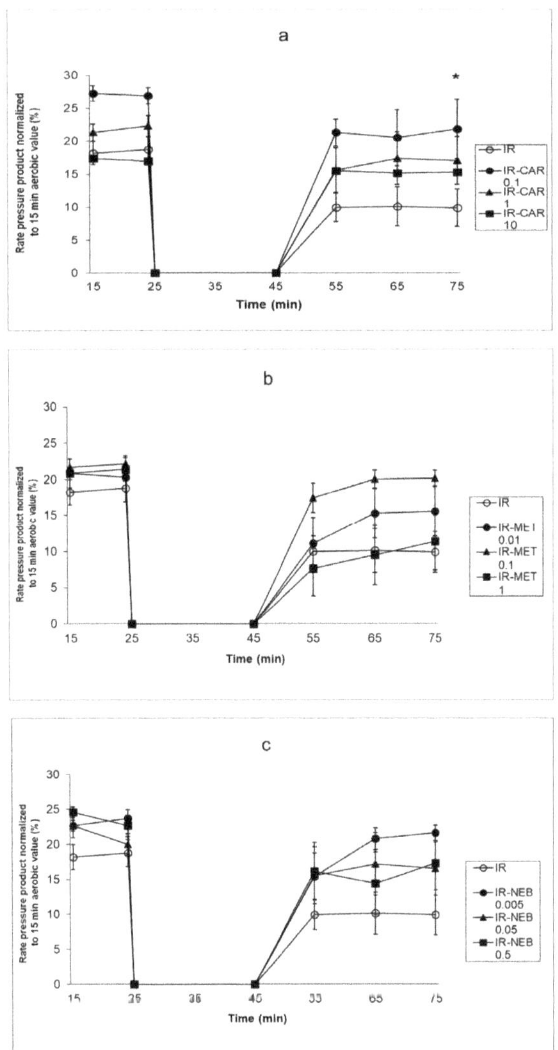

Figure 1. (**a–c**). Effect of carvedilol, metoprolol and nebivolol on cardiac mechanical function after 20 min of ischemia (**a**) Effect of carvedilol on cardiac mechanical function after 20 min of ischemia. (**b**) Effect of metoprolol on cardiac mechanical function after 20 min of ischemia. (**c**) Effect of carvedilol on cardiac mechanical function after 20 min of ischemia. The result presented as the rate-pressure product (heart rate × left ventricular developed pressure) normalized to 15 min aerobic value. C—control group, aerobically perfused hearts, IR injury—ischemia-reperfusion injury, IR-CAR 0.1– hearts from IR injury model treated with 0.1 μM carvedilol, IR-CAR 1– hearts from IR injury model treated with 1 μM carvedilol, IR-CAR 10– hearts from IR injury model treated with 10 μM carvedilol, IR-MET 0.01—hearts from IR injury model treated with 0.01 μM metoprolol, IR-MET 0.1—hearts from IR injury model treated with 0.1 μM metoprolol, IR-MET 1—hearts from IR injury model treated with 1 μM metoprolol, IR-NEB 0.005—hearts from IR injury model treated with 0.005 μM nebivolol, IR-NEB 0.05—hearts from IR injury model treated with 0.05 μM nebivolol, IR-NEB 0.5—hearts from IR injury model treated with 0.5 μM nebivolol. * IR vs IR-CAR 0.1, $p < 0.05$, $n = 6$, ANOVA.

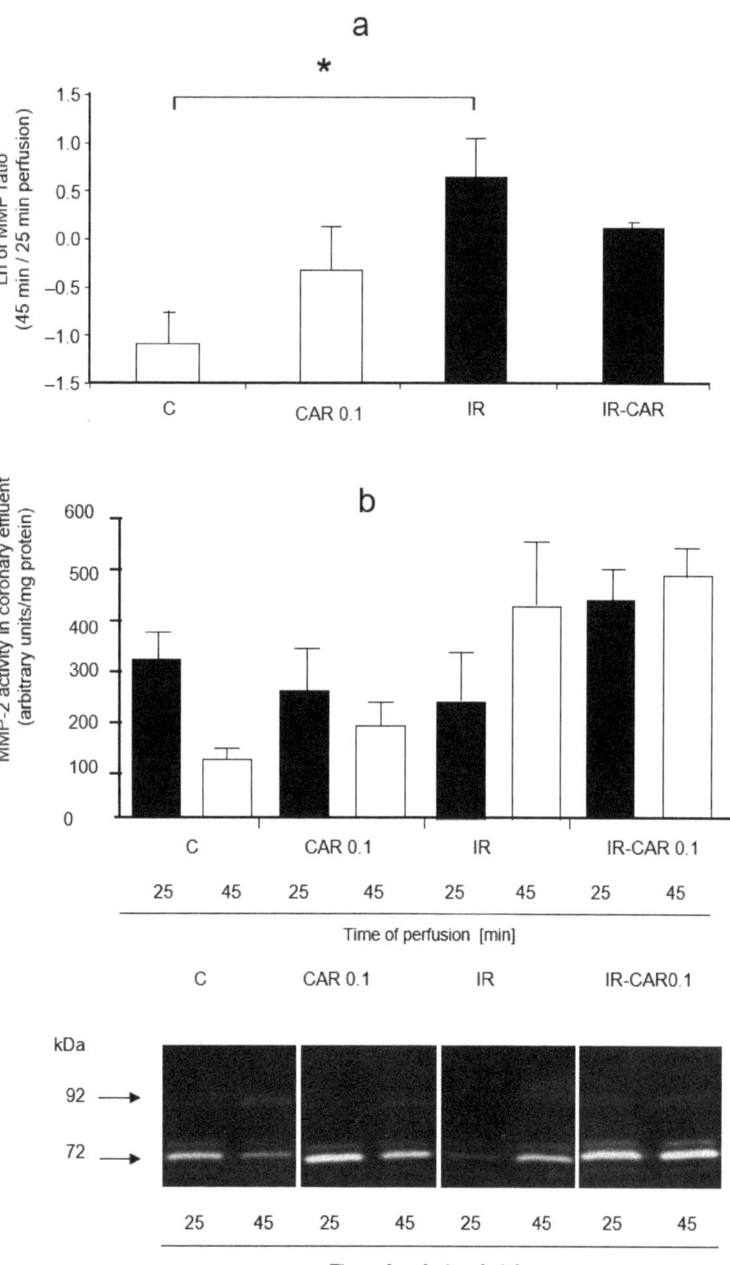

Figure 2. *Cont.*

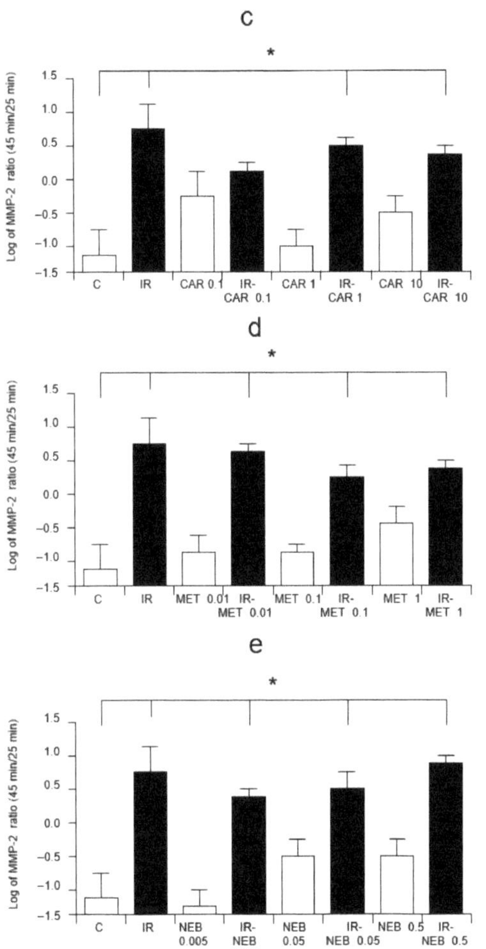

Figure 2. Influence of carvedilol, metoprolol and nebivolol on MMP-2 activity in coronary effluent. (**a**) Densitometric analysis of gelatinolytic MMP-2 activities in coronary effluent in samples from perfused hearts collected at different time points. (**b**) Representative zymogram showing gelatinolytic activities in coronary effluent samples from perfused hearts collected at different time points. The 72 kDa MMP-2 specific activity in coronary effluent samples was assessed by densitometric analysis (**c**) Log ratio of 45 vs. 25 min perfusion MMP-2 activities in coronary effluent in carvedilol treated groups. (**d**) Log ratio of 45 vs. 25 min perfusion MMP-2 activities in coronary effluent in metoprolol treated groups. (**e**) Log ratio of 45 vs. 25 min perfusion MMP-2 activities in coronary effluent in nebivolol treated groups. Only in the IR CAR 0.01 group there was no significant increase in MMP-2 activity. C—control group, aerobically perfused hearts, IR injury—ischemia-reperfusion injury, IR-CAR 0.1—hearts from IR injury model treated with 0.1 μM carvedilol, IR-CAR 1—hearts from IR injury model treated with 1 μM carvedilol, IR-CAR 10—hearts from IR injury model treated with 10 μM carvedilol, IR-MET 0.01—hearts from IR injury model treated with 0.01 μM metoprolol, IR-MET 0.1—hearts from IR injury model treated with 0.1 μM metoprolol, IR-MET 1—hearts from IR injury model treated with 1 μM metoprolol, IR-NEB 0.005—hearts from IR injury model treated with 0.005 μM nebivolol, IR-NEB 0.05—hearts from IR injury model treated with 0.05 μM nebivolol, IR-NEB 0.5—hearts from IR injury model treated with 0.5 μM nebivolol. * C vs IR, IR-CAR 1, IR-CAR 10, IR-MET 0.01, IR-MET 0.1, IR-MET 1, IR-NEB 0.005, IR-NEB 0.05, IR-NEB 0.5, $p < 0.05$, $n = 6$, ANOVA.

MMP-2 activity in the heart tissue was assessed only at the end of the experiment showing no significant changes between groups (data not shown).

2.3. Carvedilol Has No Effect on MMP-2 Activity In Vitro

Carvedilol did not inhibit the activity of MMP-2 when run out on gel zymograms incubated with 0.1 µM carvedilol (Figure 3a,b).

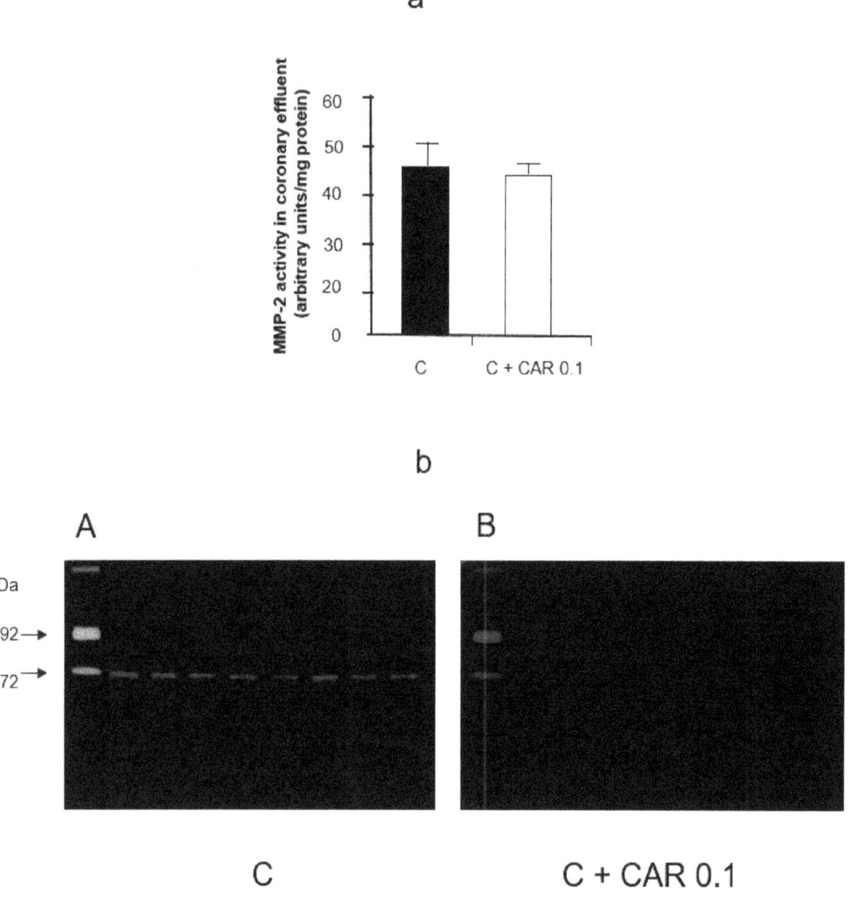

Figure 3. Effect of CAR on MMP-2 activity in vitro. (**a**) Densitometric analysis of gelatinolytic MMP-2 activities in control gel and gel incubated with addition of 0.1 µM CAR. (**b**) Representative zymograms of MMP-2 activity in control gel (**A**) and gel incubated with addition of 0.1 µM CAR (**B**). C—control group, C+ CAR 0.1—in control gel and gel incubated with addition of 0.1 µM CAR. $p > 0.05$, $n = 8$, T-test.

2.4. Carvedilol Does Not Change MMP-2 mRNA Expression in Hearts Subjected to Ischemia-Reperfusion

Real time PCR revealed no significant changes in MMP-2 mRNA expression between groups (Figure 4).

2.5. Carvedilol Does Not Affect MMP-2 Content in Coronary Effluent

IR injury caused a significant increase in MMP-2 content, assessed by western blot, in coronary effluent in the second minute of reperfusion in both IR and IR-CAR 0.1 groups, but there were no significant differences between IR and IR-CAR 0.1 groups (Figure 5a,b).

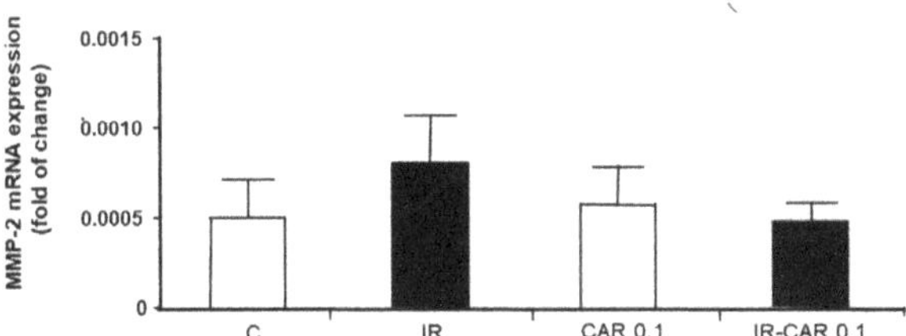

Figure 4. Effect of carvedilol on MMP-2 mRNA expression in hearts subjected to ischemia-reperfusion. Fold of change was calculated by delta-delta Ct formula. Glyceraldehyde 3-phosphate dehydrogenase served as a normalizing gene. C—control group, aerobically perfused hearts, IR injury—ischemia-reperfusion injury, CAR 0.1—aerobically perfused hearts treated with 0.1 μM carvedilol, IR-CAR 0.1—hearts from IR injury model treated with 0.1 μM carvedilol. $p > 0.05$, $n = 5$, ANOVA.

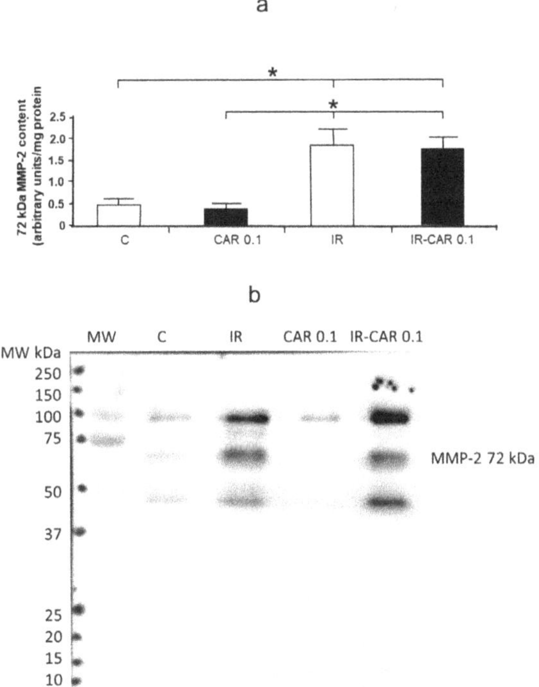

Figure 5. MMP-2 content in coronary effluents and heart tissue. (**a**) Densitometric analysis of MMP-2 content in coronary effluent collected at 2 min of reperfusion after 20 min ischemia in IR CAR 0.1 group as determined by Western blot. (**b**) Representative western blot of MMP-2 content in coronary effluents collected at 2 min of reperfusion after 20 min ischemia in IR CAR 0.1 group. C—control group, aerobically perfused hearts, IR injury—ischemia-reperfusion injury, CAR 0.1, aerobically perfused hearts treated with 0.1 μM carvedilol, IR-CAR 0.1—hearts from IR injury model treated with 0.1 μM carvedilol. * C vs IR, C vs IR-CAR 0.1, CAR 0.1 vs IR, CAR 0.1 vs IR-CAR 0.1, $p < 0.05$, $n = 5$, ANOVA.

There were no significant differences in MMP-2 content in heart tissue between groups (data not shown).

2.6. Carvedilol Abolishes Troponin I Level in Heart Tissue

Analysis of troponin I levels in hearts at the end of perfusion showed that IR injury caused a 73% decrease of the level of 31 kDa troponin I which was abolished in the IR-CAR 0.1 µM group ($p < 0.05$; Figure 6a,b). This was not observed NEB and MET in all tested concentrations (Figure 6c,d).

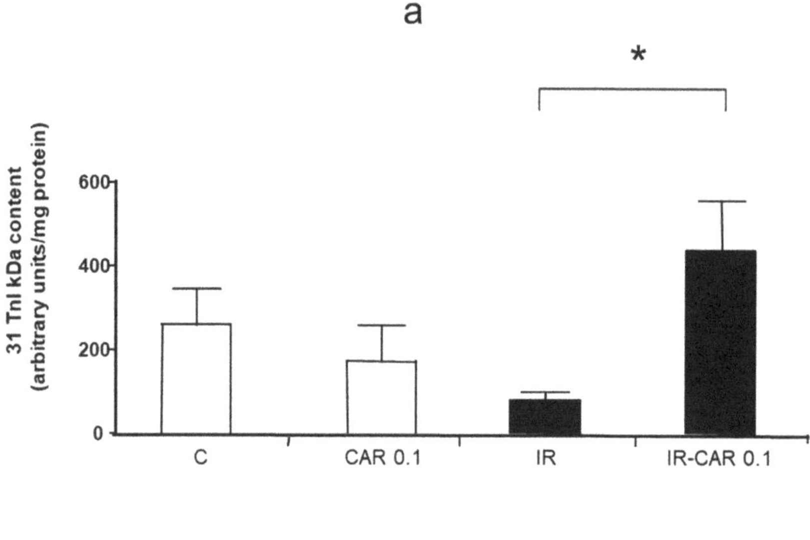

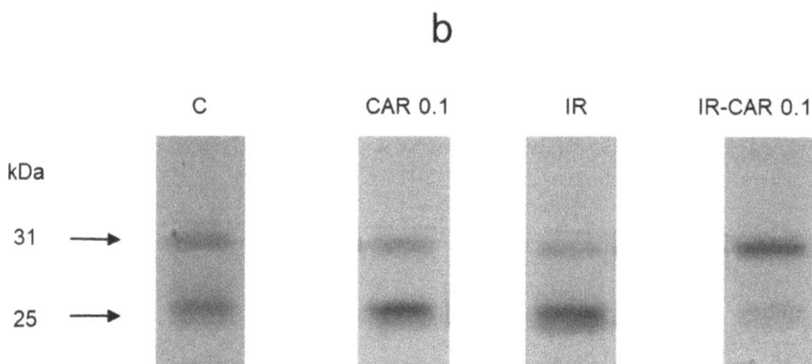

Figure 6. *Cont.*

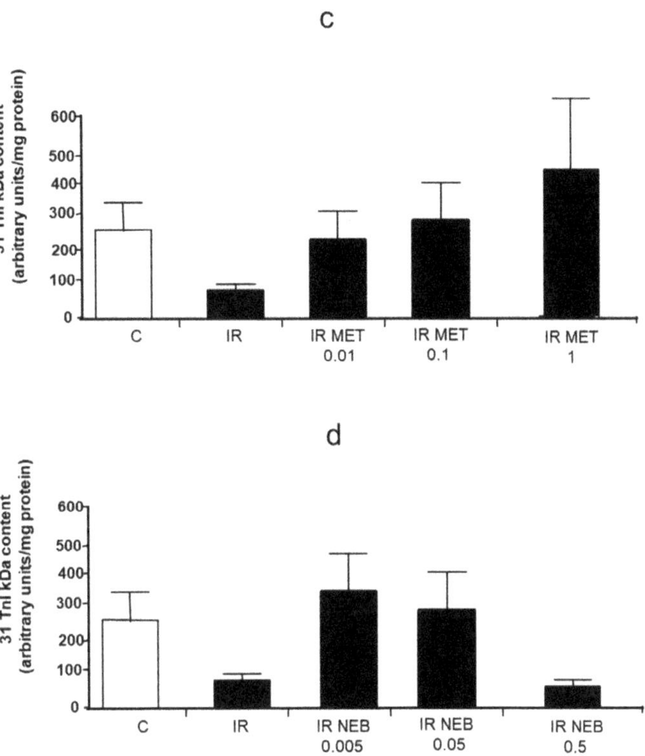

Figure 6. (**a–d**) Effect of carvedilol, metoprolol and nebivolol on troponin I content in heart tissue. (**a**) Densitometric analysis of TnI content in heart homogenates prepared at 30 min of reperfusion after 20 min ischemia in IR CAR 0.1 group as determined by Western blot. (**b**) Representative troponin I (TnI) protein level in heart homogenates prepared at the end of perfusion in IR CAR 0.1 group as determined by Western blot. (**c**) Densitometric analysis of TnI content in heart homogenates prepared at 30 min of reperfusion after 20 min ischemia in metoprolol treated hearts as determined by Western blot. (**d**) Densitometric analysis of TnI content in heart homogenates prepared at 30 min of reperfusion after 20 min ischemia in nebivolol treated hearts as determined by Western blot. Decrease of the level of 31 kDa troponin I was abolished only in the IR-CAR 0.1 μM group. C—control group, aerobically perfused hearts, IR injury—ischemia-reperfusion injury, IR-CAR 0.1—hearts from IR injury model treated with 0.1 μM carvedilol, IR-CAR 1—hearts from IR injury model treated with 1 μM carvedilol, IR-CAR 10– hearts from IR injury model treated with 10 μM carvedilol, IR-MET 0.01—hearts from IR injury model treated with 0.01 μM metoprolol, IR-MET 0.1—hearts from IR injury model treated with 0.1 μM metoprolol, IR-MET 1—hearts from IR injury model treated with 1 μM metoprolol, IR-NEB 0.005—hearts from IR injury model treated with 0.005 μM nebivolol, IR-NEB 0.05—hearts from IR injury model treated with 0.05 μM nebivolol, IR-NEB 0.5—hearts from IR injury model treated with 0.5 μM nebivolol. * IR vs IR-CAR 0.01, $p < 0.05$, $n = 6$, ANOVA.

3. Discussion

This study demonstrates for the first time that the cardioprotective role of carvedilol (CAR) in experimental IR injury may be caused in part by its ability to inhibit MMP-2 activation. These findings are of clinical importance and suggest a possible beneficial role of CAR over other β-blockers in preventing contractile dysfunction as a consequence of IR injury.

Although MMPs were first recognized for their ability to cleave substrates in the extracellular matrix, more recent studies revealed that hearts undergoing oxidative stress have increased MMP-2 activity. Moreover, it also acts intracellularly, leading to myocardial contractile dysfunction by degradation of α-actinin, myosin light chain-1, titin and troponin I [20,21]. Observations regarding MMPs function in animal hearts with induced IR injury have been verified in human studies. Lalu et al. reported an increase of both MMP-2 and MMP-9 activities in human hearts following cardiopulmonary bypass [22]. It was shown that during cardiopulmonary bypass for coronary artery graft surgery, right atrial biopsies obtained within 10 min after aortic cross-clamp release had increased activities of both MMP-2 and MMP-9 which was inversely correlated with the decrease in contractile function measured 3 h after cross-clamp release [22].

CAR and NEB were both shown to inhibit the activity of various MMPs in conditions of enhanced oxidative stress related to atherosclerosis, myocarditis, periodontitis, hypertension, or renal IR injury [11–13,15,16]. However, there were no data on the effect of these drugs on MMP activation in acute IR injury of the heart.

In our study, we showed an improvement in the recovery of cardiac contractile function in IR injury hearts along with a reduction in MMP-2 release into the coronary effluent in 0.1 μM CAR treated hearts. Interestingly, only this lowest concentration of CAR improved the recovery of mechanical function during reperfusion and attenuated the IR injury-induced MMP-2 release into the coronary effluent. The superior role of this low concentration of carvedilol in reducing MMPs activation over higher concentrations was shown in other studies. Jaggi et al. showed that only CAR- 0.1 μM attenuated IR injury in isolated rat hearts by preventing mast cell degranulation [23]. Similarly, a low concentration of CAR (0.05 μM) was shown to exert cardioprotective effects, along with reduction of creatine kinase release, during hypoxia in isolated rat hearts [24]. Our current results are also consistent with our previous findings showing that the lowest concentration of CAR was the most effective in inhibiting MMP-2 activity during acute autoimmune myocarditis [13].

NEB is also known for its anti-MMPs properties [15,16]. However, surprisingly, we did not observe cardioprotective effects and MMP inhibition with NEB despite its equivalent potency at β-receptors to CAR in the concentrations used in the experiment. The explanation of this phenomenon is not straightforward and requires further investigation. NEB is a third generation, highly selective β1-adrenoceptor antagonist endowed with the ability to induce NO release from the endothelium [14]. The protective role of NO in ischemic hearts is exerted by various mechanisms [25–27]. Nevertheless, a negative effect of NO on cardiomyocytes through the formation of peroxynitrite was when present at elevated levels also in the presence of superoxide. Mori et al. found that intracoronary administration of the precursor of nitric oxide L-arginine aggravated myocardial stunning in dogs via the biosynthesis of peroxynitrite [28]. A similar effect was achieved using an in vivo canine model with NO donor S-nitroso-N-acetylpenicillamine while NOS inhibition exerted the opposite effect [29,30]. Therefore, the lack of inhibitory action of NEB on MMPs in IR injury of the heart may be, at least in part, explained by overproduction of NO by NEB.

MMPs cause acute cardiac mechanical dysfunction by cleaving sarcomeric proteins, such as troponin I [20]. Furthermore, impaired cardiac contractile function and reduced levels of troponin I were shown in cardiomyocytes of transgenic mice with cardiac specific expression of active MMP-2 [31]. Therefore, inhibition of MMP activity should reduce troponin I degradation. In our study, we showed that in homogenates prepared from IR injury hearts that the troponin I levels were the highest in the IR-CAR 0.1 μM group. These results are in line with previous data showing the effect of CAR on troponin I in patients with adriamycin-induced cardiotoxicity [32] and our previous findings regarding the protective role of CAR on the level of troponin in acute myocarditis along with MMP-2 activity inhibition [13].

The main limitation of our study is that, based on our results, we cannot directly say what is the exact mechanism responsible for the decrease in MMP-2 activity by carvedilol. Based on the results of in vitro experiment we can exclude the direct effect of carvedilol on MMP-2 activity. We can also exclude changes at the transcriptional level as we did not observe differences in MMP-2 mRNA expression between groups, indicating that carvedilol regulates MMPs at the post-transcriptional level. Western blot analysis of MMP-2 content in coronary effluent revealed that there were no changes between the IR and IR-CAR 0.1 µM group while the increase in MMP-2 activity assessed in zymography was significantly lower in the IR injury with CAR 0.1 µM group than in the IR group. Moreover, a decrease in MMP-2 activity in the effluent from hearts treated with carvedilol was accompanied by a decrease in tissue troponin I degradation. These results indicate that the changes of MMP-2 activity resulted from its activation in heart tissue and subsequent release into the coronary effluent in the settings of IR but not from changes in enzyme localization. Further research is needed to explain the exact mechanism by which carvedilol inhibits the activity of MMP-2.

Another limitation of our study is that we did not observe significant changes in MMP-2 tissue activity. However, the tissue MMP-2 activity was assessed only at 30 min of reperfusion, but not within the first two minutes. As a result of reperfusion injury following myocardial ischemia, we showed a rapid and enhanced release of MMP-2 into the coronary effluent which peaked within the first 2 min of reperfusions shown by [19]. As the increased release of activated MMP-2 results from its intracellular activation, consequently, MMP-2 tissue activity at the beginning of reperfusion should be increased. The lack of increased MMP-2 activity in heart homogenates at the end of reperfusion is due to its release from isolated hearts as shown by Cheung et al. [33]. Again, troponin I levels in heart tissue may serve as indirect evidence of MMP activation [19]. Despite this limitation, we suggest that the effects of carvedilol on postischemic cardiac contractility and troponin levels we observe may be due to its ability to inhibit MMP-2 activation and subsequent release.

4. Materials and Methods

4.1. Animals

All β-blockers (CAR, NEB, MET) were purchased from Sigma-Aldrich (Poznan, Poland). Hearts were obtained from male Wistar rats, weighing 250–350 g, purchased from the Laboratory Animal Center, Wroclaw Medical University.

4.2. Heart Perfusion Protocol

Rats were anesthetized with thiopental (75 mg/kg) given intraperitoneally. After sternotomy, hearts were excised and immersed with ice-cold Krebs–Henseleit solution (0.5 mmol/L EDTA, 1.2 mmol/L KH_2PO_4, 1.2 mmol/L MgSO4, 3 mmol/L $CaCl_2$, 4.7 mmol/L KCl, 11 mmol/L glucose, 25 mmol/L $NaHCO_3$, and 118 mmol/L NaCl), gassed with a mixture of 95% O_2 and 5% CO_2 in pH 7.4. Then the aorta was cannulated and coronary perfusion with Krebs–Henseleit solution was initiated. Perfusion solution was delivered at constant pressure (60 mmHg) and temperature (37 °C). Left ventricular pressure was monitored by fluid filled latex balloon inserted into the left ventricle. Heart rate was also monitored during the experiments. Left ventricular developed pressure (LVDP) was calculated as the difference between systolic and diastolic pressure of the left ventricular. Cardiac mechanical function was expressed as the heart rate-pressure product (RPP) (calculated as the product of the spontaneous heart rate and LVDP).

The hearts were aerobically perfused for 25 min, then for 20 min subjected to global, no-flow normothermic ischemia and finally aerobically reperfused for 30 min (Figure 7). Control hearts were perfused for 75 min in aerobic conditions. The β-blockers (CAR, NEB, or MET) were infused into the hearts 10 min prior to the onset of ischemia and for the first 10 min of reperfusion, or for 40 min of the control aerobic perfusion (from 15 to 55 min of perfusion). The heart perfusion apparatus had separate perfusion solution reservoirs and a switching valve which allowed the selection between solutions with or without added drugs. MET was dissolved directly into Krebs–Henseleit solution. CAR and NEB were dissolved in dimethyl sulfoxide (DMSO) and subsequently diluted at the desired concentrations with Krebs–Henseleit solution. The concentration of DMSO reaching the heart was < 0.5% (*v:v*). Samples of coronary effluent for analysis were collected for an equal time of 2 min immediately prior to ischemia and after 2 and 30 min after reperfusion. At the end of the protocol, hearts were freeze clamped in liquid nitrogen and stored at −80 °C for subsequent biochemical analysis.

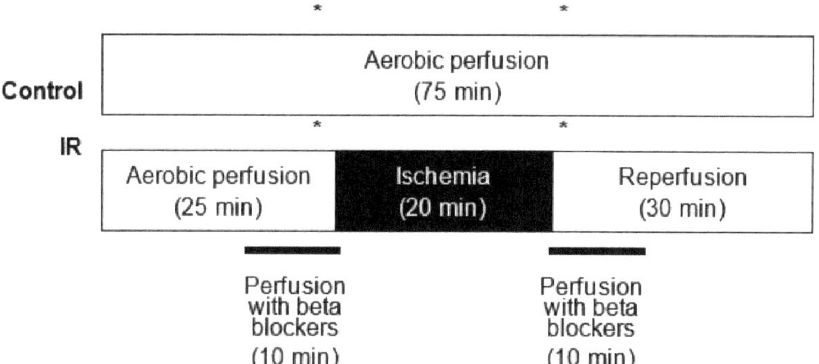

Figure 7. Experimental protocol for isolated heart perfusions. Detailed description in Material and methods part, "heart perfusion protocol" section. * show times when coronary effluent samples were collected during aerobic perfusion (23–25 min) and in the first 2 min of reperfusion (45–47 min).

4.3. Experimental Groups

The hearts were randomly divided into the following 20 experimental groups ($n = 6$ hearts per group) in aerobic or anaerobic (IR) conditions either with no drug or with three different concentrations of tested drugs: CAR (0.1, 1, 10 μM), NEB (0.005, 0.05, 0.5 nM) and MET (0.01, 0.1, 1 μM).

4.4. Preparation of Heart Extracts and Concentration of Coronary Effluent

Frozen hearts were crushed into a powder with a mortar and pestle at liquid nitrogen temperature and stored at −80 °C. Prior to the biochemical analysis, the tissue powder was homogenized in 50 mmol/L Tris-HCl (pH 7.4) containing 1 mmol/L dithiothreitol, 3.1 mmol/L sucrose, 2 μg/mL aprotinin, 10 μg/mL soybean trypsin inhibitor, 10 μg/mL leupeptin, and 0.1% Triton X-100. The samples of coronary effluent were concentrated at 4 °C 30-fold using Centricon-10 concentrating vessels purchased from Merck Millipore (Poznan, Poland). The supernatant was collected and stored at −80 °C for analysis. Protein content in both effluent and homogenates was analyzed using Bradford Protein Assay (Bio-Rad, Warszawa, Poland) and bovine serum albumin used as a protein standard.

4.5. Measurement of MMP-2 by Gelatin Zymography

MMP-2 activity was assessed in both heart extracts and concentrated coronary effluent. Equal total protein samples were applied to 8% SDS-PAGE gels. Following electrophoresis, gels were rinsed thrice for 20 min each in 2.5% Triton X-100 and afterward twice for 20 min each in incubation buffer (5 mmol/L CaCl$_2$, 50 mmol/L Tris-HCl, 150 mmol/L NaCl, and 0.05% NaN3) at room temperature. Subsequently, they were placed for 10 h at 37 °C in incubation buffer and finally were stained in 2% Coomassie Brilliant Blue G, 25% methanol, and 10% acetic acid for 2 h and destained in 30% methanol/ 10% acetic acid solution. Quantification of reactions was performed with the usage of a GS-800 Calibrated Densitometer with Quantity One v.4.6.9 software (BioRad) and the relative MMPs activity was determined and expressed in arbitrary units (AU) calculated on the basis of recombined MMP standard activity. In vitro inhibition of MMPs activity by carvedilol was evaluated by developing zymograms performed on selected heart extracts with the addition of carvedilol to the incubation buffer in comparison to control gels (with no drug added), which served as a model of 100% of MMP-2 activity.

4.6. Measurement of Troponin I and MMP-2 by Western Blot

Troponin I content was determined by Western blotting. 30 µg of protein obtained from heart extracts was applied to 15% SDS-PAGE gels. Following electrophoresis, (at 150 V, 20 °C) samples were electroblotted onto a polyvinylidene difluoride membrane (by semi-dry technique; at 25 V, 30 min). A primary monoclonal mouse antibody against cardiac troponin I as well as secondary goat-anti-mouse conjugated with horseradish peroxidase (HRP) were both used at 1:1000 dilution (Thermo Fisher Scientific, Waltham, Massachusetts, USA; BioRad, respectively). The blot was developed using a chemiluminescence assay (ClarityTM Western ECL Substrate, Biorad). Membranes were scanned using ChemiDocTM XRS+ System with Image LabTM Software v.5.2 for data analysis. Rat cardiac troponin I was used for standard curve preparation (Advanced ImmunoChemical Inc., Long Beach, California, USA). MMP-2 content was determined in coronary effluents. 30 µL of each concentrated coronary effluent was applied to 10% SDS-PAGE gels at reducing conditions. After electrophoresis samples were electroblotted for 40 min at 50 V onto a nitrocellulose membrane 0.45 µm (BioRad) by wet technique. A primary monoclonal mouse antibody against total MMP-2 ab86607 (Abcam) and secondary goat-anti-mouse conjugated with HRP (BioRad) were both used at a dilution of 1:1000. The blot was developed and scanned as described above. The 72 kDa MMP-2 was detected by comparison with Precision Plus Protein Standards (BioRad) and relative content was calculated on the basis of 75 kDa band intensity and expressed in arbitrary units.

4.7. Expression of MMP-2 Gene in Heart Tissue

Ribonucleic acid (RNA) was extracted from powdered heart tissue by phenol/chloroform technique using PureZol RNA isolation reagent (BioRad), according to the manufacturer's instruction. Briefly, 50 µg of tissue powder was mixed with 1 mL of PureZol, and immediately homogenized by Pellet Pestle® Motor (Kimble Kontes). The next steps included extraction with chloroform (Stanlab), precipitation of RNA with isopropanol (Chempur), and washing with 75% ethanol (Chempur). Purified RNA was dissolved in 50 µL of DEPC-treated water (Ambion) and its quality and concentration were assessed by NanoDrop Lite Spectrophotometer (Thermo Fisher Scientific). 500 ng of RNA was taken for reverse transcription performed using iScript™ cDNA Synthesis Kit (BioRad). Reverse transcription and subsequent real-time PCR were both performed on CFX96 Touch Real-Time PCR Detection System (BioRad). 100 ng of each cDNA template was used for both genes' real-time amplification in duplicates, using iTaq Universal SYBR® Green Supermix (BioRad) following the manufacturer's protocol. Glyceraldehyde 3-phosphate dehydrogenase (GAPDH) served as a housekeeping gene for the normalization of MMP-2 gene expression. Primers sequences were designed as follows: GAPDH F: 5′ AGTGCCAGCCTCGTCTCATA 3′, GAPDH R: 5′ GATGGTGATGGGTTTCCCGT 3′; MMP-2 F: 5′ AGCAAGTAGACGCT-GCCTTT 3′, MMP-2 R: 5′ CAGCACCTTTCTTTGGGCAC 3′. Relative fold MMP-2 gene expression was calculated according to delta-delta Ct formula.

4.8. Statistical Analysis

All data are expressed as mean +/− SEM. Comparisons between groups were assessed for significance by two-way analysis of variance (ANOVA) or T-test after assessment of normality of distribution. Logarithmic transformation was performed of data that had non-normal distribution to meet requirements of ANOVA analysis. The conclusions drawn after data transformation remain valid for the original data. If significance was established, posthoc analysis was done using Tukey's test. A value of $p < 0.05$ was considered statistically significant.

5. Conclusions

In conclusion, our results provide the first evidence that CAR reduces mechanical dysfunction of the heart muscle, along with attenuation of MMP-2 activity and degradation of troponin I in hearts subjected to acute IR injury. These data might help to differentiate carvedilol from other β-blockers and explain a peculiarity of this drug, in terms of its anti-MMP-2 properties in the settings of acute myocardial ischemia, resulting in its better cardioprotective profile compared to other β-blockers.

Author Contributions: Conceptualization, M.S.-S. and A.S.; methodology, M.S.-S. and R.S.; software, M.S.-S.; validation, M.S.-S. and A.S.; formal analysis, M.S.-S.; investigation, M.S.-S., J.U., A.S.-W., R.D., J.P., A.O. and B.K.; resources, M.S.-S.; data curation, M.S.-S., J.U., A.S.-W., R.D., J.P., A.O. and B.K.; writing—original draft preparation, M.S.-S.; writing—review and editing, I.B.-L., R.S., M.W. and A.S.; visualization, M.S.-S.; supervision, I.B.-L., R.S., M.W. and A.S.; project administration, M.S.-S.; funding acquisition, M.S.-S. All authors have read and agreed to the published version of the manuscript.

Funding: This work was supported by the grant of the Polish State Committee for Scientific Research [N401 166 32/3235].

Institutional Review Board Statement: The study protocol was approved by the Local Ethics Commission for Animal Experiments (approval No. 69/03).

Informed Consent Statement: Not applicable.

Data Availability Statement: The data used to support the findings of this study are included within the article and are available from the corresponding author.

Acknowledgments: The authors express their appreciation to Christopher Kobierzycki for his very important, critical comments to the manuscript and and to Halina Gliniak for her excellent technical assistance.

Conflicts of Interest: The authors declare no conflict of interest.

References

1. O'Rourke, S. Antianginal Actions of Beta-Adrenoceptor Antagonists. *Am. J. Pharm. Educ.* **2007**, *71*, 95. [CrossRef]
2. Heitzler, V.N.; Pavlov, M. Therapeutic approach in acute coronary syndrome focusing on oral therapy. *Acta Clin. Croat.* **2010**, *49*, 81–87.
3. Joseph, J.; Velasco, A.; Hage, F.G.; Reyes, E. Guidelines in review: Comparison of ESC and ACC/AHA guidelines for the diagnosis and management of patients with stable coronary artery disease. *J. Nucl. Cardiol.* **2018**, *25*, 509–515. [CrossRef]
4. Schwarz, E.R.; Kersting, P.H.; Reffelmann, T.; Meven, D.A.; Al-Dashti, R.; Skobel, E.C.; Klosterhalfen, B.; Hanrath, P. Cardioprotection by carvedilol: Anti-apoptosis is independent on beta-adrenoceptor blockage in the rat heart. *J. Cardiovasc. Pharm. Ther.* **2003**, *8*, 207–215. [CrossRef]
5. Lu, H.R.; Vandeplassche, G.; Wouters, L.; Borgers, M. Beta-blockade in the ischemic reperfused working rabbit heart: Dissociation of beta-adrenergic blocking and protective effects. *Arch. Int. Pharmacodyn. Ther.* **1989**, *301*, 165–181.
6. Kimura-Kurosawa, S.; Kanaya, N.; Kamada, N.; Hirata, N.; Nakayama, M.; Namiki, A. Cardioprotective effect and mechanism of action of landiolol on the ischemic reperfused heart. *J. Anesth.* **2007**, *21*, 480–489. [CrossRef]
7. Gao, F.; Chen, J.; Lopez, B.L.; Christopher, T.A.; Gu, J.; Lysko, P.; Ruffolo, R.R.; Ohlstein, E.H.; Ma, X.L.; Yue, T.-L. Comparison of bisoprolol and carvedilol cardioprotection in rabit ischemia and reperfusion model. *Eur. J. Pharmacol.* **2000**, *406*, 109–116. [CrossRef]
8. Kurosawa, S.; Kanaya, N.; Niiyama, Y.; Nakayama, M.; Fujita, S.; Namiki, A. Landiolol, esmolol and propranolol protect from ischemia/reperfusion injury in isolated guinea pig hearts. *Can. J. Anaesth.* **2003**, *50*, 489–494. [CrossRef]
9. Cargnoni, A.; Ceconi, C.; Bernocchi, P.; Boraso, A.; Parrinello, G.; Curello, S.; Ferrari, R. Reduction of oxidative stress by carvedilol: Role in maintenance of ischaemic myocardium viability. *Cardiovasc. Res.* **2000**, *47*, 556–566. [CrossRef]
10. Feuerstein, G.Z.; Ruffolo, R.R. Carvedilol, a novel vasodilating beta-blocker with the potential for cardiovascular organ protection. *Eur. Heart J.* **1996**, *17* (Suppl. B), 24–29. [CrossRef]
11. Pauschinger, M.; Rutschow, S.; Chandrasekharan, K.; Westermann, D.; Weitz, A.; Schwimmbeck, L.P.; Zeichhardt, H.; Poller, W.; Noutsias, M.; Li, J.; et al. Carvedilol improves left ventricular function in murine coxsackievirus-induced acute myocarditis association with reduced myocardial interleukin-1beta and MMP-8 expression and a modulated immune response. *Eur. J. Heart Fail.* **2005**, *7*, 444–452. [CrossRef]
12. Wu, T.-C.; Chen, Y.-H.; Leu, H.-B.; Chen, Y.-L.; Lin, F.-Y.; Lin, S.-J.; Chen, J.-W. Carvedilol, a pharmacological antioxidant, inhibits neointimal matrix metalloproteinase-2 and -9 in experimental atherosclerosis. *Free. Radic. Biol. Med.* **2007**, *43*, 1508–1522. [CrossRef]
13. Skrzypiec-Spring, M.; Haczkiewicz-Leśniak, K.; Sapa, A.; Piasecki, T.; Kwiatkowska, J.; Ceremuga, I.; Woźniak, M.; Biczysko, W.; Kobierzycki, C.; Dziegiel, P.; et al. Carvedilol inhibits matrix metalloproteinase-2 activation in experimental autoimmune myocarditis: Possibilities of cardioprotective application. *J. Cardiovasc. Pharmacol. Ther.* **2018**, *23*, 89–97. [CrossRef]

14. Kamp, O.; Metra, M.; Bugatti, S.; Bettari, L.; Cas, A.D.; Petrini, N.; Cas, L.D. Nebivolol: Haemodynamic effects and clinical significance of combined beta-blockade and nitric oxide release. *Drugs* **2010**, *70*, 41–56. [CrossRef]
15. Ceron, C.S.; Rizzi, E.; Guimarães, D.A.; Martins-Oliveira, A.; Gerlach, R.F.; Tanus-Santos, J.E. Nebivolol attenuates prooxidant and profibrotic mechanisms involving TGF-β and MMPs, and decreases vascular remodeling in renovascular hypertension. *Free. Radic. Biol. Med.* **2013**, *65*, 47–56. [CrossRef]
16. Ersan, S.; Tanrisev, M.; Cavdar, Z.; Celik, A.; Unli, M.; Kocak, A.; Kose, T. Pretreatment with nebivolol attenuates level and expression of matrix metalloproteinases in a rat model of renal ischemia-reperfusion injury. *Nephrology* **2017**, *22*, 1023–1029. [CrossRef]
17. Yasmin, W.; Strynadka, K.D.; Schulz, R. Generation of peroxynitrite contributes to ischemia-reperfusion injury in isolated rat hearts. *Cardiovasc. Res.* **1997**, *33*, 422–432. [CrossRef]
18. Viappiani, S.; Nicolescu, A.C.; Holt, A.; Sawicki, G.; Crawford, B.D.; León, H.; van Mulligen, T.; Schulz, R. Activation and modulation of 72kDa matrix metalloproteinase-2 by peroxynitrite and glutathione. *Biochem. Pharmacol.* **2009**, *77*, 826–834. [CrossRef]
19. Wang, W.; Schulze, C.J.; Suarez-Pinzon, W.L.; Dyck, J.R.B.; Sawicki, G.; Schulz, R. Intracellular action of matrix metalloproteinase-2 accounts for acute myocardial ischemia and reperfusion injury. *Circulation* **2002**, *106*, 1543–1549. [CrossRef]
20. Ali, M.A.M.; Fan, X.; Schulz, R. Cardiac sarcomeric proteins: Novel intracellular targets of matrix metalloproteinase-2 in heart disease. *Trends Cardiovasc. Med.* **2011**, *21*, 112–118. [CrossRef]
21. Sariahmetoglu, M.; Skrzypiec-Spring, M.; Youssef, N.; Jacob-Ferreira, A.L.B.; Sawicka, J.; Holmes, C.; Sawicki, G.; Schulz, R. Phosphorylation status of matrix metalloproteinase-2 in myocardial ischemia-reperfusion injury. *Heart* **2012**, *98*, 656–662. [CrossRef] [PubMed]
22. Lalu, M.M.; Pasini, E.; Schulze, C.J.; Ferrari-Vivaldi, M.; Ferrari-Vivaldi, G.; Bachetti, T.; Schulz, R. Ischaemia-reperfusion injury activates matrix metalloproteinases in the human heart. *Eur. Heart J.* **2005**, *26*, 27–35. [CrossRef]
23. Jaggi, A.S.; Singh, M.; Sharna, A.; Singh, D.; Singh, N. Cardioprotective effects of mast cell modulators in ischemia-reperfusion-induced injury in rats. *Methods Find Exp. Clin. Pharmacol.* **2007**, *29*, 593–600. [CrossRef] [PubMed]
24. Skobel, E.; Dannemann, O.; Reffelmann, T.; Böhm, V.; Weber, C.; Hanrath, P.; Uretsky, B.F.; Schwarz, E.R. Carvedilol protects myocardial cytoskeleton during hypoxia in the rat heart. *J. Appl. Res.* **2005**, *5*, 378–386.
25. Jones, S.P.; Bolli, R. The ubiquitous role of nitric oxide in cardioprotection. *J. Mol. Cell. Cardiol.* **2006**, *40*, 16–23. [CrossRef] [PubMed]
26. Shah, A.M.; MacCarthy, P.A. Paracrine and autocrine effects of nitric oxide on myocardial function. *Pharmacol. Ther.* **2000**, *86*, 49–86. [CrossRef]
27. Dawn, B.; Bolli, R. Role of nitric oxide in myocardial preconditioning. *Ann. N. Y. Acad. Sci.* **2002**, *962*, 18–41. [CrossRef] [PubMed]
28. Mori, E.; Haramaki, N.; Ikeda, H.; Imaizumi, T. Intra-coronary administration of L-arginine aggravates myocardial stunning through production of peroxynitrite in dogs. *Cardiovasc. Res.* **1998**, *40*, 113–123. [CrossRef]
29. Zhang, Y.; Bissing, J.W.; Xu, L.; Ryan, A.J.; Martin, S.M.; Miller, F.J.; Kregel, K.C.; Buettner, G.; Kerber, R.E. Nitric oxide synthase inhibitors decrease coronary sinus-free radical concentration and ameliorate myocardial stunning in an ischemia-reperfusion model. *J. Am. Coll. Cardiol.* **2001**, *38*, 546–554. [CrossRef]
30. Zhang, Y.; Davies, L.R.; Martin, S.M.; Coddington, W.J.; Miller, F.J.; Buettner, G.R.; Kerber, R.E. The nitric oxide donor S-nitroso-N-acetylpenicillamine (SNAP) increases free radical generation and degrades left ventricular function after myocardial ischemia–reperfusion. *Resuscitation* **2003**, *59*, 345–352. [CrossRef]
31. Bergman, M.R.; Teerlink, J.R.; Mahimkar, R.; Li, L.; Zhu, B.-Q.; Nguyen, A.; Dahi, S.; Karliner, J.S.; Lobertt, D.H. Cardiac transgenic matrix metalloproteinase-2 expression independently induces marked ventricular remodeling and systolic dysfunction. *Am. J. Physiol. Heart Circ. Physiol.* **2007**, *292*, 1847–1860. [CrossRef] [PubMed]
32. El-Shitany, N.A.; Tolba, O.A.; El-Shanshory, M.R.; El-Hawary, E.E. Protective effect of carvedilol on adriamycin-induced left ventricular dysfunction in children with acute lymphoblastic leukemia. *J. Card. Fail.* **2012**, *18*, 607–613. [CrossRef] [PubMed]
33. Cheung, P.-Y.; Sawicki, G.; Woźniak, M.; Wang, W.; Radomski, M.W.; Schulz, R. Matrix metalloproteinase-2 contributes to ischemia-reperfusion injury in the heart. *Circulation* **2000**, *101*, 1833–1839. [CrossRef] [PubMed]

pharmaceuticals

Article

Obeticholic Acid Reduces Kidney Matrix Metalloproteinase Activation Following Partial Hepatic Ischemia/Reperfusion Injury in Rats

Giuseppina Palladini [1,2,†], Marta Cagna [1,†], Laura Giuseppina Di Pasqua [1], Luciano Adorini [3], Anna Cleta Croce [4], Stefano Perlini [1,5], Andrea Ferrigno [1], Clarissa Berardo [1,*] and Mariapia Vairetti [1,*]

1   Department of Internal Medicine and Therapeutics, University of Pavia, 27100 Pavia, Italy;
    giuseppina.palladini@unipv.it (G.P.); marta.cagna02@universitadipavia.it (M.C.);
    lauragiuseppin.dipasqua01@universitadipavia.it (L.G.D.P.); stefano.perlini@unipv.it (S.P.);
    andrea.ferrigno@unipv.it (A.F.)
2   Internal Medicine Fondazione IRCCS Policlinico San Matteo, 27100 Pavia, Italy
3   Intercept Pharmaceuticals, San Diego, CA 92121, USA; LAdorini@interceptpharma.com
4   Institute of Molecular Genetics, Italian National Research Council (CNR), 27100 Pavia, Italy;
    annacleta.croce@igm.cnr.it
5   Emergency Department Fondazione IRCCS Policlinico San Matteo, 27100 Pavia, Italy
*   Correspondence: clarissa.berardo01@universitadipavia.it (C.B.); mariapia.vairetti@unipv.it (M.V.);
    Tel.: +39-0382-986877 (C.B.); +39-0382-986398 (M.V.)
†   These authors contributed equally to this work.

check for **updates**

Citation: Palladini, G.; Cagna, M.; Di Pasqua, L.G.; Adorini, L.; Croce, A.C.; Perlini, S.; Ferrigno, A.; Berardo, C.; Vairetti, M. Obeticholic Acid Reduces Kidney Matrix Metalloproteinase Activation Following Partial Hepatic Ischemia/Reperfusion Injury in Rats. *Pharmaceuticals* **2022**, *15*, 524. https://doi.org/10.3390/ph15050524

Academic Editor: Salvatore Santamaria

Received: 5 April 2022
Accepted: 21 April 2022
Published: 24 April 2022

Publisher's Note: MDPI stays neutral with regard to jurisdictional claims in published maps and institutional affiliations.

Abstract: We have previously demonstrated that the farnesoid X receptor (FXR) agonist obeticholic acid (OCA) protects the liver via downregulation of hepatic matrix metalloproteinases (MMPs) after ischemia/reperfusion (I/R), which can lead to multiorgan dysfunction. The present study investigated the capacity of OCA to modulate MMPs in distant organs such as the kidney. Male Wistar rats were dosed orally with 10 mg/kg/day of OCA (5 days) and were subjected to 60-min partial hepatic ischemia. After 120-min reperfusion, kidney biopsies (cortex and medulla) and blood samples were collected. Serum creatinine, kidney MMP-2, and MMP-9-dimer, tissue inhibitors of MMPs (TIMP-1, TIMP-2), RECK, TNF-alpha, and IL-6 were monitored. MMP-9-dimer activity in the kidney cortex and medulla increased after hepatic I/R and a reduction was detected in OCA-treated I/R rats. Although not significantly, MMP-2 activity decreased in the cortex of OCA-treated I/R rats. TIMPs and RECK levels showed no significant differences among all groups considered. Serum creatinine increased after I/R and a reduction was detected in OCA-treated I/R rats. The same trend occurred for tissue TNF-alpha and IL-6. Although the underlying mechanisms need further investigation, this is the first study showing, in the kidney, beneficial effects of OCA by reducing TNF-alpha-mediated expression of MMPs after liver I/R.

Keywords: kidney; obeticholic acid; ischemia/reperfusion; liver; metalloproteinases

1. Introduction

Obeticholic acid (OCA, INT-747) is a bile acid-derived FXR agonist currently in phase III trials for the treatment of NASH that has already shown its potential for treating hepatic steatosis, inflammation, and fibrosis while increasing insulin sensitivity [1]. Recently, the REGENERATE study showed that the administration of OCA in NASH patients not only ameliorates liver injury and fibrosis but also improves several health-related quality of life (HRQoL) domains [2].

Fibrosis is a type of chronic organ failure, resulting in the excessive secretion of extracellular matrix (ECM). Matrix metalloproteinases (MMPs), the main group of ECM-degrading enzymes, are considered to be a potential target for fibrosis treatment [3].

Tissue Inhibitor of Metalloproteinases (TIMP-1), the major endogenous regulator of MMP-9, plays a protective function in the control of survival and proliferation of liver cells during I/R injury [4]. MMP-2 and MMP-9 are reduced by interaction with the reversion-inducing-cysteine-rich protein with kazal motifs (RECK), a transformation-suppressor gene that regulates the expression of several MMPs and is involved in the inhibition of the tumor invasion and metastasis process [5].

We have recently shown that OCA decreases MMP-2 and MMP-9 activity in tissue and bile obtained from livers submitted to partial ischemia/reperfusion (I/R) injury [6]. This is associated with the ability of OCA to restore inhibitors of MMP-2 and MMP-9 such as RECK and TIMPs (TIMP-1 and TIMP-2) decreased by ischemic insult [6].

The I/R injury in one organ can also lead to multiorgan dysfunction, that is tissue damage in remote organs away from the body district where the I/R damage is taking place. For example, I/R intestine damage has been described to cause multiple organ dysfunction due to uncontrolled production and release of cytokines and other proinflammatory molecules [7]. Hepatic I/R injury may also cause damage to remote organs such as the kidney, heart, and lung. In addition to liver dysfunction as a consequence of hepatic reperfusion, many remote organs seem to be influenced during this process as well [8]. We have previously documented that moderate acute hepatic ischemia (30 min) followed by reperfusion (60 min) increases MMPs activity not only in the ischemic liver region but also in the lung, associated with histological damage in the liver, lung, and kidney [9]. Clinical studies have reported that, in patients with acute liver failure induced by I/R, the incidence of acute kidney injury (AKI) ranges from 40 to 85%, and up to 95% in liver transplantation [10]. Indeed, damaged liver tissue releases destructive proinflammatory cytokines and oxygen-derived radicals into the circulation that are likely causing further damage to remote organs [11]. Defined as multiple organ dysfunction syndrome (MODS) or multiple organ failure (MOF), this event is an important cause of death in surgical intensive care units. In particular, MODS includes altered organ function in sepsis, septic shock, and systemic inflammatory response syndrome [8].

Most of the current knowledge on liver I/R and remote organ injury derives from experimental studies as well as from the assessment of the efficacy of innovative therapeutic strategies. Based on the above reports, we designed a study in rats to investigate whether liver ischemia (60 min) followed by reperfusion (120 min) affect the function and the structure of remote organs such as kidney, via modulation of ECM remodeling. Using these models, we also tested if OCA treatment is able to counteract remote damage via restoration of MMPs and their specific inhibitors, TIMP-1, TIMP-2, and RECK.

2. Results

2.1. Hepatorenal Syndrome after Liver I/R

After a consensus conference in 1978, hepatorenal syndrome, renal insufficiency progressing in the presence of severe liver diseases, and the absence of recognized nephrotoxic agents were defined by plasma creatinine $\geq$1.5 mg/dL [12]. In the present study, the evaluation of serum creatinine showed a 1.8-fold increase confirming the induction of hepatorenal syndrome after liver I/R injury (Table 1). Recently, the International Club of Ascites (ICA) has adopted the concept of acute kidney injury (AKI) which was developed originally to be used in critically-ill patients. AKI is defined as an increase in creatinine of at least 0.3 mg/dL (26 µmol/L) and/or $\geq$50% from baseline, within 48 h [13].

Our results also demonstrate that rats subjected to liver I/R developed severe liver dysfunction after 1 h hepatic ischemia followed by 2 h reperfusion, with significantly higher serum levels of AST, ALT, ALP, total and direct bilirubin compared to sham-operated animals (Table 1).

Table 1. Serum biochemical parameters in sham and I/R rats and creatinine/ALT relationship.

	Sham	I/R
AST (mU/mL)	256 ± 33	9653 ± 956 *
ALT (mU/mL)	64 ± 10	8635 ± 847 *
ALP (mU/mL)	459 ± 50	803 ± 55 *
Total Bilirubin (mg/dL)	0.15 ± 0.01	0.31 ± 0.02 *
Direct Bilitubin (mg/dL)	0.05 ± 0.01	0.21 ± 0.01 *
Creatinine (mg/dL)	3.96 ± 1.10	7.35 ± 0.40 *
	r	*p*
Creatinine/ALT	0.94	0.002

n = 6 rats/group. Results are expressed as mean value ± standard error (SE). one-way ANOVA with Tukey–Kramer test. * $p < 0.05$ versus sham rats.

Moreover, we demonstrate a direct relationship between the severity of liver dysfunction (ALT) and the degree of kidney damage (creatinine) after I/R (Table 1). Our data are in agreement with those reported by Lee HT (2009) who studied renal dysfunction in mice after liver I/R injury: animals subjected to 60 min ischemia showed a direct and linear relationship between plasma ALT and creatinine levels after liver I/R [14].

2.2. Changes in MMPs, TIMPs, and RECK in Kidney Cortex and Medulla after Liver I/R

We have already reported that liver I/R injury is associated with MMP activation with profound effects on tissue integrity [15]; this event also influences the function of many remote organs [9,16]. In the present study, we evaluated the MMP-9-dimer activity in the kidney cortex after hepatic ischemia (60 min) followed by reperfusion (120 min): a significant increase in MMP-9-dimer activity was found in the I/R group compared with sham-operated rats (Figure 1a). A reduction in cortex MMP-9-dimer activity was detected in OCA-treated I/R rats compared with vehicle-treated I/R rats (Figure 1a).

The same trend in MMP-9-dimer activity occurred in the medulla obtained from the I/R group and I/R group treated with OCA (Figure 2a).

Although not significantly, the MMP-9-dimer protein trend was superimposable with MMP-9-dimer activity. In the cortex, MMP-9-dimer protein increased in I/R by 1.3-fold and in OCA-treated I/R by 0.8-fold. In the medulla, MMP-9-dimer protein increased in I/R by 1.8-fold and in OCA-treated I/R by 1.2-fold. Our data are expressed as a fold increase in their respective sham controls. Tubulin levels, which served as a loading control, remained unchanged (Figure 1a).

Although not significantly, cortex MMP-2 activity also increased after hepatic I/R compared with sham-operated rats and decreased in OCA-treated I/R rats (Figure 1a).

No changes were found in medulla MMP-2 activity in any group (Figure 2a).

In the cortex, MMP-2 protein increased in I/R by 1.2-fold and in OCA-treated I/R by 1.2-fold. In the medulla, MMP-2 protein increased in I/R by 0.8-fold and in OCA-treated I/R by 0.8-fold. Our data are expressed as a fold increase in their respective sham controls. MMP-2 protein evaluation documented no difference comparing I/R versus I/R + OCA both in the cortex and medulla. Tubulin levels, which served as a loading control, remained unchanged (Figure 2a).

Matrix metalloproteinases are inhibited by specific TIMPs; in addiction, RECK represents a novel matrix metalloproteinase regulator. The evaluation of TIMP-1 and TIMP-2 expression levels showed no significant differences among all groups considered (Figures 1b and 2b) and the same trend occurred for RECK expression levels (Figures 1c and 2c) both in the cortex and medulla.

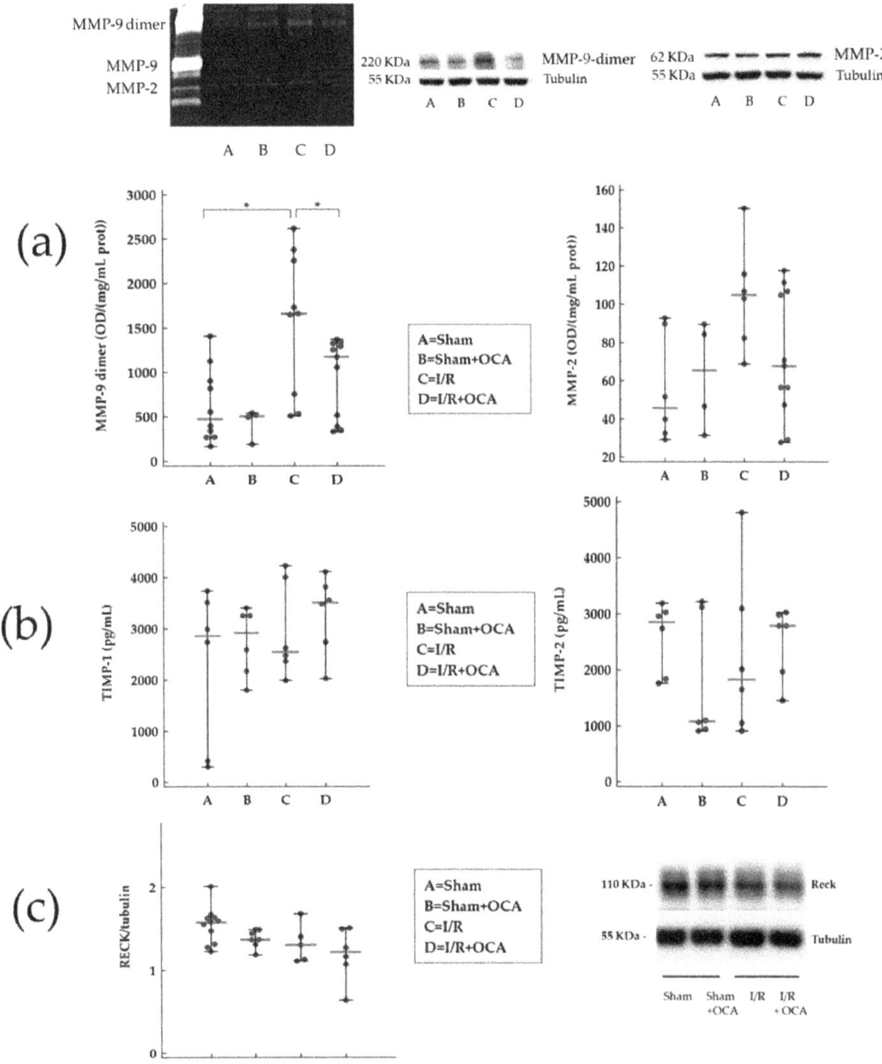

Figure 1. OCA treatment decreases MMP-9-dimer activity in kidney cortex after liver I/R. Animals were orally administered 10 mg/kg/day of OCA in methylcellulose 1% vehicle for 5 days ($n = 12$) or vehicle alone ($n = 12$). (**a**) Representative of MMP zymography and Western blot; gelatinolytic activities of MMP (MMP-9-dimer and MMP-2) quantified by densitometry. (**b**) TIMPs (TIMP-1 and TIMP-2) by Elisa kits and (**c**) RECK expression levels, by Western Blots, were determined in kidney cortex of rats submitted to partial liver ischemia (60 min) followed by reperfusion (120 min). Sham-operated animals were subjected to the same procedure without clamping the vessels. $n = 6$ rats/group. Results are expressed as median value (line red) $\pm$ error bar (1–99 percentiles) (black dot). One-way ANOVA with Tukey–Kramer test (TIMP-1 and RECK); Kruskal–Wallis test with Conover test (MMP-9-dimer, MMP-2, TIMP-2). * $p < 0.05$ versus I/R rats.

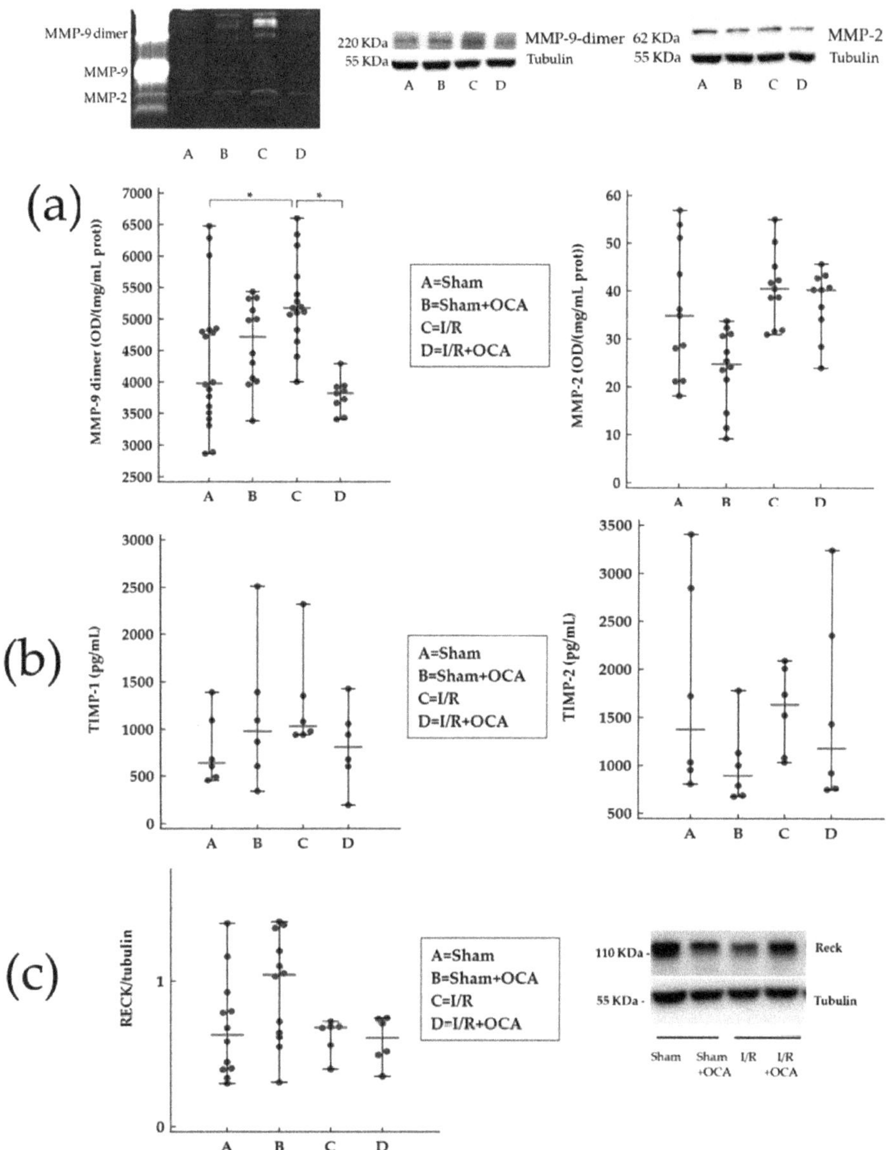

Figure 2. OCA treatment decreases MMP-9-dimer activity in kidney medulla after liver I/R. Animals were orally administered 10 mg/kg/day of OCA in methylcellulose 1% vehicle for 5 days (*n* = 12) or vehicle alone (*n* = 12). (**a**) Representative of MMP zymography and Western blot; gelatinolytic activities of MMP (MMP-9-dimer and MMP-2) quantified by densitometry. (**b**) TIMPs (TIMP-1 and TIMP-2) by Elisa kits and (**c**) RECK expression levels, by Western Blots, were determined in kidney cortex of rats submitted to partial liver ischemia (60 min) followed by reperfusion (120 min). Sham-operated animals were subjected to the same procedure without clamping the vessels. *n* = 6 rats/group. Results are expressed as median value (red line) ± error bar (1–99 percentiles) (black dot). one-way ANOVA with Tukey–Kramer test (MMP-9-dimer, MMP-2, and TIMP-1); Kruskal–Wallis test with Conover test (TIMP-2 and RECK). * $p < 0.05$ versus I/R rats.

2.3. OCA Treatment Reduces Serum Levels of Creatinine

The serum creatinine concentration is widely interpreted as a measure of the glomerular filtration rate (GFR) and is used as an index of renal function in clinical practice [14]. Serum creatinine levels were evaluated in rats submitted to liver I/R: an increase was found after I/R when compared with sham animals and a significant reduction in sham-operated levels was detected in OCA-treated I/R rats (Figure 3).

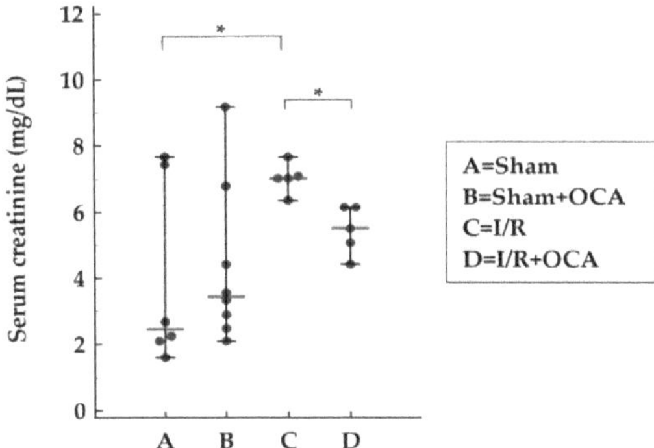

Figure 3. OCA treatment decreases serum creatinine in rats submitted to partial liver ischemia (60 min) followed by reperfusion (120 min). Animals were orally administered 10 mg/kg/day of OCA in methylcellulose 1% vehicle for 5 days (n = 12) or vehicle alone (n = 12). Sham-operated animals were subjected to the same procedure without clamping the vessels. n = 6 rats/group. Results are expressed as median value (red line) $\pm$ error bar (1–99 percentiles) (black dot). one-way ANOVA with Tukey–Kramer test. * $p < 0.05$ versus I/R rats.

2.4. OCA Treatment Reduces Kidney Cortex Levels of TNF-Alpha and IL-6

The effect of OCA on kidney TNF-alpha and IL-6 levels was evaluated by ELISA. At the end of reperfusion, OCA administration caused a significant decrease in both TNF-alpha and IL-6 in the kidney cortex of OCA-treated I/R rats compared to I/R rats (Figure 4). Undetectable levels of TNF-alpha were found in the kidney medulla in all groups considered. No changes in the IL-6 kidney medulla were found (Figure S1, supplementary materials).

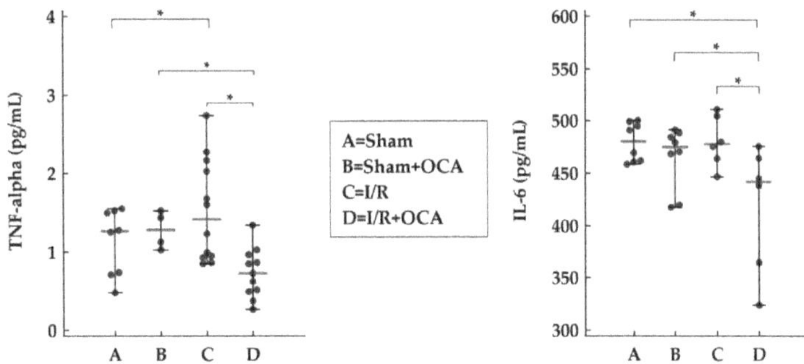

Figure 4. OCA treatment decreases cortex TNF-alpha and IL-6 in rats submitted to partial liver ischemia (60 min) followed by reperfusion (120 min). Animals were orally administered with OCA 10 mg/kg/day in methylcellulose 1% for 5 days (n = 12) or with vehicle alone (n = 12). Sham-operated animals were subjected to the same procedure without clamping the vessels. n = 6 rats/group. Results are expressed as median value (red line) $\pm$ error bar (1–99 percentiles) (black dot). Kruskal–Wallis test with Conover test. * $p < 0.05$ versus I/R rats.

2.5. FXR Expression in the Kidney

The kidney itself expresses FXR both in the cortex and in the medulla as shown in Figure 5. After OCA treatment a decrease in FXR expression occurred only in the cortex.

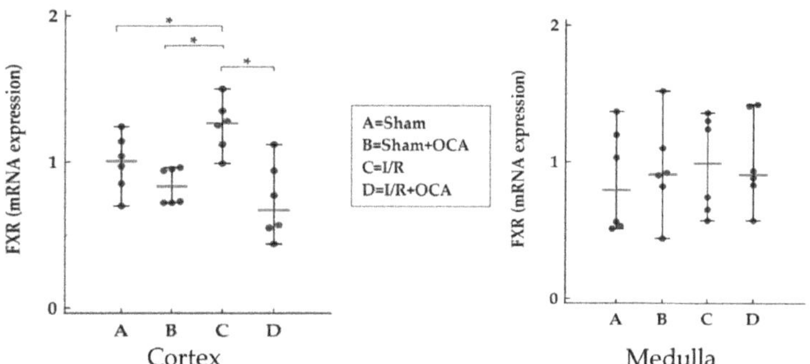

Figure 5. FXR expression in kidney cortex and medulla in rats submitted to partial liver ischemia (60 min) followed by reperfusion (120 min). Animals were orally administered with OCA 10 mg/kg/day in methylcellulose 1% for 5 days (n = 12) or with vehicle alone (n = 12). Sham-operated animals were subjected to the same procedure without clamping the vessels. n = 6 rats/group. Results are expressed as median value (red line) ± error bar (1–99 percentiles) (black dot). One-way ANOVA with Tukey–Kramer test; * p < 0.05 versus I/R rats.

2.6. TBARS Formation in the Kidney

We evaluated the kidney concentration of TBARS after 2-h reperfusion both in the cortex and in the medulla (Figure 6). No significant changes in lipid peroxidation were detected for all groups considered in the cortex (Figure 6) and no difference in I/R groups in the medulla (Figure 6).

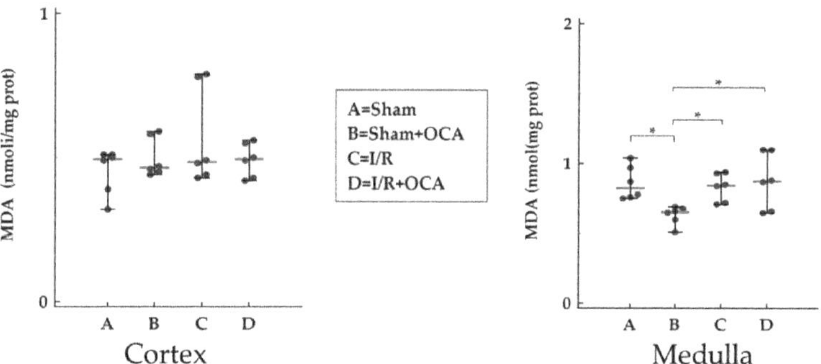

Figure 6. TBARS formation in kidney cortex and medulla in rats submitted to partial liver ischemia (60 min) followed by reperfusion (120 min). Animals were orally administered with OCA 10 mg/kg/day in methylcellulose 1% for 5 days (n = 12) or with vehicle alone (n = 12). Sham-operated animals were subjected to the same procedure without clamping the vessels. n = 6 rats/group. TBARS concentrations were calculated using malondialdehyde (MDA) as standard. Results are expressed as median value (red line) ± error bar (1–99 percentiles) (black dot). One-way ANOVA with Tukey–Kramer test (Medulla); Kruskal–Wallis test with Conover test (Cortex). * p < 0.05 versus I/R rats.

2.7. OCA Treatment and Histological Changes in the Kidney

Examples of histological patterns of kidneys collected from vehicle-treated I/R and OCA-treated I/R rats after 60 min of liver ischemia and 120 min reperfusion are shown in Figure 7. In rats submitted to liver I/R, the kidney shows dilated tubules, in particular in the cortex and outer medulla (Figure 7b,f), compared to the respective sham-operated controls (Figure 7a,e). This morphological alteration is not appreciable in kidneys from OCA-treated rats submitted to I/R (Figure 7d,h).

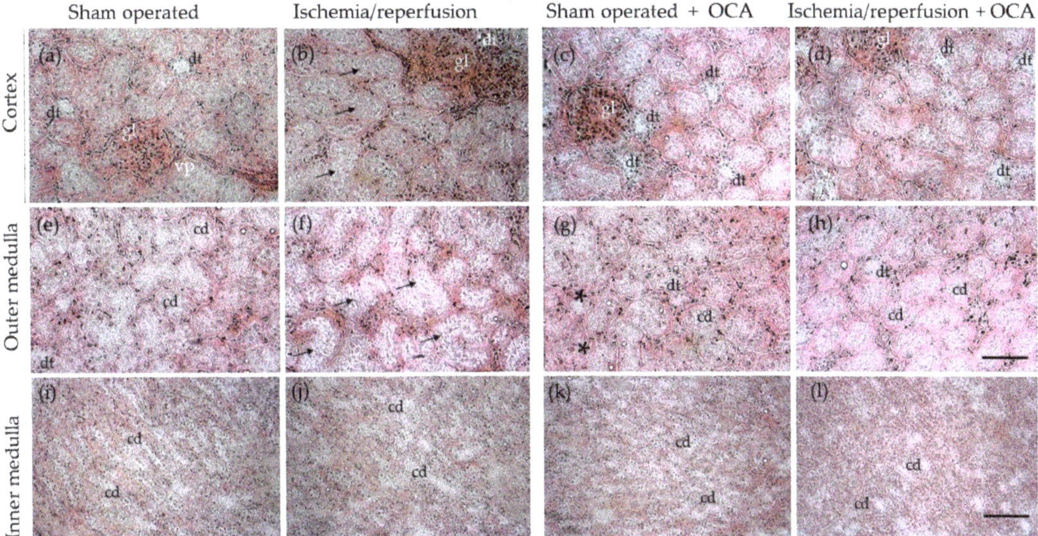

Figure 7. Histological patterns of kidneys (cortex and outer and inner medulla) collected from rats orally administered 10 mg/kg/day OCA in methylcellulose 1% for 5 days (n = 12) or vehicle alone (n = 12), and then submitted to partial liver ischemia (60 min) followed by reperfusion (120 min). Sham-operated animals were subjected to the same procedure without clamping the vessels. Cryostatic tissue sections are stained by conventional H&E procedure. (**a–d**) In the cortex the intensely stained glomeruli, or renal corpuscles, (gl) exhibit their typical rounded shape. In (**a**) the vascular pole (vp) is also easily appreciable. Distal tubules (dt) can be recognized for their much less eosinophilia than the proximal ones. (**e–h**) Cross-sections of medulla are characterized by collecting ducts (cd), larger than distal tubules (dt) and segments of the loop of Henle (asteriscs). (**i–l**) In the longitudinal sections of medulla, arrays of nuclei facing larger, empty spaces identify collecting duts (cd). Ischemia-reperfusion affects in particular the tubules in the cortex and outer medulla of rats without OCA administration, for which the lumen appears to be enlarged and filled with amorphous material (arrows). Bars: (**a–h**) 80 μm; (**i–l**) 50 μm.

Collagen I was evaluated by Western blot: collagen I was not detectable in all groups considered (Figure S2, supplementary materials).

3. Discussion

3.1. Hepatic Ischemia/Reperfusion and Renal Damage

Acute I/R-induced liver failure is a known clinical problem and often leads to remote organ dysfunction including lung, heart, and kidney [17]. Particularly, acute kidney injury (AKI), associated with liver failure, leads to hepatorenal syndrome, a serious clinical problem characterized by a high mortality rate. The pathogenesis of AKI associated with liver failure is still poorly understood, also due to the lack of reproducible and reliable animal models. In this study, we used a rat acute I/R-induced liver failure model to mimic biochemical (serum creatinine, MMP, TIMP, inflammatory markers such as TNF-alpha and IL-6, lipid peroxidation, and fibrosis) and histological (renal tubular damage as indicated by dilatation) changes observed in a hepatorenal syndrome. Through the reduction of renal matrix metalloproteinases, in particular gelatinases, our study suggests that OCA treatment can ameliorate hepatic renal syndrome following partial hepatic I/R injury in rats.

In the last decade, the Acute Kidney Injury Network (AKIN) has proposed new diagnostic criteria for AKI, based on small changes in serum creatinine levels [18]. Traditionally, kidney failure in cirrhosis has been defined as serum creatinine above 1.5 mg/dL [12]. The definition of AKI in cirrhosis has recently been changed and is based on small changes in serum creatinine associated with urinary biomarkers and pro-inflammatory cytokines [19]. In our data hepatic ischemia increased serum creatinine levels by 1.8-fold to 7.35 mg/dL, indicating acute renal failure, and thus validating our model of AKI mediated by liver I/R injury.

Molecular mechanisms underlying I/R injury involved reactive oxygen species (ROS) formation released by Kupffer cells, adherent leukocytes, or mitochondrial sources. Liver I/R injury may initiate a systemic inflammatory response that promotes remote organ dysfunctions attributed to oxidative stress mediators and other remotely released factors, including proinflammatory cytokines, tumor necrosis factor, and interleukins [8]. Oxidative stress is considered a major determinant of liver I/R induced AKI. Activated neutrophils, ROS, and cytokines release cause direct renal injury and the recruitment of monocytes and macrophages leading to further aggravation of the oxidative injury [8]. The pivotal role of ROS in the development of AKI is also demonstrated by the positive effect of the administration of free radical scavengers such as quercetin and desferrioxamine that protected the kidney by decreasing TBARS levels [20]. We evaluated the kidney concentration of TBARS after 2-h reperfusion and no changes in lipid peroxidation were detected, thus suggesting that in our model, ROS generation is not underlying the early remote damage in the kidney.

3.2. Renal Damage Is Associated to MMPs Activation

Most matrix metalloproteinases (MMPs), zinc-containing endopeptidases involved in the extracellular matrix (ECM) remodeling [21], and their specific tissue inhibitors (TIMPs) are expressed in the kidney even if their spatial expression is complex and has not been completely characterized [22]. MMPs are associated with both physiological and pathological processes in the kidney [23]. Knockout mouse models have provided insights into the cause-and-effect relationship between MMP activity and renal pathophysiology [24–26]. There is now convincing evidence that MMPs could have both pathogenic and nephroprotective effects in acute and chronic kidney diseases.

We have previously documented that moderate acute hepatic ischemia (30 min) followed by reperfusion (60 min) increases MMPs activity not only in the ischemic liver region but also in the lung, associated with histological damage in the liver, lung, and kidney [9]. Additionally, no significant difference in MMPs was observed in other distant organs such as the kidney and heart. The short duration of both the ischemic (30 min) and reperfusion periods (60 min) of our experimental model could represent a possible explanation for these results. Thus, we designed a new study to evaluate whether liver I/R may induce renal MMPs activation in rats. A liver ischemia (60 min) followed by a longer reperfusion (120 min) was associated with a significant MMP-9-dimer activity increase in the kidney cortex and medulla. The result obtained show how even the increased activity of MMP-9 dimer can be inserted in a context of renal damage characterized by a significant increase in creatinine and a histological picture of mild damage.

Zymography analysis revealed the presence of gelatinolytic activities of MMP-9 dimer, detected at about 220 kDa, and MMP-2 at 68 KDa. No MMP-9 monomers were detected in kidney tissue. MMP-9, in contrast to MMP-2, exists in two major forms: a monomeric (92 kDa) and a disulfide-bonded homodimeric (220 kDa) form. MMP-9 dimer has been identified in a variety of MMP-9-producing cells including neutrophils and normal breast epithelial cells [27]. Enzymatic activity of the monomeric and dimeric forms of MMP-9 have different biochemical and enzymatic properties. The monomer is more rapidly activated by MMP-3 and has a higher activity than the dimer [28]. The existence of the more stable, slow-activating MMP-9 dimer might serve as a regulatory mechanism during

ECM degradation [28]. In the present study, although not significantly, kidney MMP-2 activity increases. Caron et al. reported similar results in glomerular tissue in a rat model of renal ischemia-reperfusion. The most marked effect revealed by this study was a higher MMP-9 dimer induction. Authors reported that protein expression of dimeric MMP-9 forms was highly stimulated (about 8-fold) by ischemia, compared to the expression of MMP-2 (1.5-fold). It may be possible that a reduced MMP-2 expression may be accompanied by less detectable activity.

Several studies [29–32] have shown protective effects against AKI with MMP inhibitors, but there are data from knockout mouse studies where MMP-2 and -9 may show both proinjury and nephroprotective effects [24].

Loss of MMP-2 is not equivocally protective, while MMP-9 deficiency may exacerbate an injury. There is also evidence that MMP-2 may be important for recovery following injury, while MMP-9 overexpression may cause disrepair through microvascular loss. It is known that MMP-9 is an acknowledged early marker of acute renal damage [16]. These data indicate that more research remains to be performed to elucidate these pathogenic mechanisms.

3.3. OCA Treatment Reduces Kidney Gelatinases Activity following Partial Hepatic I/R Injury in Rats

In the present study, in order to characterize the mechanisms of remote organ damage, we investigated the effects of OCA on kidney levels of gelatinase, MMP-2, and MMP-9.

Very few studies have addressed the ability of OCA to modulate the activity of metalloproteases [6,33]. Recently, our group has detected the ability of OCA to limit the activation of MMP-2 and MMP-9 occurring during hepatic I/R damage, probably via a TIMP- and RECK-mediated mechanism [6].

Liver I/R injury is associated with a systemic inflammatory response that promotes remote organ dysfunctions. In the present study, a reduction in MMP-9-dimer was detected both in the kidney cortex and in the medulla of OCA-treated I/R rats. In this animal model of remote organ injury, we did not find a corresponding modulation of MMP and RECK inhibitors as in the I/R liver. It can be assumed that the modulation of OCA gelatinolytic activity is likely to be mediated by tissue TNF-alpha and IL-6. Indeed, treatment with OCA at 120 min caused a significant decrease in both TNF-alpha and IL-6 in the kidney cortex. A wealth of evidence indicates that circulatory dysfunction plays a key role in the pathophysiology of hepatorenal syndrome [34]. However, recently, evidence has emerged to support the impact of systemic inflammation on disease progression and development in extrahepatic organs, particularly renal dysfunction [35]. After liver I/R injury, circulating pro-inflammatory cytokines such as interleukin-6 (IL-6), IL-8, TNF-alpha, vascular cell adhesion protein 1 (VCAM-1), fractalkine, and macrophage inflammatory protein-1 alpha (MIP-1 alpha) have been found significantly increased [36].

Pro-inflammatory cytokines such as TNF-alpha and IL-6 are produced in response to infection. During hepatic I/R, serum TNF-alpha has been shown to act as one of the key mechanisms in modulating MMP activity in remote organs [9]. Tissue TNF-alpha may be a possible additional source for TNF-alpha traced in the serum in our previous work using the same animal model. In this study, we detected TNF-alpha in cortical renal tissue and found an increase in this proinflammatory cytokine in rats undergoing I/R, readily modulated by OCA administration. The decrease in cortex IL-6 by OCA administration supports its role in the modulation of inflammatory response.

However, sham animals reveal high levels of tissue IL-6 and TNF-alpha probably due to the operative trauma of laparotomy as already reported by Ogura et al. [37] and Wanner G et al. [38].

It is known that an excessive cytokine release enhances vascular permeability and impairs metabolic processes, thus increasing susceptibility to multiple organ dysfunction [36,39].

The evaluation of TIMP levels showed no significant differences among all groups considered and the same occurred for RECK. In our research, we also determined the content of both TIMP-1, the specific inhibitor of MMP-9, and TIMP-2, the inhibitor of MMP-2. Both inhibitors showed no significant differences among all groups tested. Usually, TIMP concentration is opposite to MMP. However, in our study, we found increased concentrations of both MMP-9 and TIMP-1, although not significant for the latter; this could result from failure of TIMP-1 inhibition. Furthermore, TIMP transcription is regulated by the same cytokines and growth factors controlling MMP expression [40]. The TIMP-2 concentration in the I/R group did not have a statistically significant influence but it showed, as expected, a negative trend in contrast with MMP2.

MMPs are also inversely regulated by RECK, a membrane-anchored glycoprotein and a key regulator of ECM integrity [6,41]. In this study, the ability of OCA to restore renal TIMP and RECK expression was lacking. Although not statistically significant, in the kidney cortex it was still possible to detect a modest increase in both TIMP-1 and TIMP-2 in OCA I/R treated group. Probably, the ability of OCA to reduce kidney matrix metalloproteinase activation following hepatic I/R injury in rats is not modulated by RECK and TIMP. Thus, we suppose that OCA protection is principally due to the reduction of the inflammatory response.

The kidney itself expresses FXR in the cortex and the medulla. No significant changes in FXR levels occurred in the sham and I/R group after 2-h reperfusion in agreement with Ogura et al.; they documented no changes in FXR expression remote damage occurred in the liver after intestinal ischemia followed 1 and 3-h reperfusion [37]. After OCA treatment a decrease in FXR expression occurred only in the cortex; this event will be further investigated in future studies with the reperfusion time prolonged.

The present study also demonstrated that OCA treatment alleviates I/R-induced kidney damage as found by a marked decrease in serum creatinine levels. Furthermore, the histological analysis in the cortex and outer medulla in rats submitted to I/R, reveals some alterations, such as dilated tubules, as compared with the sham-operated animals. These morphological changes are less noticeable in kidneys from OCA administered rats submitted to liver I/R.

Our results show that, after hepatic I/R injury, hepatorenal syndrome rapidly develops in rats, characterized by an increase in MMP-9-dimer and inflammatory changes such as an increase in renal cortex TNF-alpha, thus suggesting a key role for this pro-inflammatory cytokine in MMP activation. Although the underlying mechanism needs further investigation, this study shows, in the kidney, beneficial effects of OCA by reducing TNF-alpha-mediated expression of MMPs after liver I/R.

4. Materials

All reagents were of the highest grade of purity available and were obtained from MERCK—Life Science, 20149 Milan, Italy). The FXR agonist, OCA, was kindly provided by Intercept Pharmaceuticals, San Diego, CA, USA.

4.1. Animals and Experimental Design

The use and care of animals in this experimental study was approved by the Italian Ministry of Health and by the University of Pavia Commission for Animal Care (Document number 179/2017-PR). Male Wistar rats (200–250 g) (Charles River, 23885 Calco, LC, Italy) were used in this study. The animals (n = 24) were allowed free access to water and food in all the experiments. Animals were orally administered with 10 mg/kg/day OCA in methylcellulose 1% for 5 days (n = 12) or with vehicle alone (n = 12). The effects of I/R-induced remote damage in kidneys were studied in vivo using a partial hepatic I/R model (I/R n = 12 and I/R + OCA n = 12). The rats were anesthetized with sodium pentobarbital (40 mg/kg i.p.), the abdomen was opened via a midline incision and the bile duct was cannulated (PE-50) [17]. Ischemia to the left and the median lobe was induced for 60 min with microvascular clips by clamping the branch of portal vein and the branch of the hepatic artery after the bifurcation to the right lobe, with the abdomen temporarily closed with a suture [18]. After a 60-min ischemia, the abdomen was reopened, the clips were removed, the abdomen was closed again, and the liver was allowed to reperfuse for 120 min (Figure 1). By using partial, rather than total, hepatic ischemia, portal vein congestion, and subsequent bacterial translocation into the portal venous blood were avoided. Sham-operated animals were subjected to the same procedure without clamping the vessels (sham-operated *n* = 6 and sham-operated + OCA *n* = 6). To prevent postsurgical dehydration and hypotension, 1 mL of saline was injected into the inferior vena cava. All the animals were maintained on a warm support to prevent heat loss, rectal temperature was maintained at 37 ± 0.1 °C. Animals were sacrificed, under general anesthesia (sodium pentobarbital 40 mg/kg i.p.), by exsanguination. Each sample is derived from individual kidneys.

Blood samples were obtained after reperfusion and immediately centrifuged to isolate serum. At the end of the reperfusion period, tissue samples of the kidney (cortex and medulla) were snap-frozen in liquid nitrogen.

4.2. Biochemical Assays

Liver injury was assessed by transaminase serum release (alanine transaminase, ALT, and aspartate transaminase, AST); alkaline phosphatase (ALP), total and direct bilirubin were assayed using commercial kits, MERCK—Life Science, 20149 Milan, Italy.

Kidney injury was assessed by serum creatinine colorimetric assay kit (number 700460, Cayman Chemicals, Ann Arbor, MI 48108, USA). The kit was performed following manufacturers' instructions.

4.3. TBARS Formation

The extent of lipid peroxidation was evaluated, as previously described [42], by measuring the formation of thiobarbituric acid-reactive substances (TBARS), following the Esterbauer and Cheeseman [43] method. TBARS concentrations were calculated using malondialdehyde (MDA) as standard.

4.4. Tissue Sources for MMPs Analysis

After sacrifice, kidneys were quickly excised and placed in cold (4 °C) buffer (30 mM Histidine, 250 mM sucrose, and 2 mM EDTA, pH 7.2) to remove blood. The kidney was cleaned of external tissue; the renal cortex and medulla were separated and subsequently frozen in liquid nitrogen and stored at −80 °C, until use. Fifty milligrams of cortex and medulla were homogenized in a dissociation buffer containing 10 mmol/L cacodylic acid, 0.15 mmol/L NaCl, 1 mmol/L $ZnCl_2$, 20 mmol/L $CaCl_2$, 1.5 mmol/L NaN_3, and 0.01% Triton X-100, pH 5.0 [25]. The homogenate was then shaken at 4 °C for 24 h and the protein concentration of the supernatant was measured with the colorimetric Lowry method [24]. Samples were stored at −20 °C before use.

4.5. MMP Zymography

In order to detect MMPs activity present in the samples, the homogenate protein content was normalized by a final concentration of 400 µg/mL in sample loading buffer (0.25 M Tris-HCl, 4% sucrose *w/v*, 10% SDS *w/v*, and 0.1% bromophenol blue *w/v*, pH 6.8). After dilution, the samples and purified MMP-9 dimer, MMP-9, and MMP-2 (Enzo Lifescience and Calbiochem-Merck Life Science, 20149 Milan, Italy) were loaded onto electrophoretic gels (SDS-PAGE) containing 1 mg/mL of gelatin under nonreducing conditions [27,28], followed by zymography as described previously [29]. Zymogram experiment was replicated 3 times for each sample.

The zymograms were analyzed by densitometer (GS 710 Densitometer BIORAD, Hercules, CA, USA) and data were expressed as optical density (OD), reported to 1 mg/mL protein content.

4.6. Western Blots

CelLytic Buffer and Protease Inhibitor Cocktail were purchased from Sigma-Aldrich (Milan, Italy), as well as the monoclonal antibody anti-alpha-tubulin (DM1A). Rabbit monoclonal antibody against RECK (D8C7) was purchased from Cell Signaling Technology (Euroclone, Milan, Italy). Rabbit polyclonal antibody against Collagen I alpha 1 (NBP1-30054) was purchased from Novus Biological (Bio-techne, Milan, Italy). Mouse monoclonal antibody against MMP-2 and MMP-9 were purchased by Thermo Fisher Scientific (Monza, Italy).

Kidney tissue samples were homogenized in an ice-cold CelLytic Buffer supplemented with Protease Inhibitor Cocktail and centrifuged at $15,000 \times g$ for 10 min. The collected supernatant was divided into aliquots containing the same amount of proteins and stocked at $-80\ ^\circ$C. Samples of kidney extracts containing the same amount of proteins were diluted with $2\times$ Laemmli sample buffer purchased by BIO-RAD (Segrate, Italy) added with 0.1 M DTT (Sigma-Aldrich, Milan, Italy). Samples were incubated for 10 min at 70 $^\circ$C and then put for 1 min on ice to cause thermal shock. Samples used for the evaluation of MMP-9 dimer were loaded under unreducing conditions: they were diluted with $2\times$ Laemmli sample buffer without reducing agent, and were not heated at 70 $^\circ$C. No thermal shock was caused to them. Thus, samples of kidney extracts containing the same amount of proteins were separated in SDS-PAGE on 7.5% acrylamide gels and transferred to PVDF membrane. Unspecific sites were blocked for 2 h with 5% Bovine Serum Albumin (BSA) in TBST (20 mM Tris/Base, 150 mM NaCl, 7.4 pH, 0.1% Tween 20) at 4 $^\circ$C. The membranes were incubated with primary antibodies overnight at 4 $^\circ$C, under gentle agitation. Primary antibodies against alpha-tubulin and RECK were used at a dilution of 1:1000. Membranes were washed in TBST (20 mM Tris/Base, 150 mM NaCl, pH 7.4, 0.1% Tween 20) and incubated with peroxidase-conjugated secondary antibody at a 1:2000 dilution for both tubulin and RECK. Primary antibodies against alpha-tubulin and Collagen I Alpha 1 antibody were used at a dilution of 1:1000 and membranes were washed in TBST (20 mM Tris/Base, 150 mM NaCl, pH 7.4, 0.1% Tween 20) and incubated with peroxidase-conjugated secondary antibody at a 1:5000 diluition for both tubulin and Collagen I Alpha 1. Immunostaining was revealed with BIO-RAD Chemidoc XRS+. Bands intensity quantification was performed by BIO-RAD Image Lab software, 20090 Segrate, MI, Italy.

4.7. TIMP-1, TIMP-2, TNF-Alpha, and IL-6 Enzyme-Linked Immunosorbent Assay

Kidney Cortex and Medulla were homogenized with suitable lysis buffer as indicated by the collection procedures of the kit suppliers. After centrifuging the tissue homogenates for 5 min at 10.000 rpm, supernatants were collected and the concentration of TIMP-1 (Abnova), TIMP-2 (Abnova), and TNF-alpha and IL-6 (Antibodies-online GmbH, Aachen, Germany) was immediately measured by ELISA kit.

4.8. FXR Expression

FXR mRNA expression was analyzed by a real-time polymerase chain reaction (RT-PCR). Total RNA was isolated from both kidney cortex and medulla homogenized in TRI reagent (Sigma-Aldrich, St. Louis, MO, USA), in accordance with the method described by Chomczynski et al. (1995) [44]. RNA quantification was evaluated by measuring the absorbance at 260 nm with a T92$^+$ UV Spectrophotometer (PG Instruments, Lutterworth, UK); the RNA purity was also evaluated by calculating the 260/280 nm ratio. iScript Supermix (Bio-Rad, Milan, Italy) was employed to generate the cDNA [45]. mRNA expression was analyzed using the SsoAdvancedTM SYBR® Green Supermix (Bio-Rad, Milan, Italy) and the amplification was performed through two-step cycling (60–95 $^\circ$C) for 40 cycles in a CFX96TMReal-TimeSystem (Bio-Rad, Milan, Italy). FXR, ubiquitin C (UBQ), and glyceraldehyde 3-phosphate dehydrogenase (GAPDH) gene amplification efficiency (99.8%, 100%, 102.8%, respectively), was established by means of calibration curves, in a cDNA concentration range of 10–0.625 ng/µL. As regards housekeeping genes, UBC and GAPDH were used. The sequence of forward and reverse primers used in the experiments is reported in Table 2. All samples were assayed in duplicate and the expression of the reference genes remained constant in all the experimental groups. Gene expression was calculated using the ΔCt method. Comparison between groups was calculated using the $\Delta\Delta$Ct method.

Table 2. List of forward and reverse primers used in the experiments.

Gene	Sequence
rat FXR	Forward: CGCCTCATCGGCGGGAAGAA Reverse: TCACGCAGTTGCCCCCGTTC
rat GAPDH	Forward: AACCTGCCAAGTATGATGAC Reverse: GGAGTTGCTGTTGAAGTCA
rat UBQ	Forward: CACCAAGAAGGTCAAACAGGAA Reverse: AAGACACCTCCCCATCAAACC

4.9. Tissue Morphology

Kidney tissue samples collected after animal sacrifice were frozen in liquid nitrogen and stored at $-80\,^{\circ}$C until they were cut at cryostat (CM1850, Leica Microsystems, GmbH, Wetlzar, Germany). Tissue sections (12 µm thick) were air-dried for at least 24 h and stained following conventional hematoxylin eosin (H&E) procedure. Tissue morphology was observed using an Olympus BX51 microscope (Olympus Optical Co. GmBH, Hamburg, Germany) equipped with a Canon EOS 1300D (Canon Inc. Ōta, Tokyo, Japan) digital photo camera, and an Olympus 20× UplanFl objective (NA 0.50 Olympus Optical Co. GmBH, Hamburg, Germany).

4.10. Statistical Analysis

Statistical analysis was performed using MedCalc Statistical Software version 18.11.3 (MedCalc Software bvba, Ostend, Belgium; https://www.medcalc.org; accessed on 1 March 2022). Statistical analysis was performed with one-way ANOVA with Tukey's test, as post hoc test, or Kruskall–Wallis with Conover's test when data were not normally distributed. To assess normality of distribution, Kolmogorov–Shapiro normality test was used. Results are expressed as mean value $\pm$ standard error (SE) or median value with range percentiles. The value of $p = 0.05$ was considered the criterion for statistical significance.

5. Conclusions

This is the first study reporting that the pretreatment with OCA resulted in attenuation of renal injury after liver I/R by reducing TNF-alpha-and IL6 mediated expression of MMPs.

Supplementary Materials: The following supporting information can be downloaded at: https://www.mdpi.com/article/10.3390/ph15050524/s1. Figure S1. IL-6 in rats submitted to partial liver ischemia (60 min) followed by reperfusion (120 min). Animals were orally administered with OCA 10 mg/kg/day in methylcellulose 1% for 5 days ($n = 12$) or with vehicle alone ($n = 12$). Sham-operated animals were subjected to the same procedure without clamping the vessels. $n = 6$ rats/group. Results are expressed as median value $\pm$ error bar (1–99 percentiles). Kruskal-Wallis test with Conover test. Figure S2. Collagen I expression levels, by Western Blots, were determined in kidney cortex (a) and medulla (b) of rats submitted to partial liver ische-mia (60 min) followed by reperfusion (120 min). Sham-operated animals were subjected to the same procedure with-out clamping the vessels. $n = 6$ rats/group.

Author Contributions: Conceptualization, G.P., A.F. and M.V.; methodology, M.C., C.B., L.G.D.P. and A.C.C.; formal analysis, G.P.; investigation, M.C.; data curation, A.C.C. and C.B.; writing—original draft preparation, G.P. and M.V.; writing—review and editing, C.B. and A.F.; visualization, L.A. and S.P. All authors have read and agreed to the published version of the manuscript.

Funding: This work was funded by the University of Pavia.

Institutional Review Board Statement: The use and care of animals in this experimental study were approved by the Italian Ministry of Health and by the University Commission for Animal Care, University of Pavia (Document number 179/2017-PR).

Data Availability Statement: Data is contained within the article and its supporting information.

Conflicts of Interest: The authors declare no conflict of interest. The authors declare that Luciano Adorini is a part-time consultant of Intercept Pharmaceuticals. The specific roles of this author are articulated in the 'author contributions' section. However, the Intercept Pharmaceuticals did not provide support in the form of salaries for authors and did not have any additional role in the design of the study; in the collection, analysis, or interpretation of data; in the writing of the manuscript, or in the decision to publish the results.

References

1. Makri, E.; Cholongitas, E.; Tziomalos, K. Emerging role of obeticholic acid in the management of nonalcoholic fatty liver disease. *World J. Gastroenterol.* **2016**, *22*, 9039–9043. [CrossRef] [PubMed]
2. Younossi, Z.M.; Stepanova, M.; Nader, F.; Loomba, R.; Anstee, Q.M.; Ratziu, V.; Harrison, S.; Sanyal, A.J.; Schattenberg, J.M.; Barritt, A.S.; et al. Obeticholic Acid Impact on Quality of Life in Patients With Nonalcoholic Steatohepatitis: REGENERATE 18-Month Interim Analysis. *Clin. Gastroenterol. Hepatol.* **2021**. [CrossRef] [PubMed]
3. Chuang, H.M.; Chen, Y.S.; Harn, H.J. The Versatile Role of Matrix Metalloproteinase for the Diverse Results of Fibrosis Treatment. *Molecules* **2019**, *24*, 4188. [CrossRef] [PubMed]
4. Duarte, S.; Hamada, T.; Kuriyama, N.; Busuttil, R.W.; Coito, A.J. TIMP-1 deficiency leads to lethal partial hepatic ischemia and reperfusion injury. *Hepatology* **2012**, *56*, 1074–1085. [CrossRef]
5. Meng, N.; Li, Y.; Zhang, H.; Sun, X.F. RECK, a novel matrix metalloproteinase regulator. *Histol. Histopathol.* **2008**, *23*, 1003–1010. [CrossRef] [PubMed]
6. Ferrigno, A.; Palladini, G.; Di Pasqua, L.G.; Berardo, C.; Richelmi, P.; Cadamuro, M.; Fabris, L.; Perlini, S.; Adorini, L.; Vairetti, M. Obeticholic acid reduces biliary and hepatic matrix metalloproteinases activity in rat hepatic ischemia/reperfusion injury. *PLoS ONE* **2020**, *15*, e0238543. [CrossRef] [PubMed]
7. Clichici, S.; David, L.; Moldovan, B.; Baldea, I.; Olteanu, D.; Filip, M.; Nagy, A.; Luca, V.; Crivii, C.; Mircea, P.; et al. Hepatoprotective effects of silymarin coated gold nanoparticles in experimental cholestasis. *Mater. Sci. Eng. C* **2020**, *115*, 111117. [CrossRef]
8. Nastos, C.; Kalimeris, K.; Papoutsidakis, N.; Tasoulis, M.K.; Lykoudis, P.M.; Theodoraki, K.; Nastou, D.; Smyrniotis, V.; Arkadopoulos, N. Global consequences of liver ischemia/reperfusion injury. *Oxid. Med. Cell. Longev.* **2014**, *2014*, 906965. [CrossRef] [PubMed]
9. Palladini, G.; Ferrigno, A.; Rizzo, V.; Tarantola, E.; Bertone, V.; Freitas, I.; Perlini, S.; Richelmi, P.; Vairetti, M. Lung matrix metalloproteinase activation following partial hepatic ischemia/reperfusion injury in rats. *Sci. World J.* **2014**, *2014*, 867548. [CrossRef]
10. Barri, Y.M.; Sanchez, E.Q.; Jennings, L.W.; Melton, L.B.; Hays, S.; Levy, M.F.; Klintmalm, G.B. Acute kidney injury following liver transplantation: Definition and outcome. *Liver Transpl.* **2009**, *15*, 475–483. [CrossRef] [PubMed]
11. Ozsoy, M.; Gonul, Y.; Bal, A.; Ozkececi, Z.T.; Celep, R.B.; Adali, F.; Hazman, O.; Koçak, A.; Tosun, M. Effect of IL-18 binding protein on hepatic ischemia-reperfusion injury induced by infrarenal aortic occlusion. *Ann. Surg. Treat. Res.* **2015**, *88*, 92–99. [CrossRef] [PubMed]
12. Arroyo, V.; Ginès, P.; Gerbes, A.L.; Dudley, F.J.; Gentilini, P.; Laffi, G.; Reynolds, T.B.; Ring-Larsen, H.; Schölmerich, J. Definition and diagnostic criteria of refractory ascites and hepatorenal syndrome in cirrhosis. International Ascites Club. *Hepatology* **1996**, *23*, 164–176. [CrossRef]
13. Angeli, P.; Ginès, P.; Wong, F.; Bernardi, M.; Boyer, T.D.; Gerbes, A.; Moreau, R.; Jalan, R.; Sarin, S.K.; Piano, S.; et al. Diagnosis and management of acute kidney injury in patients with cirrhosis: Revised consensus recommendations of the International Club of Ascites. *J. Hepatol.* **2015**, *62*, 968–974. [CrossRef] [PubMed]
14. Lee, H.T.; Park, S.W.; Kim, M.; D'Agati, V.D. Acute kidney injury after hepatic ischemia and reperfusion injury in mice. *Lab. Investig.* **2009**, *89*, 196–208. [CrossRef] [PubMed]
15. Palladini, G.; Ferrigno, A.; Rizzo, V.; Boncompagni, E.; Richelmi, P.; Freitas, I.; Perlini, S.; Vairetti, M. Lobe-specific heterogeneity and matrix metalloproteinase activation after ischemia/reperfusion injury in rat livers. *Toxicol. Pathol.* **2012**, *40*, 722–730. [CrossRef]
16. Teng, L.; Yu, M.; Li, J.M.; Tang, H.; Yu, J.; Mo, L.H.; Jin, J.; Liu, X.Z. Matrix metalloproteinase-9 as new biomarkers of severity in multiple organ dysfunction syndrome caused by trauma and infection. *Mol. Cell. Biochem.* **2012**, *360*, 271–277. [CrossRef]
17. Davis, C.L.; Gonwa, T.A.; Wilkinson, A.H. Pathophysiology of renal disease associated with liver disorders: Implications for liver transplantation. Part I. *Liver Transpl.* **2002**, *8*, 91–109. [CrossRef] [PubMed]
18. Solé, C.; Pose, E.; Solà, E.; Ginès, P. Hepatorenal syndrome in the era of acute kidney injury. *Liver Int.* **2018**, *38*, 1891–1901. [CrossRef] [PubMed]
19. Manyalich, M.; Nelson, H.; Delmonico, F.L. The need and opportunity for donation after circulatory death worldwide. *Curr. Opin. Organ Transplant.* **2018**, *23*, 136–141. [CrossRef] [PubMed]
20. Polat, C.; Tokyol, Ç.; Kahraman, A.; Sabuncuoğlu, B.; Yılmaz, S. The effects of desferrioxamine and quercetin on hepatic ischemia-reperfusion induced renal disturbance. *Prostaglandins. Leukot. Essent. Fatty Acids* **2006**, *74*, 379–383. [CrossRef] [PubMed]

21. Raeeszadeh-Sarmazdeh, M.; Do, L.D.; Hritz, B.G. Metalloproteinases and Their Inhibitors: Potential for the Development of New Therapeutics. *Cells* **2020**, *9*, 1313. [CrossRef]
22. Tan, R.J.; Liu, Y. Matrix metalloproteinases in kidney homeostasis and diseases. *Am. J. Physiol. Renal Physiol.* **2012**, *302*, F1351–F1361. [CrossRef]
23. Catania, J.M.; Chen, G.; Parrish, A.R. Role of matrix metalloproteinases in renal pathophysiologies. *Am. J. Physiol. Renal Physiol.* **2007**, *292*, F905–F911. [CrossRef] [PubMed]
24. Bengatta, S.; Arnould, C.; Letavernier, E.; Monge, M.; De Préneuf, H.M.; Werb, Z.; Ronco, P.; Lelongt, B. MMP9 and SCF protect from apoptosis in acute kidney injury. *J. Am. Soc. Nephrol.* **2009**, *20*, 787–797. [CrossRef]
25. Kunugi, S.; Shimizu, A.; Kuwahara, N.; Du, X.; Takahashi, M.; Terasaki, Y.; Fujita, E.; Mii, A.; Nagasaka, S.; Akimoto, T.; et al. Inhibition of matrix metalloproteinases reduces ischemia-reperfusion acute kidney injury. *Lab. Investig.* **2011**, *91*, 170–180. [CrossRef] [PubMed]
26. Lee, S.Y.; Hörbelt, M.; Mang, H.E.; Knipe, N.L.; Bacallao, R.L.; Sado, Y.; Sutton, T.A. MMP-9 gene deletion mitigates microvascular loss in a model of ischemic acute kidney injury. *Am. J. Physiol. Renal Physiol.* **2011**, *301*, F101–F109. [CrossRef]
27. Roy, R.; Louis, G.; Loughlin, K.R.; Wiederschain, D.; Kilroy, S.M.; Lamb, C.C.; Zurakowski, D.; Moses, M.A. Tumor-specific urinary matrix metalloproteinase fingerprinting: Identification of high molecular weight urinary matrix metalloproteinase species. *Clin. Cancer Res.* **2008**, *14*, 6610–6617. [CrossRef]
28. Olson, M.W.; Bernardo, M.M.; Pietila, M.; Gervasi, D.C.; Toth, M.; Kotra, L.P.; Massova, I.; Mobashery, S.; Fridman, R. Characterization of the monomeric and dimeric forms of latent and active matrix metalloproteinase-9. Differential rates for activation by stromelysin 1. *J. Biol. Chem.* **2000**, *275*, 2661–2668. [CrossRef] [PubMed]
29. Parrish, A.R. Matrix Metalloproteinases in Kidney Disease: Role in Pathogenesis and Potential as a Therapeutic Target. *Prog. Mol. Biol. Transl. Sci.* **2017**, *148*, 31–65. [CrossRef] [PubMed]
30. Sutton, T.A.; Kelly, K.J.; Mang, H.E.; Plotkin, Z.; Sandoval, R.M.; Dagher, P.C. Minocycline reduces renal microvascular leakage in a rat model of ischemic renal injury. *Am. J. Physiol. Renal Physiol.* **2005**, *288*, F91–F97. [CrossRef] [PubMed]
31. Ihtiyar, E.; Yaşar, N.F.; Erkasap, N.; Köken, T.; Tosun, M.; Öner, S.; Erkasap, S. Effects of doxycycline on renal ischemia reperfusion injury induced by abdominal compartment syndrome. *J. Surg. Res.* **2011**, *167*, 113–120. [CrossRef] [PubMed]
32. Chang, M.W.; Chen, C.H.; Chen, Y.C.; Wu, Y.C.; Zhen, Y.Y.; Leu, S.; Tsai, T.H.; Ko, S.F.; Sung, P.H.; Yang, C.C.; et al. Sitagliptin protects rat kidneys from acute ischemia-reperfusion injury via upregulation of GLP-1 and GLP-1 receptors. *Acta Pharmacol. Sin.* **2015**, *36*, 119–130. [CrossRef] [PubMed]
33. Anfuso, B.; Tiribelli, C.; Adorini, L.; Rosso, N. Obeticholic acid and INT-767 modulate collagen deposition in a NASH in vitro model. *Sci. Rep.* **2020**, *10*, 1699. [CrossRef] [PubMed]
34. Ginès, P.; Schrier, R.W. Renal failure in cirrhosis. *N. Engl. J. Med.* **2009**, *361*, 1279–1290. [CrossRef] [PubMed]
35. Simbrunner, B.; Trauner, M.; Reiberger, T.; Mandorfer, M. Recent advances in the understanding and management of hepatorenal syndrome. *Fac. Rev.* **2021**, *10*. [CrossRef]
36. Cai, D.; Duan, H.; Fu, Y.; Cheng, Z. Renal Tissue Damage Induced by Acute Kidney Injury in Sepsis Rat Model Is Inhibited by Cynaropicrin via IL-1β and TNF-α Down-Regulation. *Dokl. Biochem. Biophys.* **2021**, *497*, 151–157. [CrossRef]
37. Ogura, J.; Terada, Y.; Tsujimoto, T.; Koizumi, T.; Kuwayama, K.; Maruyama, H.; Fujikawa, A.; Takaya, A.; Kobayashi, M.; Itagaki, S.; et al. The decrease in farnesoid X receptor, pregnane X receptor and constitutive androstane receptor in the liver after intestinal ischemia-reperfusion. *J. Pharm. Pharm. Sci.* **2012**, *15*, 616–631. [CrossRef]
38. Wanner, G.A.; Ertel, W.; Müller, P.; Höfer, Y.; Leiderer, R.; Menger, M.D.; Messmer, K. Liver ischemia and reperfusion induces a systemic inflammatory response through Kupffer cell activation. *Shock* **1996**, *5*, 34–40. [CrossRef]
39. Burdon, D.; Tiedje, T.; Pfeffer, K.; Vollmer, E.; Zabel, P. The role of tumor necrosis factor in the development of multiple organ failure in a murine model. *Crit. Care Med.* **2000**, *28*, 1962–1967. [CrossRef]
40. Norman, J.T.; Lewis, M.P. Matrix metalloproteinases (MMPs) in renal fibrosis. *Kidney Int. Suppl.* **1996**, *54*, S61–S63.
41. Kim, C.W.; Oh, E.T.; Kim, J.M.; Park, J.S.; Lee, D.H.; Lee, J.S.; Kim, K.K.; Park, H.J. Hypoxia-induced microRNA-590-5p promotes colorectal cancer progression by modulating matrix metalloproteinase activity. *Cancer Lett.* **2018**, *416*, 31–41. [CrossRef] [PubMed]
42. Palladini, G.; Di Pasqua, L.G.; Berardo, C.; Siciliano, V.; Richelmi, P.; Mannucci, B.; Croce, A.C.; Rizzo, V.; Perlini, S.; Vairetti, M.; et al. Fatty acid desaturase involvement in non-alcoholic fatty liver disease rat models: Oxidative stress versus metalloproteinases. *Nutrients* **2019**, *11*, 799. [CrossRef] [PubMed]
43. Esterbauer, H.; Cheeseman, K.H. Determination of aldehydic lipid peroxidation products: Malonaldehyde and 4-hydroxynonenal. *Methods Enzymol.* **1990**, *186*, 407–421. [CrossRef] [PubMed]
44. Chomczynski, P.; Mackey, K. Substitution of chloroform by bromo-chloropropane in the single-step method of RNA isolation. *Anal. Biochem.* **1995**, *225*, 163–164. [CrossRef]
45. Berardo, C.; Siciliano, V.; Di Pasqua, L.G.; Richelmi, P.; Vairetti, M.; Ferrigno, A. Comparison between lipofectamine RNAiMAX and genmute transfection agents in two cellular models of human hepatoma. *Eur. J. Histochem.* **2019**, *63*, 189–194. [CrossRef]

pharmaceuticals

Review

Natural Marine and Terrestrial Compounds as Modulators of Matrix Metalloproteinases-2 (MMP-2) and MMP-9 in Alzheimer's Disease

Lidia Ciccone [1,†], Jennifer Vandooren [2,†], Susanna Nencetti [1,3] and Elisabetta Orlandini [4,5,*]

1   Department of Pharmacy, University of Pisa, via Bonanno 6, 56126 Pisa, Italy; lidia.ciccone@unipi.it (L.C.); susanna.nencetti@unipi.it (S.N.)
2   Laboratory of Immunobiology, Department of Microbiology, Immunology and Transplantation, Rega Institute for Medical Research, University of Leuven, KU Leuven—Herestraat 49—Box 1044, 3000 Leuven, Belgium; jennifer.vandooren@kuleuven.be
3   Interdepartmental Research Centre "Nutraceuticals and Food for Health (NUTRAFOOD), University of Pisa, 56126 Pisa, Italy
4   Department of Earth Sciences, University of Pisa, via Santa Maria 53, 56126 Pisa, Italy
5   Research Center "E. Piaggio", University of Pisa, 56122 Pisa, Italy
*   Correspondence: elisabetta.orlandini@unipi.it; Tel.: +39-050-221-9581
†   These authors contributed equally to the manuscript.

Abstract: Several studies have reported neuroprotective effects by natural products. A wide range of natural compounds have been investigated, and some of these may play a beneficial role in Alzheimer's disease (AD) progression. Matrix metalloproteinases (MMPs), a family of zinc-dependent endopeptidases, have been implicated in AD. In particular, MMP-2 and MMP-9 are able to trigger several neuroinflammatory and neurodegenerative pathways. In this review, we summarize and discuss existing literature on natural marine and terrestrial compounds, as well as their ability to modulate MMP-2 and MMP-9, and we evaluate their potential as therapeutic compounds for neurodegenerative and neuroinflammatory diseases, with a focus on Alzheimer's disease.

Keywords: MMP-2; MMP-9; Alzheimer's disease; AD; neurodegeneration; neuroinflammation; natural compounds; marine compounds; terrestrial compounds; nutraceuticals

check for updates

Citation: Ciccone, L.; Vandooren, J.; Nencetti, S.; Orlandini, E. Natural Marine and Terrestrial Compounds as Modulators of Matrix Metalloproteinases-2 (MMP-2) and MMP-9 in Alzheimer's Disease. *Pharmaceuticals* **2021**, *14*, 86. https://doi.org/10.3390/ph14020086

Academic Editor: Thierry Besson
Received: 24 November 2020
Accepted: 19 January 2021
Published: 24 January 2021

Publisher's Note: MDPI stays neutral with regard to jurisdictional claims in published maps and institutional affiliations.

1. Introduction

Most neurodegenerative diseases are characterized by an incurable loss of neurons in the brain and spinal cord, leading to impaired movement and/or mental functioning. In 2019, an estimated 50 million individuals were suffering from dementia worldwide, and this number is projected to increase to 152 million cases by 2050. The worldwide cost of dementia exceeds one trillion dollars per year and is set to double by 2030 [1]. While Alzheimer's disease (AD) is the most common neurodegenerative disorder, other neurodegenerative disorders resulting in cognitive defects include vascular dementia, frontotemporal dementia, mixed dementia, and dementia with Lewy bodies. Furthermore, neurodegenerative diseases affecting the motor system include amyotrophic lateral sclerosis (ALS), Huntington's disease (HD), Parkinson's disease (PD), and spinocerebellar ataxias [2]. Despite the high burden of these pathologies, current therapies mostly manage symptoms but do not prevent progressive deterioration [3,4].

Brain tissue from patients with AD is characterized by the presence of extracellular amyloid-β (Aβ) plaques and intracellular neurofibrillary tangles of hyperphosphorylated tau proteins [5]. The most known molecular pathway in AD is the generation of Aβ peptides, which are released after the consecutive cleavage of membrane-associated amyloid precursor protein (APP) by α and γ-secretases (a process referred to as the amyloidogenic pathway; Figure 1). After release, Aβ peptides induce the formation of protein aggregates.

While the main component is Aβ, up to 488 other proteins that also influence the process of aggregation have been detected [6]. For example, a cross-reaction with apolipoprotein A1, cystatin C, or transthyretin inhibits amyloid formation [7–12]. However, the formation of pathogenic Aβ peptides can be avoided through the cleavage of APP by α-secretase and the release of soluble APPα (sAPPα) (a process referred to as the non-amyloidogenic pathway; Figure 1) [13]. Though tau and Aβ aggregates are hallmarks of AD [14], the failure of clinical trials targeting Aβ (resulting in adverse effects on cognition) has sparked a debate regarding whether the production of Aβ is the primary underlying cause of AD [15].

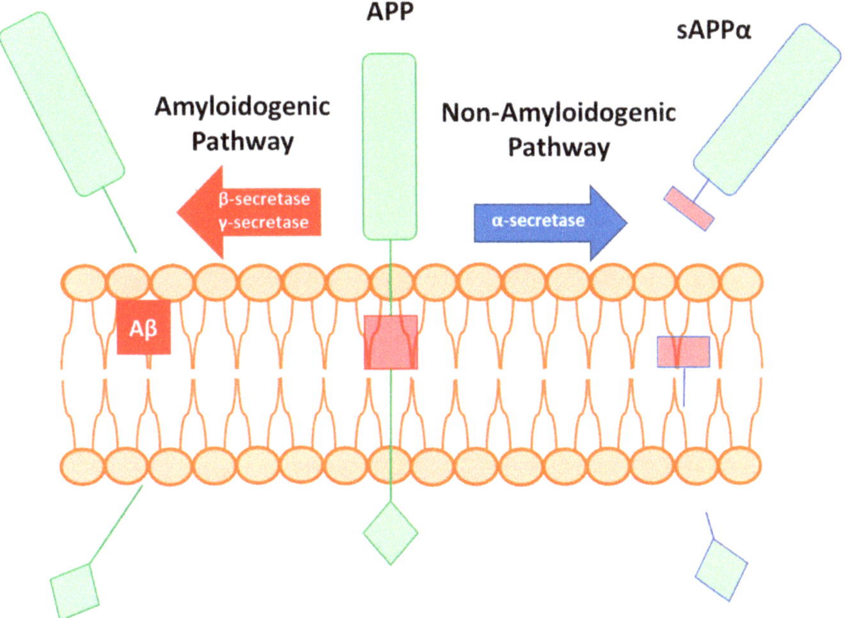

Figure 1. Graphical representation of amyloid precursor protein (APP) cleavage and the amyloidogenic and non-amyloidogenic pathways. sAPPα: soluble APPα.

One family of proteins that has gained attention in neurodegeneration and neuroinflammation are the matrix metalloproteinases (MMPs). MMPs are endopeptidases that are able to modify a broad range of proteins with key functions in the extracellular and pericellular environment (discussed in Section 1.1.). MMPs have been studied as therapeutic targets for several pathologies, in particular cancer. However, early clinical trials in which broad spectrum metalloproteinase inhibitors were used for the treatment of cancer failed due to off-target effects. Since then, selectivity has become one of the principal aspects of MMP inhibitor design. In addition, various molecules acting against two or more MMPs have recently been investigated, suggesting that carefully selected multitarget approaches also represent a promising strategy for targeting MMPs [16].

Interesting, both the beneficial and detrimental functions of MMPs in neurodegeneration and neuroinflammation have been described [17]. For example, MMPs cleave myelin basic protein (MBP), thereby contributing to the demyelination of neurons and neurodegeneration. In contrast, several MMPs are able to degrade Aβ aggregates [18,19]. Though this suggests a potential role for MMPs in aggregate clearance, it remains to be seen whether this is clinically significant. Nevertheless, given the many pathological functions of MMPs, they remain considered as potential targets for neurodegenerative diseases such as AD. An overview of the roles of MMPs in neuropathology and discussion on their suitability as targets in AD is given in Section 1.2.2.

Better knowledge on products that modulate MMPs could inspire new approaches for treatment of neurodegenerative and neuroinflammatory diseases such as AD. Natural bioactive products extracted from plants, minerals, animals, and microorganism are common claimed nutraceuticals, and they have been used to fight many diseases [20]. Knowledge on medicinal use of natural products is often handed over from one generation to another, and it stems from thousands of years ago when medicinal plants, rich in phytochemicals and microorganisms, were a major source of medicines.

Recently, many natural products were placed under investigation in pre-clinical and clinical trials in the treatment of AD [21], confirming the relevance of natural compounds. Several studies report that nutraceuticals and phytochemicals can have a crucial role in cell survival, neuron function, synaptic plasticity, and memory formation, thus contributing to prevention of neurodegenerative diseases onset [22–24]. Furthermore, in addition to other functions such as antioxidant activity and anti-inflammatory activity, several of these compounds are able to modulate the expression or proteolytic activity of MMPs.

In this manuscript, we briefly discuss MMPs and their known roles in neurodegeneration and neuroinflammation, followed by an overview of natural compounds derived from marine and terrestrial sources that can modulate the expressions level and/or activity of MMP-2 and MMP-9. The aim of this manuscript is to function as a reference for medicinal chemists who wish to develop new molecules that combine the beneficial actions of natural compounds with the ability to modulate MMPs, perhaps being useful in the treatment of neuropathologies such as AD.

1.1. Introduction to MMPs

Matrix metalloproteinases (MMPs), also called matrixins, are a family of zinc-dependent endopeptidases [25]. MMPs have a modular domain structure and share a similar catalytic domain and zinc-binding domain, together forming the active site for substrate catalysis (Figure 2). MMPs are secreted by cells such as inactive pro-enzymes (proMMPs) that require the proteolytic removal (e.g., by other proteases) or chemical modification (e.g., by reactive oxygen species) of a self-inhibitory prodomain in order to become catalytically active proteases [26]. The prodomain, catalytic domain, and zinc-binding domain form the structural basis of all MMPs. Several MMPs also share a hemopexin domain that is involved in binding to substrates [27], inhibitors [28], and cell surface receptors [29]. MMP-2 and MMP-9 (gelatinase A and gelatinase B) both have three fibronectin repeats that give them the ability to bind large substrates such as collagens and to efficiently cleave gelatins [30]. In addition, MMP-9 has a unique linker sequence (64 amino acids) that connects the active site to the hemopexin domain and is rich in O-glycans; hence, it is named the O-glycosylated domain. This domain is indispensable for MMP-9 functions including cell migration [31], substrate catalysis [32], and the regulation of bioavailability [33]. Furthermore, this domain allows MMP-9 to form higher order multimers that are differentially regulated by natural MMP inhibitors than MMP-9 monomers [34]. There are several MMPs that are bound to the cell surface through a membrane anchoring domain such as a transmembrane domain or glycosylphosphatidylinositol anchor, thereby executing cell-surface associated functions [25]. Finally, MMP-23 is an atypical MMP, having a transmembrane domain attached to the N-terminus and an immunoglobulin-like domain and cysteine-rich domain attached to the C-terminus [35].

MMPs are mainly known for their ability to cleave components of the extracellular matrix (ECM); hence, their original nomenclature was based on substrate specificity, e.g., collagenases, gelatinases, and matrilysins. The cleavage of ECM components by MMPs contributes to processes such as tissue remodeling, cell migration, and the release of growth factors from the ECM [36]. However, many other (even intracellular [37]) MMP substrates have been found, and intracellular roles of MMPs are being discovered [38,39]. For example, the nuclear localization of MMP-2 was reported in cigarette smoke-exposed endothelial cells and associated with cell apoptosis [40]. MMP-12 is transported to the nucleus, where it binds to the NF-kappa-B inhibitor alpha NFKBIA promoter and mediates NFKBIA transcription, leading to Interferonα secretion and protection against viral infections [41]. Finally, functions unrelated to the proteolytic actions of MMPs have also been found. These effects mainly rely on the cell-surface association of MMPs and the subsequent activation of signal transduction pathways, e.g., by binding of cell surface receptors such as CD44 and integrins [42].

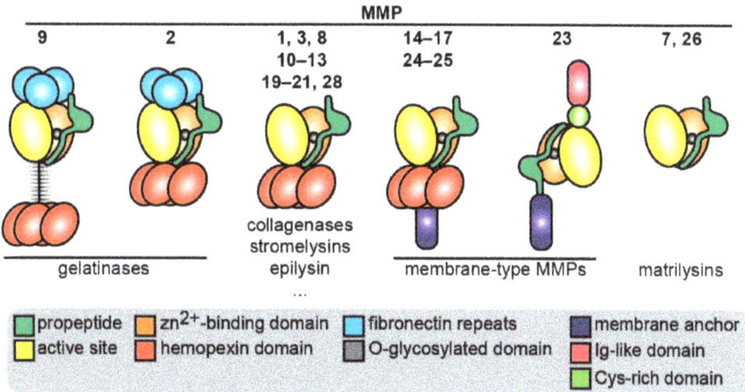

Figure 2. Overview of matrix metalloproteinases (MMPs) and their functional domain organization. Cys: cysteine; Ig: immunoglobulin.

1.2. MMPs in Neurodegeneration and Neuroinflammation

A common feature of several neurodegenerative disorders is the presence and dispersal of anomalous protein aggregates throughout the brain. For example, the formation of amyloid fibrils and Aβ peptide aggregates in AD, tau tangles in tauopathies and α-synuclein aggregates in PD. The formation of these anomalies is associated with neurotoxicity and the progressive loss of neurons. As an additional factor, innate immune mechanisms are also implicated in the pathogenesis of neurodegenerative diseases. The transitioning of the innate immune response into chronic inflammation results in the proliferation of glial cells (gliosis) and elevated levels of proinflammatory cytokines [2]. For a comprehensive overview of the actions of all MMPs in neurodegeneration and neuroinflammation, we refer the reader to several recent reviews [17,43–45]. In this manuscript, we focus on the actions of the two gelatinases—MMP-2 and MMP-9.

1.2.1. Localization and Origin of MMP-2 and MMP-9 in the Nervous System

Both MMP-2 and MMP-9 can be expressed by cells of the nervous system. While expression of MMP-9 is most often induced, MMP-2 is more constitutively present and less influenced by damaging factors [46]. The basal expression of MMP-9 in healthy brain tissue is low [47] but increases significantly in disease models and/or patient samples with neurological damage. The disease- or damage-mediated induction of MMP-9 is found in the endothelial cells of cerebral vasculature [47], astrocytes surrounding amyloid plaques or in injured nerves [18,48,49], areas of astrogliosis [50], meninges and neurons of the injured spinal cord [51], human pyramidal neurons [52], and Schwann cells stimulated with proinflammatory cytokines (e.g., tumor necrosis factor-α and interleukin-1β) [53]. In patients with secondary progressive multiple sclerosis (MS), MMP-9 is expressed at the rim of plaques in chronic active lesions, suggestive for the expression of MMP-9 by activated microglia [54]. Furthermore, all forms of MMP-9 (monomers, multimers, and charge variants) are increased in serum from patients with MS [55]. Interestingly, in a rat model for HD, MMP-9 immunoreactivity was found in the nuclei of the neurons of a healthy rat striatum [50]. In a mouse model for peripheral nerve injury, MMP-2 was found to be constitutively present in nerve tissue [56]. However, in a mouse model for AD, MMP-2 was also increased in astrocytes surrounding amyloid plaques [18].

Infiltrating immune cells are a second source of MMP-2 and MMP-9. Polymorphonuclear leukocytes (PMNs) are the first cells to arrive in damaged tissue and PMN-derived MMPs have neurotoxic activity [57]. PMN infiltration is followed by monocyte infiltration, and tissue-differentiated monocytes secrete higher levels of MMP-9 compared undifferentiated monocytes, thereby contributing to neurotoxicity [58]. In animal models for peripheral nerve injury and spinal cord injury, MMP-9 is associated with infiltrating macrophages [56,59]. In HIV-induced dementia, proMMP-2 is secreted by HIV-infected macrophages [56], and in a model for neuroinflammation, macrophage-derived gelatinase (MMP-2 and MMP-9) activity was found to be crucial for leukocyte infiltration into the central nervous system (CNS) [60]. In conclusion, there are several origins of MMP-2 and MMP-9 in CNS pathology; both resident and infiltrating immune cells are able to supply these proteases or induce their production, e.g., upon stimulation with pro-inflammatory molecules.

1.2.2. Mechanisms of MMP-2 and MMP-9 in CNS Pathology

MMPs have a potential role in the turnover of protein aggregates (Figure 3). Several MMPs, including MMP-2 and MMP-9, are able to cleave Aβ monomers and oligomers, but MMP-9 is unique in its ability to also cleave Aβ fibrils and clear plaques from amyloid-laden brains. For a detailed overview of MMP cleavage sites on APP, the reader is referred to the review manuscript by Cauwe et al. [61]. A recent overview of APP processing by MMPs is also available in the review manuscript by Zipfel et al. [16]. Though mainly relying on in vitro studies, ex vivo studies, and steady state mouse models, a positive cross-interaction between Aβ, MMP-2, and MMP-9 has been suggested [18,48]. Moreover, several studies have shown that Aβ can also induce MMP-9 expression and activity in vitro, e.g., in astrocytes [48] and in THP-1 cells (a monocytic cell line) [62]. Hence, Aβ-induced MMP-2 and MMP-9 expression might also enhance MMP-associated neurotoxicity and outweigh the Aβ-clearing effect. Interestingly, tau, the aggregating protein associated with tauopathies, including AD, is also an MMP-2, MMP-3, and MMP-9 substrate. However, proteolysis by MMP-9, not MMP-3, induces tau oligomer formation [63], whereas a physiological function of MMP-2 in normal tau proteolysis has been suggested [64].

MMP-2 and MMP-9 can cleave several soluble factors such as chemokines and growth factors, thereby altering their functional properties [65]. Several MMPs (including MMP-2 and MMP-9) are able to cleave C-X-C motif chemokine Ligand 12 (CXCL12) stromal cell-derived factor-1 (SDF-1), thereby either degrading it or converting it to a neurotoxic protein that activates the intrinsic apoptotic pathway (Figure 3) [66]. In HIV-associated neurodegeneration, proMMP-2 is secreted by HIV-infected macrophages, and after activation by neuronal MMP-14/ membrane-type 1 (MT1)-MMP, MMP-2 converts astrocyte-derived SDF-1 into neurotoxic SDF-1 (5–67) [67]. Another crucial substrate of MMPs is MBP (Figure 3). Many in vitro and in vivo studies have shown that MMPs promote demyelination by degrading MBP, a major constituent of myelin sheets supporting neuronal signals [53,68–70].

MMPs also have a fundamental role in leukocyte migration (Figure 3). During neuroinflammation, leukocytes migrate from the circulation into the CNS. This process requires their migration through a layer of vascular endothelial cells, two layers of basement membranes (the endothelial basement membrane and the parenchymal basement membrane), and a layer of astrocytes, together forming the blood brain barrier (BBB) [36]. Both MMP-2 and MMP-9 are important players in this process. In mouse experimental autoimmune encephalomyelitis (EAE), a model for CNS inflammation, leukocytes were found to use MMP-2 and MMP-9 to migrate through the parenchymal basement membrane, specifically through the cleavage of dystroglycan, a transmembrane receptor that connects astrocyte endfeet with parenchymal basement membrane BM [60]. For a detailed overview on the roles of MMP-2 and MMP-9 in neuroinflammation, we refer the reader to a recent review by Hannocks et al. [45].

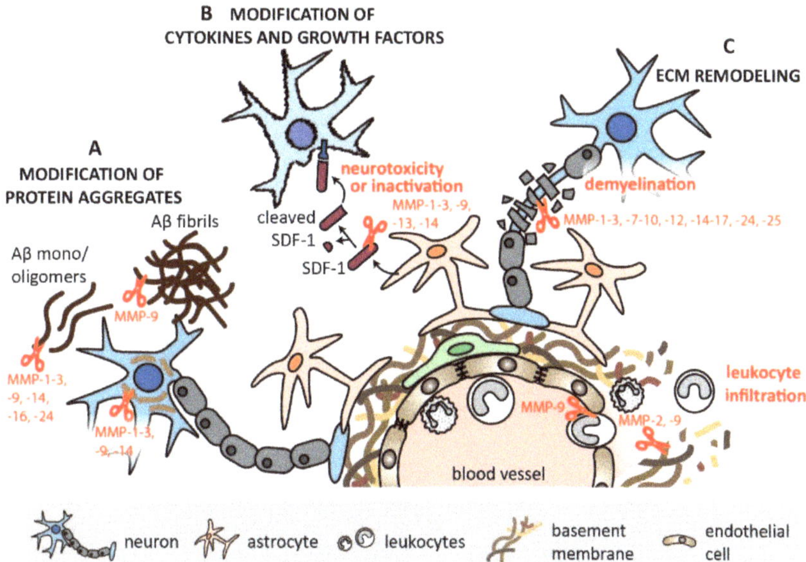

Figure 3. Overview of MMP functions in the nervous system. (**A**) Several MMPs can cleave protein aggregates such as Aβ and tau, with MMP-9 being unique in its ability to cleave Aβ fibrils. (**B**) MMPs can modify several cytokines and growth factors. This is exemplified by the cleavage of stromal cell-derived factor-1 (SDF-1)/ C-X-C motif chemokine Ligand 12 (CXCL12) and either its inactivation or its conversion into a neurotoxic peptide. (**C**) MMPs cleave several components of the extracellular matrix (ECM) such as tight junction proteins and components of the basement membranes, thereby allowing for immune cell migration and contributing to neuroinflammation.

Several models for neurodegeneration and neuroinflammation in animals deficient in MMP-2 and/or MMP-9 have been used to show beneficial effects on disease progression and outcome. Upon spinal cord injury, MMP-9-knockout (KO) mice have less disruption of the blood–spinal cord barrier, reduced neutrophil infiltration, and improved locomotor recovery [51]. The deletion of the MMP-9 gene protects nerve fibers from demyelination and reduces neuropathic pain after injury [53]. While MMP-9 KO mice are protected in traumatic brain injury [71], in ischemia, the knock-out of MMP-2 does not alter acute brain injury [72]. The genetic ablation of both MMP-2 and MMP-9 in mice results in resistance to EAE by inhibiting dystroglycan cleavage and preventing leukocyte infiltration [60]. Finally, the beneficial effects of MMP inhibitors (minocycline, simvastatin, and GM6001) have been described in a model for cerebral amyloid angiopathy [73]. Overall, these studies justify the inhibition of MMP-2/-9 in neurodegenerative and neuroinflammatory diseases such as AD and highlight the potential of new inhibitory compounds.

2. Natural Products from Marine Source That Modulate MMP-2 and/or MMP-9

Marine organisms such as sponges, macroalgae, microalgae, and bacteria are considered effective biological sources of new bioactive drugs. These organisms are usually rich in nutraceuticals and pharmaceuticals that are secreted to survive in the hostile environment where they live. The bioactive molecules are metabolites such as small chemical molecules, as well as short peptides and enzymes. One critical point associated with marine compounds is that only limited amounts are produced by the natural source. In addition, their chemical structures are often too complex to be synthetized in vitro. In recent years, several products with anti-MMP activity have been identified from marine sources. Most investigations have been focused on their effect on MMPs during inflammatory diseases and/or cancer. Due to the relevance of MMPs (especially MMP-2 and MMP-9) in neurodegenerative diseases, marine products could also be useful for the inhibition of MMPs in AD [74].

In the following paragraphs, several natural MMP modulators of marine origin are reported. For more detailed information regarding the literature prior to 2018, we refer the reader to previous reviews cited below [75–77]. The active products are divided in two groups: protein/peptides and small molecules (Figure 4).

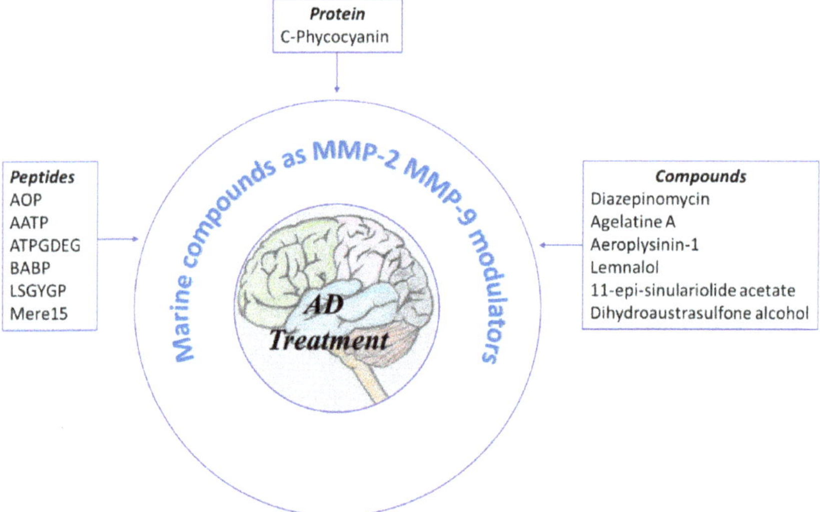

Figure 4. Overview of marine compounds able to modulate MMP-2 and MMP-9. AOP: abalone oligopeptide; AATP: abalone anti-tumor peptide; ATPGDEG: Ala-Thr-Pro-Gly-Asp-Glu-Gly; BABP: boiled abalone byproduct peptide; LSGYGP: Leu-Ser-Gly-Tyr-Gly-Pro; Mere15: Mere Meretrix 15 kDa polypeptide.

Of note, many manuscripts reporting on the analysis of MMPs (in particular MMP-2 and MMP-9) miss a clear distinction between differences in mRNA levels, protein levels, and proteolytic activity. This problem stems from the fact that several of the standard methods to evaluate MMP-2 and MMP-9 levels and activity are prone to misinterpretation (as discussed in more detail by Vandooren et al., 2013) [78]. In this review manuscript, a clear distinction between these different levels of regulation was made based on the methodology used in each of the discussed manuscripts.

2.1. Protein and Peptides

C-phycocyanin (C-PC) is a deep blue colored pigment protein that can be isolated and purified from several seaweeds. C-PC is largely found in *Spirulina*, a microalgae used in many countries as dietary supplement and whose nutritional benefits have been well-described [79]. The structure of C-PC is characterized by a heterodimeric monomer ($\alpha\beta$) formed by the α and β subunits. Usually, the $\alpha\beta$ monomers of C-PC further polymerize into higher order multimers; $(\alpha\beta)_n$ $n = 1 \sim 6$ [80]; see Figure 5.

C-PC has several biological activities such as improving wound healing, antioxidant activity, pro-apoptotic effects, and antitumor activity [81]. The beneficial effects of C-PC have also been observed in various models for degenerative diseases such as PD, MS and ALS [82–84].

Recently, C-PC was reported as an inhibitor of MMP-2 and MMP-9. In a vasculogenic mimicry assay with breast cancer cells (MDA-MB 231 cell line), treatment with C-PC resulted in a drastic reduction in the number of vascular channels formed compared to a control. A real-time quantitative reverse transcription-PCR (qPCR) analysis recorded a decrease in the mRNA levels of both vascular endothelial growth factor receptor-2 (VEGFR2) and MMP-9, two key regulators of cancer-associated angiogenesis [85].

In a 1,2-dimethylhydrazine-induced colon cancer rat model, treatment with piroxicam and C-PC resulted in a lower tumor expression of MMP-2 and MMP-9 compared to a control. Both pro- and activated forms of MMP-2 and MMP-9 were reduced [86]. In the HepG2 cell line (hepatocellular carcinoma cell line), C-PC could also inhibit the mRNA and protein expression of both investigated MMPs [87]. Moreover, it has been reported that C-PC might cross the BBB in a model for tributyltin chloride (TBTC)-induced neurotoxicity, adding to its neuroprotective effects and making C-PC a potential drug candidate in neurodegenerative diseases [88].

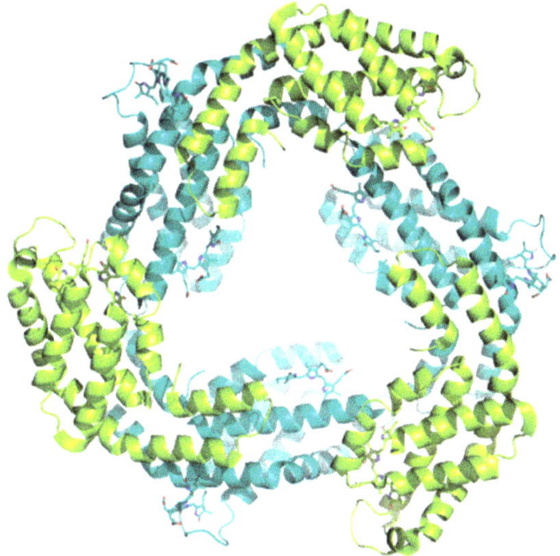

Figure 5. Graphical representation of the C-phycocyanin (C-PC) crystal structure PDB code 3O18 [80]. In this structure, C-PC forms trimers $(\alpha\beta)_3$. The α subunit is colored in lemon, and the β subunit is colored in teal.

Abalone, *Haliotis discus hannai*, is a marine univalve gastropod that is predominantly cultured on the southwestern coast of Korea and considered as a precious food at Asian markets. Several studies have found that abalone is a source of nutraceuticals with anti-microbial, anti-oxidant, anti-thrombotic, anti-inflammatory, and anti-cancer activities [89].

The digestion of abalone intestine with an in vitro gastrointestinal (GI) digestion system resulted in the identification of two peptides with anti-MMP-2 and anti-MMP-9 activity in human fibrosarcoma cells (HT1080 cells) (Table 1), namely abalone oligopeptide (AOP) and abalone anti-tumor peptide (AATP). AOP (Ala-Glu-Leu-Pro-Ser-Leu-Pro-Gly) was first characterized in 2013 by Nguyen et al. [90], while AATP (Lys-Val-Asp-Ala-Gln-Asp-Pro-Ser-Glu-Trp) was described in 2019 by Gong at al. [91]. In addition to inhibitory activity against MMPs, the AATP peptide also reduces the level of mRNA expression of both gelatinases.

Other peptides were also identified from boiled abalone such as ATPGDEG (Ala-Thr-Pro-Gly-Asp-Glu-Gly) and BABP (Glu-Met-Asp-Glu-Ala-Gln-Asp-Gly-Asp-Pro-Lys) (Table 1). In a human keratinocyte cell line (HaCaT cells), treatment with ATPGDEG resulted in a reduction of MMP-9 secretion. Moreover, molecular docking analysis suggested that ATPGDEG interacts with the MMP-9 active site, thereby blocking the catalytic activity [92].

The treatment of HT1080 cells with BABP also resulted in reduced levels of MMP-9 but not of MMP-2. BABP was able to suppress both MMP-9 protein levels (as determined by gelatin zymography and Western blot analysis) and mRNA expression in a dose-dependent manner [93].

Despite several studies, it is still not fully known how these peptides interact with MMPs. One hypothesis is that the amino acids Glu, Asp, Pro, and Lys could be involved [91].

The LSGYGP (Leu-Ser-Gly-Tyr-Gly-Pro), a peptide isolated from tilapia fish skin gelatin hydrolysates (TGHs) (Table 1), showed a high hydroxyl radical scavenging activity in an in vivo study. It has been suggested that TGHs could protect mouse skin collagen fibers against UV irradiation damage. In UVB-stimulated mouse embryonic fibroblasts (MEFs), LSGYGP significantly decreased the levels of MMP-9 in a dose-dependent manner. Furthermore, after using molecular modeling simulation, it was suggested that LSGYGP can enter into the catalytic site of MMP-9, thereby inhibiting its proteolytic activity [94]; however, in vitro or in vivo evidence of its effect on MMP-9 proteolytic activity is not yet available.

The Mere Meretrix 15 kDa polypeptide (Mere15) peptide was isolated from *Meretrix meretrix Linnaeus*, a mollusk used in traditional Chinese medicine as an anticancer molecule (Table 1). In human lung adenocarcinoma A549 cells, Mere15 downregulated the secretion of proteins and the expression of the mRNA of MMP-2 and MMP-9. Moreover, MMP-9 expression was lower than that of MMP-2, suggesting that MMP-9 is more sensitive to Mere15 inhibition [95].

Table 1. Peptides inhibitors of MMP-2 and MMP-9 from marine source.

Compound	Source	Sequence	MMP-2	MMP-9	Model	Ref.
AOP	*Haliotis discus hannai*	Ala-Glu-Leu-Pro-Ser-Leu-Pro-Gly	Inhibition [2]	Inhibition [2]	HT1080 cells	[90]
AATP	*Haliotis discus hannai*	Lys-Val-Asp-Ala-Gln-Asp-Pro-Ser-Glu-Trp	Inhibition [2,3]	Inhibition [2,3]	HT1080 cells	[91]
ATPGDEG	*Haliotis discus hannai*	Ala-Thr-Pro-Gly-Asp-Glu-Gly	n.d.	Inhibition [1,2]	HaCaT cells	[92]
BABP	*Haliotis discus hannai*	Glu-Met-Asp-Glu-Ala-Gln-Asp-Gly-Asp-Pro-Lys	Inhibition [2,3]	Inhibition [2,3]	HT1080 cells	[93]
LSGYGP	*Tilapia fish skin gelatin hydrolysate*	Leu-Ser-Gly-Tyr-Gly-Pro	n.d.	Inhibition [1,2]	mice	[94]
Mere15	*Meretrix meretrix Linnaeus*	unknown	Inhibition [2,3]	Inhibition [2,3]	A549 cells	[95]

AOP: abalone oligopeptide; AATP: abalone anti-tumor peptide; ATPGDEG: Ala-Thr-Pro-Gly-Asp-Glu-Gly; BABP: boiled abalone byproduct peptide; LSGYGP: Leu-Ser-Gly-Tyr-Gly-Pro; Mere15: Mere Meretrix 15 kDa polypeptide. [1] Inhibition of proteolytic activity (e.g., substrate degradation assays and molecular docking); [2] inhibition of protein expression (e.g., in gel zymography, Western-blot, and ELISA); [3] inhibition of mRNA expression (e.g., reverse transcription polymerase chain reaction, RT-PCR). n.d.: not defined.

2.2. Compounds

Diazepinomicin (BU-4664L) is a terpenoid firstly identified in the extract of marine *Micromonospora* sp. [96]. BU-4664L has mainly been studied for its anti-invasive and anti-migratory effects on cancer cells. In murine colon 26-L5 carcinoma cells, BU-4664L inhibited the proteolytic activities of MMP-2 and MMP-9 with IC_{50} values of 0.46 and 0.60 μg/mL, respectively (Table 2). Furthermore, it was confirmed that BU-4664L is able to influence both extracellular matrix degradation and cell migration [97].

In 2008, a Phase II clinical trial with BU-4664L was started in patients affected by glioblastoma. Unfortunately, due to a lack of efficacy, the trail failed. Nevertheless, BU-4664L is a bioactive, farnesylated, dibenzodiazepinone alkaloid that is able to cross the BBB and provide a potential scaffold for further optimization and for investigation in the study of the relationship between AD and MMPs.

Ageladine A, isolated from the extract of the marine sponge *Agelas nakamurai*, possesses antiangiogenic activity. An in vitro study on isolated enzymes showed that ageladine A inhibits both MMP-2 and MMP-9 proteolytic activity (Table 2), whereas the N-methylated derivatives did not inhibit MMP-9 [98]. Since several MMP inhibitors exhibit their inhibitory activity by interaction with the Zn^{+2}-ion of the catalytic domain (chelation), the chelation power of ageladine A was investigated but could not be established. Therefore, ageladine A is likely to inhibit through a yet-to-be-determined mechanism [98].

Aeroplysinin-1 is a brominated antibiotic secreted by *Aplysina aerophoba* sponges as a chemical defense response triggered by tissue injury. Aeroplysinin-1 is thought to have anti-tumor and anti-angiogenic actions. A study conducted in different human endothelial cell lines reported a decrease of MMP-2 expression [99]; see Table 2. Interestingly, aqueous extracts of *Aplysina aerophoba* were able to reduce the protein and mRNA expression of both MMP-2 and MMP-9 in rat astrocyte cultures [101].

Lemnalol (Table 2), isolated from soft coral (*Lemnalia cervicornis* and *Lemnalia tenuis Verseveldt*), has anti-inflammatory effects on mast cells (MCs) and osteoclasts activity in rats with monosodium urate (MSU) crystal-induced gouty arthritis. Though immunohistochemical analysis, it was shown that lemnalol decreases the infiltration and degranulation of MCs, and it was suggested that this effect is partially related to reduced expression of MMP-9 [100].

11-*Epi*-sinularoide acetate (11-*epi*-SA) was isolated from the soft coral *Sinularia flexibilis* (Table 2). In hepatocellular carcinoma cells (HA22T cells), 11-*epi*-SA was found to inhibit cell migration and invasion in a concentration-dependent manner. These activities remained at low, non-toxic doses (<7.98 μM) and also reduced the expression and activity of MMP-2 and MMP-9, suggesting that the effect could be associated with the modulation of MMPs or their endogenous inhibitors [102].

Recently, 11-*epi*-SA was also studied in a human bladder cancer cell line (TSGH-8301 cells), again showing a relevant effect against cell migration and invasion. Similarly, these effects were associated with decreased levels of MMP-2 and MMP-9 protein secretion [103]. Considering the reported results in cancer models, 11-*epi*-SA might also be a promising candidate for further development as a new modulator of MMPs in other pathologies such as AD.

Dihydroaustrasulfone alcohol (DA), isolated from marine coral, has antioxidant and anti-cancer activity (Table 2). Furthermore, DA has a concentration-dependent inhibitory effect on the migration and motility of human non-small cell lung carcinoma cells (NSCLC A549 cells), as determined by trans-well and wound healing assays. Gelatin zymography analysis, a standard method to detect MMP-2 and MMP-9 levels and proteoforms, showed that DA also significantly inhibited the presence of MMP-2 and MMP-9. These results proved that the anti-metastatic effect of DA was associated with the suppression of enzymes involved in cancer cell migration [104].

DA has also been proposed as an anti-restenosis molecule. Restenosis is characterized by the abnormal proliferation and migration of vascular smooth muscle cells (VSMCs) and the stimulation of platelet-derived growth factor (PDGF)-BB. Based on gelatin zymography data, it was suggested that DA dose-dependently decreased the pro-forms of all gelatinases, as well as the active form of MMP-9, in comparison with the control group (PDGF-BB alone). These results showed that DA decreased the activation and expression levels of both MMP-2 and MMP-9, which are involved in cell migration [105].

Table 2. Compound inhibitors of MMP-2 and MMP-9 from marine sources.

Structure	Source	MMP-2	MMP-9	Model	Ref
BU-4664L (Diazepinomicin)	*Micromonospora* sp.	Inhibition [1]	Inhibition [1]	26-L5 cells	[97]
Ageladine A	*Agelas nakamurai*	Inhibition [1]	Inhibition [1]	Isolated enzymes	[98]
Aeroplysinin-1	*Aplysina aerophoba*	Inhibition [2]	n.d.	Endothelial cells	[99]
Lemnalol	*Lemnalia* sp.	n.d.	Inhibition [2]	rats	[100]
11-epi-sinulariolide acetate	*Sinularia flexibilis*	Inhibition [2]	Inhibition [2]	HA22T cells	[101] [102]
Dihydroaustrasulfone alcohol	*Cladiella australis*	Inhibition [2] Inhibition [2]	Inhibition [2] Inhibition [2]	A549 cells VSMC (vascular smooth muscle cells)	[103] [104]

[1] Inhibition of proteolytic activity (e.g., substrate degradation assays, and molecular docking); [2] inhibition of protein expression (e.g., in gel zymography, Western blot, ELISA, and immunohistochemistry); n.d.: not defined.

3. Natural Products from Terrestrial Source That Modulate MMP-2 and/or MMP-9

Natural bioactive products, phytochemicals, and nutraceuticals extracted from plants, minerals, animals, and microorganism have been the source of most of the bioactive molecules used in traditional medicine. Several in vitro and in vivo studies have demonstrated the therapeutic potential of natural products in various pathologies including degenerative and neurodegenerative diseases such as AD, PD, HD, ALS, and MS [21,106–111].

Due to their multifunctional properties, natural products may interfere with AD progression at all stages including the formation and clearance of pathological aggregates, the release of damaging reactive oxygen species, and even neuroinflammation. Recently, Andrade et al. generated a detailed overview of the state of the art of natural molecules and natural extracts currently studied in AD [21]. Interesting, the majority of natural molecules reported as a potential drug candidate against AD also influence MMP-2 and MMP-9. In Figure 6, an overview is given of flavonoids with MMP-2/-9 modulatory activity, including their confirmed effects on pathological processes in AD. Below, we further discuss the effects of these flavonoid on MMP-2 and MMP-9 mRNA expression, protein production, and proteolytic activity, as reported in several different disease models, in vitro cell-based assays, and biochemical assays.

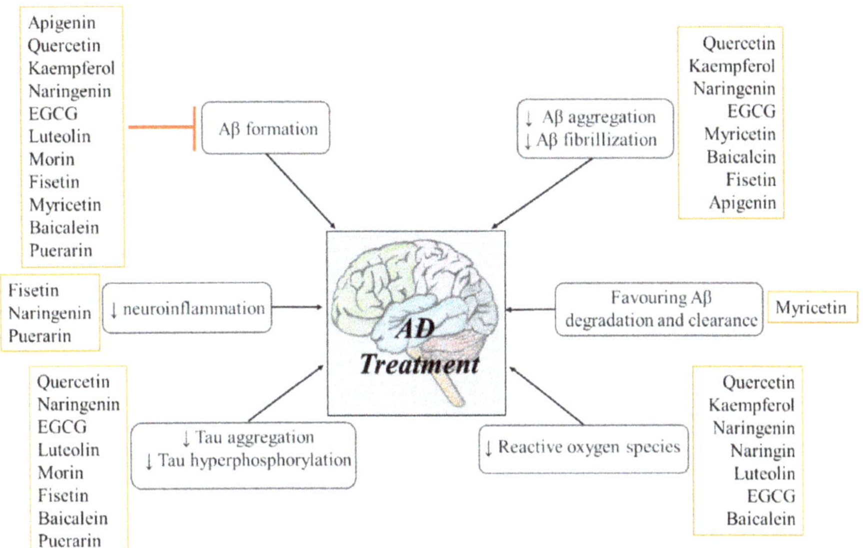

Figure 6. Overview of flavonoids with modulatory activity on MMP-2/-9 and with confirmed effects on different processes involved in Alzheimer's disease (AD) pathology. EGCG: epigallocatechin gallate.

Flavonoids

Flavonoids are phenolic compounds that can be isolated from a wide range of plants. Several of these compounds have been attributed beneficial actions in health and disease. Their main effects include anti-carcinogenic, anti-inflammatory, antiviral, antioxidant, and psychostimulant activities. In the following section, we report on flavonoids with known positive effects in AD [21] and that have the ability to module MMP-2 and MMP-9; see Table 3.

Quercetin (Que), Table 3, has been largely studied for its anti-inflammatory and anti-cancer activity.

In the human fibrosarcoma cells line (HT1080), Que was found to inhibit both MMP-2 and MMP-9 protein levels were in a dose-dependent manner [112]. In a human hepatocarcinoma cell line (HCCLM3 cells), Que inhibits cell migration and invasion in vitro, and it has been suggested that these anti-migratory and anti-invasive effects are due to the ability of Que to downregulate the protein expression of MMP-2 and MMP-9 [113]. Moreover, the effect of Que was also studied in human oral cancer cell lines (HSC-6 and SCC-9 cells), where Que also decreased the abundances of MMP-9 and MMP-2 [114].

In an asphyxia-based rat model for cardiopulmonary resuscitation (CPR), rats treated with treated by intragastric injection of 50 mg/kg quercetin once a day for five days had significantly less reactive oxygen species (ROS) generation, inflammation, and MMP-2 protein expression [147]. Que was also studied in the two-kidney one-clip rat model for hypertension. Animals treated with quercetin (10 mg/kg/day for three weeks by gavage) had reduced aortic ROS levels and MMP-2 activity, as determined by situ zymography and immunofluorescence [115].

Table 3. Flavonoid compounds able to modulate MMP-2 and MMP-9 levels.

Structure	Source	MMP-2	MMP-9	Model	Ref
Quercetin	Fruit, vegetables, seeds, and grains	Inhibition [1,2]	Inhibition [2]	HT1080 cells HCCLM3 cells HSC-6 cells SCC-9 cells 2K1C rats	[112] [113] [114] [114] [115]
Kaempferol	Tea, cabbage broccoli, kale, beans, endive, tomato, strawberries grapes, and endive	Inhibition [2] Inhibition [2]	n.d. Inhibition [2]	SCC-4 cells MCF-7	[116] [117]
Naringenin	Grapefruit, bergamot, orange, tomatoes, and cocoa	Inhibition [2]	Inhibition [2]	A549 cells	[118]
Epigallocatechin Gallate	Green tea	n.d. Inhibition [2] n.d. Inhibition [2]	Inhibition [1] n.d. Inhibition [2,3] Inhibition [2]	Biochemical assay MCF-7 cells MDA-MB-231 AsPC-1 cells	[32] [119] [120] [121]
Luteolin	Salvia, broccoli, parsley, thyme, green pepper, and artichoke	Inhibition [2] Inhibition [2] Inhibition [2] Inhibition [2,3] Inhibition [2]	Inhibition [2] Inhibition [2] Inhibition [2] Inhibition [2,3] Inhibition [2]	mice A2780 cells mice A375 cells mice	[122] [123] [123] [124] [124]

Table 3. *Cont.*

Structure	Source	MMP-2	MMP-9	Model	Ref
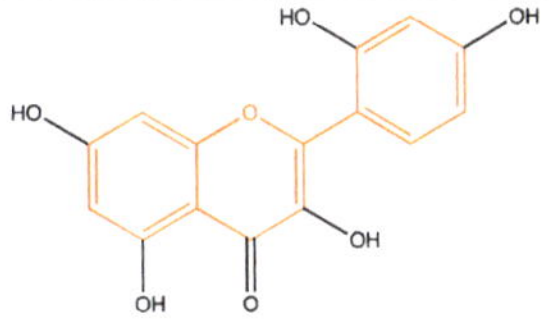 Morin	Osage orange, common guava, and old fustic	Inhibition [2] Inhibition [2] Inhibition [2] n.d.	Inhibition [2] Inhibition [2] Inhibition [2] Inhibition [2,3]	rats LX-2 cells rats MCF-7	[125] [126] [126] [127]
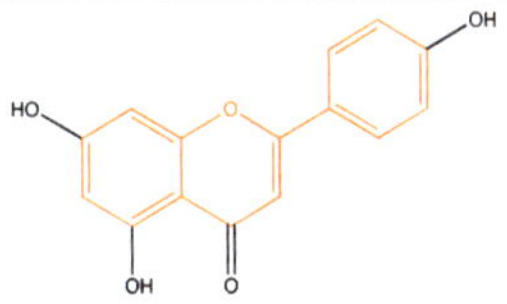 Apigenin	Chamomile grapefruit, parsley, celery, celeriac, and onion	n.d. n.d. Inhibition [2,3]	Inhibition [2] Inhibition [3] Inhibition [2]	SW480 cells U87 A375 cells	[128] [129] [130]
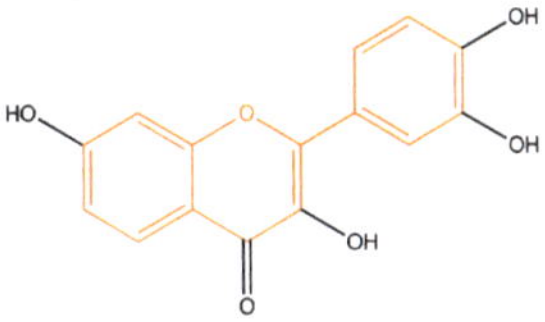 Fisetin	Kiwifruit, tomato, strawberries, apples, persimmons, onions, and cucumbers	n.d. Inhibition [2] Inhibition [2]	Inhibition [2,3] Inhibition [2] Inhibition [2]	AsPC-1 cells 4T1 cells JC cells	[131] [132] [132]
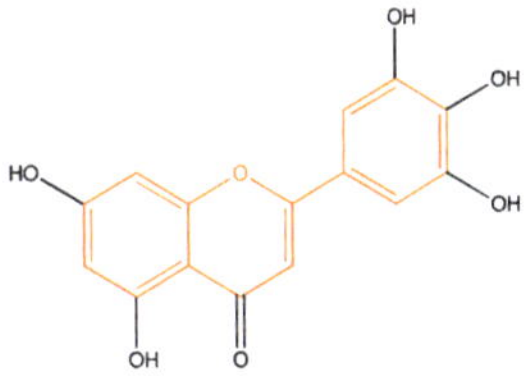 Myricetin	Tomatoes, oranges, nuts, berries, tea, and red wine	Inhibition [1,2] Inhibition [1,2] Inhibition [2] Inhibition [2,3] Inhibition [2,3]	n.d. n.d. Inhibition [2] Inhibition [2,3] Inhibition [2,3]	Isolated MMP-2 COLO 205 cells MDA-Mb-231Br cells A549-IR C57BL/6 mice	[133] [133] [134] [135] [136]
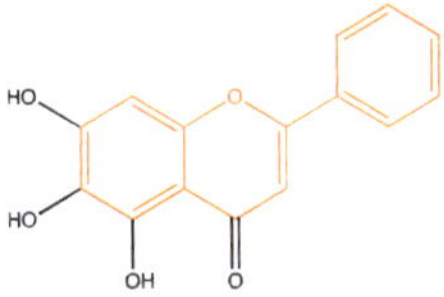 Baicalein	Root of *Scutellaria baicalensis*	Inhibition [2,3] Inhibition [2] Inhibition [2] Inhibition [2] Inhibition [2]	Inhibition [2,3] n.d. n.d. Inhibition [2] Inhibition [2]	mice A375 SK-MEL-28 BON1 cells CRL-1427 cells	[137] [138] [138] [139] [140]
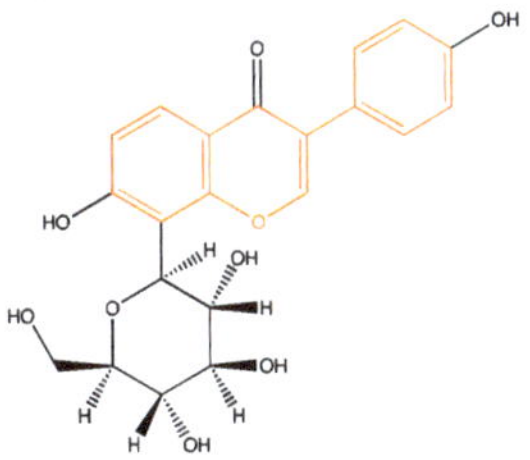 Puerarin	Root of *Pueraria thomsonii*, *Pueraria tuberosa*, and *Pueraria lobate*	Inhibition [2] Inhibition [2,3] Inhibition [2,3]	Inhibition [2] Inhibition [2,3] Inhibition [2,3]	rats MCF-7 cells MDA-MB-231 cells	[141] [142] [142]

Table 3. *Cont.*

Structure	Source	MMP-2	MMP-9	Model	Ref
Rutin	Capes, olive buckwheat, asparagus, red-raspberry, tomato, prune, fenugreek, zucchini, and apricot	n.d. Inhibition [2]	Inhibition [2] Inhibition [2]	rats rats	[143] [144]
Naringin	citrus fruits, grapefruit, artichokes, brussels sprouts, strawberries, rosemary, oregano, and tomato	Inhibition [2] Inhibition [2]	Inhibition [2] Inhibition [2]	U251 cells U87 cells	[145] [146]

Flavonoid scaffold is highlighted in orange. [1] Inhibition of proteolytic activity (e.g., substrate degradation assays, molecular docking, and in situ zymography); [2] inhibition of protein expression (e.g., in gel zymography, Western blot, ELISA, and immunohistochemistry); [3] inhibition of mRNA expression (e.g., RT-PCR). n.d.: not defined.

Kaempferol (Kae), Table 3, is a bioactive substance that possesses several properties such as anti-cancer, anti-diabetic, anti-inflammatory, anti-aging, anti-allergic, and cardio-protective activities [148]. In human tongue squamous cell carcinoma cells (SCC4 cells), Kae inhibited migration and invasion, reduced the protein expression of MMP-2, and decreased the nuclear translocation of the transcription factor AP-1 to the MMP-2 promoter [116].

As a phytoestrogen, Kae is known to play a chemopreventive role inhibiting carcinogenesis and cancer progression. In the MCF-7 breast cancer cell line, Kae inhibits the protein expression of epithelial-mesenchymal transition-related markers and suppresses metastasis-related markers such as MMP-2 and MMP-9 [117].

Naringenin (Nar), Table 3, is a bioactive compound found in several fruits that has anti-inflammatory and antitumor effects. One study investigated the effect of Nar on the migration of lung cancer cells (A549 cells) and found a significant alteration in A549 cell proliferation in response to Nar treatment. Gelatin zymography revealed that Nar reduces MMP-2 and MMP-9 levels in a concentration-dependent manner [118].

Epigallocatechin Gallate (EGCG), Table 3, is the most abundant catechin found in green tea. EGCG has various biological effects such as antioxidant, radical scavenging, antimicrobial, anti-inflammatory, anticarcinogenic, antiapoptotic, and metal-chelating activities [149]. Moreover, several studies have reported that EGCG offers potential protection from neurodegeneration [150] or can be considered an inhibitor of cancer cell metastasis via the inhibition of the expression and activity of several proteins such as MMP-2 and MMP-9. MCF-7 cells treated with EGCG suppress the expression of pro-MMP-2 [119]. In a human breast cancer cell line (MDA-MB-231) with high metastatic properties, treatment with EGCG resulted in the inhibition of MMP-9 mRNA and protein expression [120]. Additionally in human pancreatic cancer cells (AsPC-1 cells), EGCG inhibits the expression of MMP-2 and MMP-9 [121]. Furthermore, in biochemical assays with recombinant MMP-9, the direct inhibition of MMP-9 gelatinolytic activity by EGCG has also been established [32].

Luteolin (Lut), Table 3, is largely present in herbs, vegetables, and fruits, and it exhibits anti-inflammatory and antioxidant activities. The effect of Lut on MMP-2 and MMP-9 in azoxymethane (AOM)-induced colon carcinogenesis in BALB/c mice was investigated. The expression of MMP-2 and MMP-9 was increased during AOM induction, whereas treatment with Lut (15 mg/kg, intraperitoneally, once a week for three weeks) reduced their expressions reduced their expressions [122]. Recently, a study showed that Lut inhibits the metastasis of ovarian cancer cells (A2780 cells) by downregulating the expression of MMP-2 and MMP-9 both in vitro (A2780 cells) and in vivo in a tumor model comprising subcutaneous injection of A2780 cells in nude mice [123]. Lut exhibited a similar behavior in human melanoma cells (A375 cells), where it inhibits proliferation, induces apoptosis, and reduces the expression of MMP-2 and MMP-9 (in vitro and in vivo) [124].

Morin (Mor), Table 3, is a flavonol with various bioactive properties including neuroprotection, the suppression of inflammation, and anticancer activity [151]. An in vivo study was performed to evaluate the role of Mor in diethylnitrosamine (DEN)-induced hepatocarcinogenesis in Wistar albino rats. Both MMP-2 and MMP-9 levels were increased in DEN-induced animals when compared to a control. In animals treated with Mor, MMP-2, and MMP-9 levels were decreased [125]. Next, Mor was also tested in cultured LX-2 cells (hepatic stellate cells; HSCs) and diethylnitrosamine-induced fibrotic rats. Again, significantly decreased levels of MMP-2 and MMP-9 were found upon Mor treatment (200 mg/kg in drinking water) when compared to untreated cells and DEN-induced fibrotic rats [126].

Recently, Mor hydrate was studied in the metastasis of MCF-7 human breast cancer cells, where Mor hydrate suppressed 12-*O*-tetradecanoylphorbol-13-acetate (TPA)-induced cell migration and invasion via the inhibition of MMP-9 mRNA and protein expression [127].

Apigenin (Api), Table 3, a flavonoid present in vegetables, fruits, and herbs, possesses several bioactive properties, including anti-inflammatory, neuroprotective [152], and anticancer activities [153].

In colorectal adenocarcinoma cell lines (SW480 cells), Api reduced MMP-9 protein levels in a dose-dependent manner correlating with anti-metastasis and antitumor effects [128].

The positive action of Api was also evaluated in U87 glioma cells, where it was found to reduce tumor cell metastasis and invasion, inhibit MMP-9 mRNA levels, and downregulate nuclear factor-$_K$B (NF-$_K$B) [129], a known regulator of MMP-9 expression under inflammatory conditions. The anti-metastatic action of Api was also found in human melanoma cells (A375 cells), where Api reduces the MMP-2 and MMP-9 in a dose-dependent manner [130].

Fisetin (Fis), Table 3, is a bioactive flavonol found in several fruits and vegetables, and it is recognized for its anti-inflammatory, anti-proliferative, and neuroprotective effects [154,155].

Fis is able to dose-dependently inhibit MMP-9 protein and mRNA expression in a pancreatic cancer cell line (AsPC-1 cells) [131]. Recently, the effect of Fis on MMP-2 and MMP-9 expression in triple breast cancer cells (4T1 and JC cells) was investigated, with the findings that Fis reduced cell motility and that this phenomenon was partially associated with a reduction of MMP-2 and MMP-9 expressions [132].

Myricetin, (Myr), Table 3, is abundantly found in vegetables, fruits, teas, and some medicinal plants. Several studies have illustrated that Mir can exert anti-oxidant, anti-inflammatory, anti-cancer, and neuroprotective effects [156]. Myr acts as an anti-cancer agent through different mechanisms including the modulation of MMP-2 and MMP-9. It inhibits MMP-2 protein expression in colorectal carcinoma cells (COLO 205). Furthermore, one study reported that purified MMP-2 incubated with Myr had reduced activity when analyzed by gelatin gel zymography [133], which would suggest a direct and strong Myr/MMP-2 interaction. Recently, it has been reported that Myr suppresses breast cancer metastasis through the downregulation of the activity of MMP-2 and MMP-9 (MDA-Mb-231Br cells) [134]. In another study, the effect of Myr on the migration and invasion of radioresistant lung cancer cells (A549-IR cells) was investigated. Experimental evidence showed that Myr can inhibit the invasion and migration of A549-IR cells by suppressing the expression of MMP-2 and MMP-9 through the inhibition of the focal adhesion kinase FAK-ERK signaling pathway [135].

Recently, Myr has been studied in a pentylenetetrazole (PTZ)-induced mouse model of epilepsy (C57BL/6 male mice). It is known that PTZ-induction increases the expression of MMP-9 and that the selective inhibition of MMP-9 confers neuroprotection in patients with epilepsy. Interestingly, treatment with Myr (100–200 mg/kg, orally for 26 days, 30 min prior to each PTZ injection) reduced the mRNA and protein levels of MMP-9 in a dose-dependent manner, thus confirming the neuroprotective role of Myr [136].

Baicalein (Bai), Table 3, is produced by the root of Chinese skullcap, *Scutellaria baicalensis* Georgi (Lamiaceae), and it is a bioactive substance part of traditional Chinese medicine. Bai has several beneficial effects conferred through its anti-oxidant, anti-viral, anti-inflammatory, anti-angiogenic, and anti-cancer activities [157]. Moreover, studies have shown that Bas exerts a neuroprotective role in AD [158].

Several studies have reported that Bai acts as an anticancer agent through several pathways including the modulation of MMP-2 and MMP-9 expression. In a benzo(a)pyrene-induced pulmonary carcinogenesis mouse model, animals treated with Bai (12 mg/kg) had the significantly reduced mRNA and protein expression of MMP-2 and MMP-9 [137].

The anti-proliferative potential of Bai was also studied in melanoma cell lines (A375 and SK-MEL-28), where MMP-2 expression was significantly reduced in cells treated with Bai [138]. Moreover, Bai was tested in a pancreatic neuroendocrine tumor cell line (BON1 cells), and the observed reduction of tumor migration and invasion related to the decreased MMP-2 and MMP-9 [139]. Recently, it has been reported that Bai contributes to reduced metastasis in osteosarcoma. Upon treatment with Bai, the invasive capacity of human osteosarcoma cells (CRL-1427 cells) was reduced. This result was attributed to a series of enzymes modulated by Bai, including the reduced production of MMP-9 and MMP-2 [140].

Puerarin (Pue), Table 3, is part of the isoflavone glycoside group, and it is extracted from *Pueraria lobate*, *Pueraria thomsonii*, and *Pueraria tuberosa*. Pue was approved by the Chinese Ministry of Health for clinical treatment in 1993. It was primarily used for the treatment of cardiovascular diseases and later also reported to have anticancer activity [159]. Recently, studies have reported the ability of Pue to protect against AD [160]. Pue is known for its properties including bone-sparing, anti-inflammatory, and anti-proteinase properties. Rats with periodontitis that were treated with Pue showed a reduction of MMP-2 and MMP-9 expression [141]. Pue also significantly inhibited lipopolysaccharide (LPS)-induced MCF-7 and MDA-MB-231 cell migration, invasion, and adhesion. The mRNA and protein levels showed that Pue treatment effectively negated the expression of several proteins including MMP-2 and MMP-9 in LPS-activated cells [142].

Rutin, (Rut), Table 3, is a polyphenolic natural flavonoid known as quercetin-3-*O*-rutinoside and vitamin P, that is found in vegetables, citrus fruits, and plant-derived beverages. Rut has been largely studied for its several bioactive properties [161]. Rut was studied in vivo in a rat photothrombotic cerebral ischemic model, and the administration of Rut reduced BBB disruption via the downregulation of MMP-9 protein expression [143]. In mice with LPS-induced heart injury, Rut mitigated fibrosis-related genes, reduced MMP-2 and MMP-9 levels in the heart, and prevented LPS-induced cardiac fibrosis [144].

Naringin (Nar), Table 3, is a bioflavonoid compound especially found in grapefruit and is related to citrus herb species. It has an extensive spectrum of pharmacological activities such as anti-inflammatory, anti-oxidant, antitumor, and neuroprotective effects [162].

In a glioma cell line (U251 cells), Nar inhibited invasion and migration at several concentrations. In addition, a decrease in the levels of MMP-2 and MMP-9 was measured [145]. Similar results were obtained in human glioblastoma cells (U87 cells), where Nar exhibited inhibitory effects on the invasion and adhesion of U87 cells and reduced the protein levels of both MMP-2 and MMP-9 [146].

4. Conclusions and Future Prospective

AD is a chronic crippling disease for which the approved drugs are only palliative. In this manuscript, we briefly discuss the complex role that MMPs, specifically the gelatinases MMP-2 and MMP-9, have in neurodegeneration and neuroinflammation. It is clear that by modifying structural proteins and altering the functions of cytokines, MMPs contribute to the progression of neuropathology. In addition, in several in vivo studies, the beneficial effects of targeting MMPs in neuroinflammation and neurodegeneration were confirmed.

Both in clinical and preclinical studies, it was shown that many natural products from marine and terrestrial sources are promising bioactive substances for AD treatment. Interestingly, many of these natural products also have the ability to modulate MMP-2 and MMP-9. In this manuscript, we summarized the MMP-2 and MMP-9-modulatory activities of marine and terrestrial compounds with known beneficial effects on processes involved in AD pathology. Many natural compounds appear to regulate signal transduction pathways, thus leading to the simultaneous downregulation of MMP-2 and MMP-9 gene and protein expression. While such effects might correspond with general anti-oxidant and anti-inflammatory properties, for some compounds, direct anti-proteolytic activity on MMP-2 and/or MMP-9 has been established (e.g., ATPGDEG, LSGYGP, BU-4664L, ageladine A, quercetin and myricetin). In addition, several compounds require further research to decipher their exact regulator mechanisms. For example, it is imperative to determine whether their activity is due to the direct inhibition of the proteolytic mechanisms of MMPs or due to the direct or indirect modulation of signaling pathways upstream of MMP production.

In conclusion, natural compounds might represent a pipeline or 'blueprint' to develop new molecules that can modulate MMPs. Furthermore, the inhibition of MMPs in combination with other properties such as anti-inflammatory and anti-oxidant activity, as well as abilities such as diffusion across the blood–brain barrier, might provide valuable for treatments against AD progression.

Author Contributions: L.C. and J.V. writing—original draft preparation, L.C., J.V., S.N., and E.O. review and editing. All authors have read and agreed to the published version of the manuscript.

Funding: J.V. is a postdoctoral fellow, funded by the Fund for Scientific Research of Flanders (FWO-Vlaanderen, Grants 12Z0920N and G0A3820N) and the Belgian Charcot Foundation. This research was funded by the Italian *Ministero dell'Istruzione, dell'Università e della Ricerca* (PRIN 2017SNRXH3) and by the University of Pisa (PRA_2018_20).

Acknowledgments: The authors thank G. Opdenakker for useful suggestions and comments on the manuscript.

Conflicts of Interest: The authors declare no conflict of interest.

References

1. Alzheimer's Disease International World Alzheimer Report 2019. Available online: https://www.alz.co.uk/research/WorldAlzheimerReport2019.pdf (accessed on 20 September 2020).
2. Gan, L.; Cookson, M.R.; Petrucelli, L.; La Spada, A.R. Converging pathways in neurodegeneration, from genetics to mechanisms. *Nat. Neurosci.* **2018**, *21*, 1300–1309. [CrossRef]
3. Vaz, M.; Silvestre, S. Alzheimer's disease: Recent treatment strategies. *Eur. J. Pharmacol.* **2020**, *887*, 173554. [CrossRef]
4. Stanciu, G.D.; Luca, A.; Rusu, R.N.; Bild, V.; Chiriac, S.B.; Solcan, C.; Bild, W.; Ababei, D.C. Bild Alzheimer's Disease Pharmacotherapy in Relation to Cholinergic System Involvement. *Biomolecules* **2019**, *10*, 40. [CrossRef]
5. Ittner, L.M.; Götz, J. Amyloid-β and Tau—A Toxic Pas de Deux in Alzheimer's Disease. *Nat. Rev. Neuro Sci.* **2011**, *12*, 67–72. [CrossRef] [PubMed]
6. Liao, L.; Cheng, D.; Wang, J.; Duong, D.M.; Losik, T.G.; Gearing, M.; Rees, H.D.; Lah, J.J.; Levey, A.I.; Peng, J. Proteomic Characterization of Postmortem Amyloid Plaques Isolated by Laser Capture Microdissection. *J. Biol. Chem.* **2004**, *279*, 37061–37068. [CrossRef] [PubMed]
7. Levy, E.; Jaskolski, M.; Grubb, A. The Role of Cystatin C in Cerebral Amyloid Angiopathy and Stroke: Cell Biology and Animal Models. *Brain Pathol.* **2006**, *16*, 60–70. [CrossRef] [PubMed]
8. Wisniewski, T.; Golabek, A.A.; Kida, E.; Wisniewski, K.E.; Frangione, B. Conformational Mimicry in Alzheimer's Disease. Role of Apolipoproteins in Amyloidogenesis. *Am. J. Pathol.* **1995**, *147*, 238–244. [PubMed]

60. Agrawal, S.; Anderson, P.; Durbeej-Hjalt, M.; Van Rooijen, N.; Ivars, F.; Opdenakker, G.; Sorokin, L.M. Dystroglycan is selectively cleaved at the parenchymal basement membrane at sites of leukocyte extravasation in experimental autoimmune encephalomyelitis. *J. Exp. Med.* **2006**, *203*, 1007–1019. [CrossRef]

61. Cauwe, B.; Van den Steen, P.E.; Opdenakker, G. The Biochemical, Biological, and Pathological Kaleidoscope of Cell Surface Substrates Processed by Matrix Metalloproteinases. *Crit. Rev. Biochem. Mol. Biol.* **2007**, *42*, 113–185. [CrossRef]

62. Chong, Y.H.; Sung, J.H.; Shin, S.A.; Chung, J.H.; Suh, Y.H. Effects of the β-Amyloid and Carboxyl-terminal Fragment of Alzheimer's Amyloid Precursor Protein on the Production of the Tumor Necrosis Factor-α and Matrix Metalloproteinase-9 by Human Monocytic THP-1. *J. Biol. Chem.* **2001**, *276*, 23511–23517. [CrossRef] [PubMed]

63. Nübling, G.; Levin, J.; Bader, B.; Israel, L.; Bötzel, K.; Lorenzl, S.; Giese, A. Limited cleavage of tau with matrix-metalloproteinase MMP-9, but not MMP-3, enhances tau oligomer formation. *Exp. Neurol.* **2012**, *237*, 470–476. [CrossRef] [PubMed]

64. Terni, B.; Ferrer, I. Abnormal Expression and Distribution of MMP2 at Initial Stages of Alzheimer's Disease-Related Pathology. *J. Alzheimer's Dis.* **2015**, *46*, 461–469. [CrossRef] [PubMed]

65. Proost, P.; Struyf, S.; Van Damme, J.; Fiten, P.; Ugarte-Berzal, E.; Opdenakker, G. Chemokine isoforms and processing in inflammation and immunity. *J. Autoimmun.* **2017**, *85*, 45–57. [CrossRef] [PubMed]

66. Adelita, T.; Stilhano, R.S.; Han, S.W.; Justo, G.Z.; Porcionatto, M. Proteolytic processed form of CXCL12 abolishes migration and induces apoptosis in neural stem cells in vitro. *Stem Cell Res.* **2017**, *22*, 61–69. [CrossRef]

67. Zhang, K.; McQuibban, G.A.; Silva, C.; Butler, G.S.; Johnston, J.B.; Holden, J.; Clark-Lewis, I.; Overall, C.M.; Power, C. HIV-induced metalloproteinase processing of the chemokine stromal cell derived factor-1 causes neurodegeneration. *Nat. Neurosci.* **2003**, *6*, 1064–1071. [CrossRef]

68. Proost, P.; Vandamme, J.; Opdenakker, G. Leukocyte Gelatinase B Cleavage Releases Encephalitogens from Human Myelin Basic Protein. *Biochem. Biophys. Res. Commun.* **1993**, *192*, 1175–1181. [CrossRef]

69. Chandler, S.; Coates, R.; Gearing, A.; Lury, J.; Wells, G.; Bone, E. Matrix metalloproteinases degrade myelin basic protein. *Neurosci. Lett.* **1995**, *201*, 223–226. [CrossRef]

70. Kobayashi, H.; Chattopadhyay, S.; Kato, K.; Dolkas, J.; Kikuchi, S.-I.; Myers, R.R.; Shubayev, V.I. MMPs initiate Schwann cell-mediated MBP degradation and mechanical nociception after nerve damage. *Mol. Cell. Neurosci.* **2008**, *39*, 619–627. [CrossRef]

71. Wang, X.; Jung, J.; Asahi, M.; Chwang, W.; Russo, L.; Moskowitz, M.A.; Dixon, C.E.; Fini, M.E.; Lo, E.H. Effects of Matrix Metalloproteinase-9 Gene Knock-Out on Morphological and Motor Outcomes after Traumatic Brain Injury. *J. Neurosci.* **2000**, *20*, 7037–7042. [CrossRef]

72. Asahi, M.; Sumii, T.; Fini, M.E.; Itohara, S.; Lo, E.H. Matrix metalloproteinase 2 gene knockout has no effect on acute brain injury after focal ischemia. *NeuroReport* **2001**, *12*, 3003–3007. [CrossRef] [PubMed]

73. Garcia-Alloza, M.; Prada, C.; Lattarulo, C.; Fine, S.; Borrelli, L.A.; Betensky, R.; Greenberg, S.M.; Frosch, M.P.; Bacskai, B.J. Matrix Metalloproteinase Inhibition Reduces Oxidative Stress Associated with Cerebral Amyloid Angiopathy in Vivo in Transgenic Mice. *J. Neurochem.* **2009**, *109*, 1636–1647. [CrossRef] [PubMed]

74. Martins, M.; Silva, R.; Pinto, M.; Sousa, E. Marine Natural Products, Multitarget Therapy and Repurposed Agents in Alzheimer's Disease. *Pharmaceuticals* **2020**, *13*, 242. [CrossRef] [PubMed]

75. Thomas, N.V.; Kim, S.-K. Metalloproteinase Inhibitors: Status and Scope from Marine Organisms. *Biochem. Res. Int.* **2010**, *2010*, 845975. [CrossRef] [PubMed]

76. Kumar, G.B.; Nair, B.G.; Perry, J.J.P.; Martin, D.B.C. Recent insights into natural product inhibitors of matrix metalloproteinases. *MedChemComm* **2019**, *10*, 2024–2037. [CrossRef] [PubMed]

77. Gentile, E.; Liuzzi, G.M. Marine pharmacology: Therapeutic targeting of matrix metalloproteinases in neuroinflammation. *Drug Discov. Today* **2017**, *22*, 299–313. [CrossRef]

78. Vandooren, J.; Geurts, N.; Martens, E.; Van den Steen, P.E.; Opdenakker, G. Zymography Methods for Visualizing Hydrolytic Enzymes. *Nat. Methods* **2013**, *10*, 211. [CrossRef]

79. De Morais, M.G.; da Fontoura Prates, D.; Moreira, J.B.; Duarte, J.H.; Costa, J.A.V. Phycocyanin from Microalgae: Properties, Extraction and Purification, with Some Recent Applications. *Ind. Biotechnol.* **2018**, *14*, 30–37. [CrossRef]

80. David, L.; Marx, A.; Adir, N. High-Resolution Crystal Structures of Trimeric and Rod Phycocyanin. *J. Mol. Biol.* **2011**, *405*, 201–213. [CrossRef]

81. Jiang, L.; Wang, Y.; Yin, Q.; Liu, G.; Liu, H.; Huang, Y.; Liangqian, J. Phycocyanin: A Potential Drug for Cancer Treatment. *J. Cancer* **2017**, *8*, 3416–3429. [CrossRef]

82. Pabón, M.M.; Jernberg, J.N.; Morganti, J.; Contreras, J.; Hudson, C.E.; Klein, R.L.; Bickford, P.C. A Spirulina-Enhanced Diet Provides Neuroprotection in an α-Synuclein Model of Parkinson's Disease. *PLoS ONE* **2012**, *7*, e45256. [CrossRef] [PubMed]

83. Pentón-Rol, G.; Martínez-Sánchez, G.; Cervantes-Llanos, M.; Lagumersindez-Denis, N.; Acosta-Medina, E.F.; Falcón, V.; Alonso-Ramirez, R.; Valenzuelasilva, C.M.; Rodríguez-Jiménez, E.; Llópiz-Arzuaga, A.; et al. C-Phycocyanin ameliorates experimental autoimmune encephalomyelitis and induces regulatory T cells. *Int. Immunopharmacol.* **2011**, *11*, 29–38. [CrossRef] [PubMed]

84. Garbuzova-Davis, S.; Bickford, P.C. Short Communication: Neuroprotective Effect of Spirulina in a Mouse Model of ALS. *Open Tissue Eng. Regen. Med. J.* **2010**, *3*, 36–41. [CrossRef]

85. Ravi, M.; Tentu, S.; Baskar, G.; Prasad, S.R.; Raghavan, S.; Jayaprakash, P.; Jeyakanthan, J.; Rayala, S.K.; Venkatraman, G. Molecular mechanism of anti-cancer activity of phycocyanin in triple-negative breast cancer cells. *BMC Cancer* **2015**, *15*, 768. [CrossRef] [PubMed]

86. Saini, M.K.; Sanyal, S.N. Targeting angiogenic pathway for chemoprevention of experimental colon cancer using C-phycocyanin as cyclooxygenase-2 inhibitor. *Biochem. Cell Biol.* **2014**, *92*, 206–218. [CrossRef]

87. Kunte, M. The Inhibitory Effect of C-Phycocyanin Containing Protein Extract (C-PC Extract) on Human Matrix Metalloproteinases (MMP-2 and MMP-9) in Hepatocellular Cancer Cell Line (HepG2). *Protein J.* **2017**, *36*, 186–195. [CrossRef] [PubMed]

88. Mitra, S.; Siddiqui, W.A.; Khandelwal, S. C-Phycocyanin protects against acute tributyltin chloride neurotoxicity by modulating glial cell activity along with its anti-oxidant and anti-inflammatory property: A comparative efficacy evaluation with N-acetyl cysteine in adult rat brain. *Chem. Interact.* **2015**, *238*, 138–150. [CrossRef]

89. Kim, Y.-S.; Kim, E.-K.; Tang, Y.; Hwang, J.-W.; Natarajan, S.B.; Kim, W.-S.; Moon, S.-H.; Jeon, B.-T.; Park, P.-J. Antioxidant and anticancer effects of extracts from fermented Haliotis discus hannai with Cordyceps militaris mycelia. *Food Sci. Biotechnol.* **2016**, *25*, 1775–1782. [CrossRef]

90. Nguyen, V.-T.; Qian, Z.-J.; Ryu, B.; Kim, K.-N.; Kim, D.; Kim, Y.-M.; Jeon, Y.-J.; Park, W.S.; Choi, I.-W.; Kim, G.H.; et al. Matrix metalloproteinases (MMPs) inhibitory effects of an octameric oligopeptide isolated from abalone Haliotis discus hannai. *Food Chem.* **2013**, *141*, 503–509. [CrossRef]

91. Gong, F.; Chen, M.-F.; Zhang, Y.-Y.; Li, C.-Y.; Zhou, C.-X.; Hong, P.-Z.; Sun, S.-L.; Qian, Z.-J. A Novel Peptide from Abalone (*Haliotis discus hannai*) to Suppress Metastasis and Vasculogenic Mimicry of Tumor Cells and Enhance Anti-Tumor Effect In Vitro. *Mar. Drugs* **2019**, *17*, 244. [CrossRef]

92. Chen, J.; Liang, P.; Xiao, Z.; Chen, M.-F.; Gong, F.; Li, C.; Zhou, C.; Hong, P.; Jung, W.-K.; Qian, Z.-J. Antiphotoaging effect of boiled abalone residual peptide ATPGDEG on UVB-induced keratinocyte HaCaT cells. *Food Nutr. Res.* **2019**, *63*. [CrossRef] [PubMed]

93. Gong, F.; Chen, M.-F.; Chen, J.; Li, C.; Zhou, C.; Hong, P.; Sun, S.; Qian, Z.-J. Boiled Abalone Byproduct Peptide Exhibits Anti-Tumor Activity in HT1080 Cells and HUVECs by Suppressing the Metastasis and Angiogenesis in Vitro. *J. Agric. Food Chem.* **2019**, *67*, 8855–8867. [CrossRef] [PubMed]

94. Ma, Q.; Liu, Q.; Yuan, L.; Zhuang, Y. Protective Effects of LSGYGP from Fish Skin Gelatin Hydrolysates on UVB-Induced MEFs by Regulation of Oxidative Stress and Matrix Metalloproteinase Activity. *Nutrients* **2018**, *10*, 420. [CrossRef] [PubMed]

95. Wang, H.; Wei, J.; Wu, N.; Liu, M.; Wang, C.; Zhang, Y.; Wang, F.; Liu, H.; Lin, X. Mere15, a novel polypeptide from Meretrix meretrix, inhibits adhesion, migration and invasion of human lung cancer A549 cells via down-regulating MMPs. *Pharm. Biol.* **2012**, *51*, 145–151. [CrossRef]

96. Igarashi, Y.; Miyanaga, S.; Onaka, H.; Takeshita, M.; Furumai, T. Revision of the Structure Assigned to the Antibiotic BU-4664L from Micromonopora. *J. Antibiot.* **2005**, *58*, 350–352. [CrossRef] [PubMed]

97. Miyanaga, S.; Sakurai, H.; Saiki, I.; Onaka, H.; Igarashi, Y. Anti-invasive and anti-angiogenic activities of naturally occurring dibenzodiazepine BU-4664L and its derivatives. *Bioorganic Med. Chem. Lett.* **2010**, *20*, 963–965. [CrossRef]

98. Fujita, M.; Nakao, Y.; Matsunaga, S.; Seiki, M.; Itoh, Y.; Yamashita, J.; Van Soest, R.W.M.; Fusetani, N. Ageladine A: An Antiangiogenic Matrixmetalloproteinase Inhibitor from the Marine SpongeAgelasnakamurai1. *J. Am. Chem. Soc.* **2003**, *125*, 15700–15701. [CrossRef]

99. Martínez-Poveda, B.; García-Vilas, J.A.; Cárdenas, C.; Melgarejo, E.; Quesada, A.R.; Medina, M.A. The Brominated Compound Aeroplysinin-1 Inhibits Proliferation and the Expression of Key Pro-Inflammatory Molecules in Human Endothelial and Monocyte Cells. *PLoS ONE* **2013**, *8*, e55203. [CrossRef]

100. Lee, H.-P.; Lin, Y.-Y.; Duh, C.-Y.; Huang, S.-Y.; Wang, H.-M.; Wu, S.-F.; Lin, S.-C.; Jean, Y.-H.; Wen, Z.-H. Lemnalol attenuates mast cell activation and osteoclast activity in a gouty arthritis model. *J. Pharm. Pharmacol.* **2015**, *67*, 274–285. [CrossRef]

101. Di Bari, G.; Gentile, E.; Latronico, T.; Corriero, G.; Fasano, A.; Marzano, C.N.; Liuzzi, G.M. Inhibitory Effect of Aqueous Extracts from Marine Sponges on the Activity and Expression of Gelatinases A (MMP-2) and B (MMP-9) in Rat Astrocyte Cultures. *PLoS ONE* **2015**, *10*, e0129322. [CrossRef]

102. Lin, J.-J.; Su, J.-H.; Tsai, C.-C.; Chen, Y.-J.; Liao, M.-H.; Wu, Y.-J. 11-epi-Sinulariolide Acetate Reduces Cell Migration and Invasion of Human Hepatocellular Carcinoma by Reducing the Activation of ERK1/2, p38MAPK and FAK/PI3K/AKT/mTOR Signaling Pathways. *Mar. Drugs* **2014**, *12*, 4783–4798. [CrossRef] [PubMed]

103. Cheng, T.-C.; Din, Z.-H.; Su, J.-H.; Wu, Y.-J.; Liu, C.-I. Sinulariolide Suppresses Cell Migration and Invasion by Inhibiting Matrix Metalloproteinase-2/-9 and Urokinase through the PI3K/AKT/MTOR Signaling Pathway in Hu-man Bladder Cancer Cells. *Mar. Drugs* **2017**, *15*, 238. [CrossRef] [PubMed]

104. Chen, S.-C.; Chien, Y.-C.; Pan, C.-H.; Sheu, J.-H.; Chen, C.-Y.; Wu, C.-H. Inhibitory Effect of Dihydroaustrasulfone Alcohol on the Migration of Human Non-Small Cell Lung Carcinoma A549 Cells and the Antitumor Effect on a Lewis Lung Carcinoma-Bearing Tumor Model in C57BL/6J Mice. *Mar. Drugs* **2014**, *12*, 196–213. [CrossRef] [PubMed]

105. Li, P.-C.; Sheu, M.-J.; Ma, W.-F.; Pan, C.-H.; Sheu, J.-H.; Wu, C.-H. Anti-Restenotic Roles of Dihydroaustrasulfone Alcohol Involved in Inhibiting PDGF-BB-Stimulated Proliferation and Migration of Vascular Smooth Muscle Cells. *Mar. Drugs* **2015**, *13*, 3046–3060. [CrossRef]

106. Ciccone, L.; Tonali, N.; Nencetti, S.; Orlandini, E. Natural compounds as inhibitors of transthyretin amyloidosis and neuroprotective agents: Analysis of structural data for future drug design. *J. Enzym. Inhib. Med. Chem.* **2020**, *35*, 1145–1162. [CrossRef]

107. Ortore, G.; Orlandini, E.; Braca, A.; Ciccone, L.; Rossello, A.; Martinelli, A.; Nencetti, S. Targeting Different Transthyretin Binding Sites with Unusual Natural Compounds. *ChemMedChem* **2016**, *11*, 1865–1874. [CrossRef]

156. Gupta, G.; Siddiqui, M.A.; Khan, M.M.; Ajmal, M.; Ahsan, R.; Rahaman, M.A.; Ahmad, M.A.; Arshad, M.; Khushtar, M. Current Pharmacological Trends on Myricetin. *Drug Res.* **2020**, *70*, 448–454. [CrossRef] [PubMed]
157. Tuli, H.S.; Aggarwal, V.; Kaur, J.; Aggarwal, D.; Parashar, G.; Parashar, N.C.; Tuorkey, M.; Kaur, G.; Savla, R.; Sak, K.; et al. Baicalein: A metabolite with promising antineoplastic activity. *Life Sci.* **2020**, *259*, 118183. [CrossRef] [PubMed]
158. Sonawane, S.K.; Balmik, A.A.; Boral, D.; Ramasamy, S.; Chinnathambi, S. Baicalein suppresses Repeat Tau fibrillization by sequestering oligomers. *Arch. Biochem. Biophys.* **2019**, *675*, 108119. [CrossRef]
159. Ahmad, B.; Khan, S.; Liu, Y.; Xue, M.; Nabi, G.; Kumar, S.; Alshwmi, M.; Qarluq, A.W. Molecular Mechanisms of Anticancer Activities of Puerarin. *Cancer Manag. Res.* **2020**, *12*, 79–90. [CrossRef]
160. Yao, Y.; Chen, X.; Bao, Y.; Wu, Y. Puerarin inhibits β-amyloid peptide 1-42-induced tau hyperphosphorylation via the Wnt/β-catenin signaling pathway. *Mol. Med. Rep.* **2017**, *16*, 9081–9085. [CrossRef]
161. Negahdari, R.; Bohlouli, S.; Sharifi, S.; Dizaj, S.M.; Saadat, Y.R.; Khezri, K.; Jafari, S.; Ahmadian, E.; Jahandizi, N.G.; Raeesi, S. Therapeutic benefits of rutin and its nanoformulations. *Phytother. Res.* **2020**, 6904. [CrossRef]
162. Ahmed, S.; Khan, H.; Aschner, M.; Hasan, M.M.; Hassan, S.T. Therapeutic potential of naringin in neurological disorders. *Food Chem. Toxicol.* **2019**, *132*, 110646. [CrossRef]

pharmaceuticals

Communication

(2-Aminobenzothiazole)-Methyl-1,1-Bisphosphonic Acids: Targeting Matrix Metalloproteinase 13 Inhibition to the Bone

Antonio Laghezza [1], Luca Piemontese [1], Leonardo Brunetti [1], Alessia Caradonna [1], Mariangela Agamennone [2], Fulvio Loiodice [1,*] and Paolo Tortorella [1,*]

[1] Department of Pharmacy and Pharmaceutical Sciences, University of Bari "A. Moro", via E. Orabona 4, 70125 Bari, Italy; antonio.laghezza@uniba.it (A.L.); luca.piemontese@uniba.it (L.P.); leonardo.brunetti@uniba.it (L.B.); a.caradonna@hotmail.it (A.C.)

[2] Department of Pharmacy, University "G. d'Annunzio" of Chieti-Pescara, Via Dei Vestini, 31, 66100 Chieti, Italy; magamennone@unich.it

* Correspondence: fulvio.loiodice@uniba.it (F.L.); paolo.tortorella@uniba.it (P.T.)

Abstract: Matrix Metalloproteinases (MMPs) are a family of secreted and membrane-bound enzymes, of which 24 isoforms are known in humans. These enzymes degrade the proteins of the extracellular matrix and play a role of utmost importance in the physiological remodeling of all tissues. However, certain MMPs, such as MMP-2, -9, and -13, can be overexpressed in pathological states, including cancer and metastasis. Consequently, the development of MMP inhibitors (MMPIs) has been explored for a long time as a strategy to prevent and hinder metastatic growth, but the important side effects linked to promiscuous inhibition of MMPs prevented the clinical use of MMPIs. Therefore, several strategies were proposed to improve the therapeutic profile of this pharmaceutical class, including improved selectivity toward specific MMP isoforms and targeting of specific organs and tissues. Combining both approaches, we conducted the synthesis and preliminary biological evaluation of a series of (2-aminobenzothiazole)-methyl-1,1-bisphosphonic acids active as selective inhibitors of MMP-13 via in vitro and in silico studies, which could prove useful for the treatment of bone metastases thanks to the bone-targeting capabilities granted by the bisphosphonic acid group.

Keywords: Matrix Metalloproteinase inhibitors; bone targeting; bisphosphonates; antitumor agents; skeletal malignancies

Citation: Laghezza, A.; Piemontese, L.; Brunetti, L.; Caradonna, A.; Agamennone, M.; Loiodice, F.; Tortorella, P. (2-Aminobenzothiazole)-Methyl-1,1-Bisphosphonic Acids: Targeting Matrix Metalloproteinase 13 Inhibition to the Bone. *Pharmaceuticals* **2021**, *14*, 85. https://doi.org/10.3390/ph14020085

Academic Editor: Salvatore Santamaria

Received: 18 December 2020
Accepted: 20 January 2021
Published: 24 January 2021

Publisher's Note: MDPI stays neutral with regard to jurisdictional claims in published maps and institutional affiliations.

1. Introduction

The remodeling of the extracellular matrix (ECM) is an essential process for the development and maintenance of numerous organs and tissues, such as bone. This process is regulated by a variety of mediators and enzymes, among which Matrix Metalloproteinases (MMPs) are widely considered to be the most important players [1–3]. Their expression is finely regulated, and their activity is also modulated by tissue inhibitors of MMPs (TIMPs) [3].

In bone, MMPs are physiologically involved in several processes, including cellular differentiation of osteoblasts, bone formation, bone resorption, and osteoclast recruitment and migration. During the metastatic process, MMPs are involved in the so-called "vicious cycle" established by malignant cells. In order to carve out the metastatic niche, these cells trigger RANK-Ligand (RANKL) production by osteoblasts, which promotes osteoclast differentiation and bone matrix resorption; the degradation of the extracellular matrix results in the release of growth factors, driving forward tumor growth and fueling the vicious cycle [4–6].

Bone metastases are linked to so-called skeletal-related events, which include spinal cord compression and bone fractures. Moreover, they cause chronic pain and bone marrow aplasia and impair patient mobility, resulting in a severe reduction in quality of life. A wide range of treatments is available for bone metastasis (e.g., radiotherapy, chemotherapy, endocrine treatments, and orthopedic intervention). In this context, bisphosphonates, such as zoledronic acid, were standard-of-care for more than a decade and currently share the spotlight with RANKL-targeted monoclonal antibodies, such as Denosumab, for the treatment of this condition. Regrettably, although these therapeutic options significantly reduce the morbidity related to bone metastases, they are generally palliative and do not lead to tumor eradication [7–10].

Likewise, the ongoing research efforts toward the development of Matrix Metalloproteinase Inhibitors (MMPIs) have not yet resulted in satisfying clinical applications, mostly due to a lack of selectivity caused by the hydroxamic zinc-binding group (ZBG), typical of the first generation of MMPIs. Indeed, indiscriminate inhibition of MMPs and other zinc proteases leads to severe adverse reactions, ranging from pain to musculoskeletal syndromes, since zinc ions are essential to an enormous variety of biological processes [1,11]. Therefore, in the last decade, most research efforts focused on curbing the adverse effects of these compounds by acting on their selectivity or by targeting them to specific tissues, such as bone. This was achieved either by designing nonzinc-binding MMPIs, which target allosteric sites, or by developing compounds containing novel, more selective ZBGs [1,3,12].

Bisphosphonic acids belong to the latter class of compounds due to the bone-seeking nature of the bisphosphonic moiety itself. In the last few years, they proved to be a reliable and versatile scaffold for the development of novel bone-seeking MMPIs [13–16]. Lead compound ML115 showed high potency and was also endowed with antiosteoclastic activity, which could reduce skeletal disease burden in patients with conditions involving abnormal bone resorption [8,17]. Further developments along this line of research showed that modifications on the arylbisphosphonic scaffold could afford compounds which selectively inhibit MMP-2 [8,15,17] and MMP-9 [18], and that also exhibit significant potential for bone malignancy therapy, being superior at promoting cancer apoptosis than standard-of-care bisphosphonates, such as zoledronic acid [8,17]. Moreover, these bisphosphonic MMPIs (BMMPIs) showed no particular side effects in vivo at therapeutic dosages [8].

The significant inhibitory potency of ML115 toward MMP-2 and MMP-8 can be explained by the increased hydrophobic interactions that the biphenyl group forms with the deep S1$'$ sites of such MMP isoforms [19]. However, inhibition of MMP-13, an enzyme with an equally deep S1$'$ pocket [20], has not yet been investigated as part of the activity spectrum of bisphosphonic acids. This MMP isoform is strongly overexpressed in the metastatic microenvironment, where it plays a central role by activating MMP-9 and osteolysis, resulting in the release of growth factors that further stimulate cancer cell proliferation. Indeed, in recent years, various studies validated MMP-13 as a therapeutic target for the treatment of bone metastases [21–24]. Following this particular line of research, we report the development of a series of (2-aminobenzothiazole)methyl-1,1-bisphosphonic acids with particular attention to their activity as MMP-13 inhibitors, and provide a rationalization of their activity profile via computational studies.

2. Results and Discussion

Compounds **1–12** (reported in Table 1) were specifically developed to evaluate how the substitution of the biphenylsulfonamide of ML115 with a 2-aminobenzothiazole moiety could modify the MMP inhibition profile of these compounds, and whether substitutions on the benzothiazole scaffold could be used to further modulate their activity.

Table 1. Compounds 1–12.

ML115

1: R = H
2: R = F
3: R = Cl
4: R = NO$_2$

5: R = H
6: R = Br
7: R = NO$_2$

8

9

10

11

12

Treatment of the appropriate commercially available 2-aminobenzothiazole with triethyl orthoformate and diethyl phosphite afforded tetraethyl-1,1-bisphosphonates 1a–4a. These bisphosphonic esters were dealkylated with trimethylsilyl bromide (TMBS) in anhydrous acetonitrile, affording bisphosphonic acids 1–4. Intermediate 4a, bearing a nitro group, was also separately reduced to compound **5b**, which was the key to the synthesis of compounds **5–11**. For compounds **5–8**, the precursor was acylated with the appropriate acyl chloride and subsequently dealkylated with TMBS. Compounds **9–11** were instead obtained via condensation with phenyl isocyanate, phthalic anhydride or Boc-L-Ala and subsequent dealkylation of the bisphosphonic ester intermediates with TMBS (Figure 1).

Compound **12** required a different synthetic route, involving the preparation of a tetraethyl ethyleniden-bisphosphonate (**12b**), which was condensed with 2-amino-6-chlorobenzothiazole, affording intermediate **12a**, whose dealkylation in acidic conditions resulted in the desired product **12** (Figure 2).

chromatography was performed using Fluka silica gel 60 Å (63–200 μm) or silica gel Si 60 (40–63 μm). Mass spectra were recorded on an HP MS 6890-5973 MDS spectrometer, electron impact 70 eV, equipped with an HP ChemStation or with an Agilent 6530 Series Accurate-Mass Quadrupole Time-of-FLIGHT (Q-TOF) LC/MS. High-resolution mass spectrometry (HRMS) analyses were performed using a Bruker microTOF QII mass spectrometer equipped with an electrospray ion source (ESI). ^{1}H NMR was recorded using the suitable deuterated solvent on a Varian Mercury 300 or 500 NMR Spectrometer. Chemical shift (δ) was expressed as parts per million (ppm) and the coupling constant (J) in Hertz (Hz). Melting points were determined in open capillaries on a Gallenkamp electrothermal apparatus and were uncorrected.

3.2.1. General Procedure for the Preparation of Tetraethyl Bisphosphonates (1a–4a)

Triethyl orthoformate, diethyl phosphite and the opportune 2-aminobenzothiazole were added to a round-bottom flask fitted with a distillation apparatus in a 1.2:3:1 stoichiometric ratio. The resulting mixture was then heated to 160 °C until all EtOH was distilled away; the residue was dissolved in ethyl acetate, and evaporated away under vacuum, affording a crude yellow oil, which was further purified via column chromatography over silica gel (eluent EtOAc/MeOH 9:1). The desired compounds were obtained as white solids in 15–53% yields.

Tetraethyl [(benzo[*d*]thiazol-2-amino)methyl]-1,1-bisphosphonate (1a): White solid, 42% yield. 1H NMR (500 MHz, CDCl$_3$): δ = 1.26–1.32 (m, 12H, CH$_3$), 4.17–4.30 (m, 8H, CH$_2$), 5.26 (t, J$_{HP}$ = 21.52, 1H, PCHP), 6.17 (bs, 1H, NH), 7.12 (t, J = 7.83,1H, aromatic), 7.30 (t, J = 7.83,1H, aromatic), 7.55–7.59 (m, 2H, aromatics). MS (ESI): *m/z*: 437[M+H]+; MS2 *m/z* (%): 299(46), 271 (88), 243(100), 225(66), 161 (39).

Tetraethyl [(6-fluorobenzo[*d*]thiazol-2-amino)methyl]-1,1-bisphosphonate (2a): White solid, 53% yield. 1H NMR (500 MHz, CDCl$_3$): δ = 1.26–1.31 (m, 12H, CH$_3$), 4.18–4.27 (m, 8H, CH$_2$), 5.19 (t, 1H, PCHP), 5.95 (bs, 1H, NH), 7.00–7.03 (m, 1H, aromatic), 7.26–7.29 (m, 1H, aromatic), 7.45–7.48 (m, 1H, aromatic). MS(ESI): *m/z*: 455 [M+H]+; MS2: *m/z* (%): 317(38), 289(89), 261(100), 243 (75), 179(39).

Tetraethyl [(6-chlorobenzo[*d*]thiazol-2-amino)methyl]-1,1-bisphosphonate (3a): White solid, 15% yield. 1H NMR(500 MHz, CDCl$_3$): δ = 1.26–1.31 (m, 12 H, CH$_3$), 4.16–4.29 (m, 8H, CH$_2$), 5.19 (t, J$_{HP}$ = 21.53, 1H, PCHP), 5.80 (bs, 1H, NH), 7.24–7.27 (dd, J$_1$ = 8.81, J$_2$ = 2.45, 1H, aromatic), 7.44 (d, J = 8.81, 1H, aromatic), 7.54 (d, J = 2.45, 1H, aromatic). MS(ESI): *m/z*: 495 [M+2+Na]+, 493[M+Na]+, 473[M+2+H]+, 471[M+H]+; MS2(471): *m/z* (%): 335 (27), 333(43), 307(49), 305(86), 279 (45), 277(100), 197 (18), 195 (41).

Tetraethyl [(6-nitrobenzo[d]thiazol-2-amino)methyl]-1,1-bisphosphonate (4a): Yellow solid, 40% yield. 1H NMR(500 MHz, CDCl3): δ = 1.26 (t, J = 7.095, 6H, CH3), 1.33 (t, J = 7.095, 6H, CH3), 1.7(bs, 1H, NH), 4.18–4.32 (m, 8H, CH$_2$), 5.29 (t, J$_{HP}$ = 21.53, 1H, PCHP), 7.55 (d, J = 8.81, 1H, aromatic), 8.19–8.21(dd, J$_1$ = 8.81, J$_2$ = 2.45, 1H, aromatic), 8.48 (d, J = 2.45, 1H, aromatic). MS(ESI): *m/z*: 482[M+H]+; MS2: *m/z* (%): 344(18), 316.01(69), 288 (100), 272 (12), 206 (25), 139 (16).

Tetraethyl [(6-aminobenzo[d]thiazol-2-amino)methyl]-1,1-bisphosphonate (5b): Ten percent Pd/C (0.20 mmol) was added to the solution of the nitro compound (0.532 mmol) in 8.5 mL EtOH, and the mixture was hydrogenated at room temperature at a pressure of 3 bar for 12 h. The reaction mixture was filtered through a pad of celite and the filtrate was evaporated in vacuo to give an oil. The residue was purified by chromatography on silica gel (eluent: CHCl$_3$/ MeOH 9.5: 0.5 *v/v*) to give the desired amino derivate. White solid, 77% yield. 1H NMR (CDCl$_3$, 500 MHz): δ = 1.25–1.30 (m, 12H, CH$_3$), 3.64 (b, 2H, NH$_2$), 4.09–4.27 (m, 8H, CH$_2$), 5.15 (t, J$_{HP}$ = 21.53, 1H, PCHP), 5.52 (bs, 1H, NH), 6.67 (dd, J$_1$ = 8.32, J$_2$ = 2.45, 1H, aromatic), 6.91 (d, J = 2.45, 1H, aromatic), 7.35 (d, J = 8.32, 1H, aromatic). MS(ESI): *m/z*: 474[M+Na]$^+$; MS$_2$: *m/z* (%): 336 (100), 308 (73), 262 (32), 174 (24), 146 (22).

3.2.2. General Procedure for N-Acylation of Compound 5b

The appropriate acyl chloride (1.1–2 mmol) and triethylamine (2 mmol) were added to the solution of amino derivate (5b) (1 mmol) in anhydrous THF and the mixture was stirred at room temperature on nitrogen or argon for 3–12 h. After the given time, the eluent was evaporated in vacuo and the residue was partitioned between EtOAc and $NaHCO_3$ and the layers were separated. The organic phase was washed with HCl 1N, NH_4Cl ss, brine, dried over anhydrous Na_2SO_4, filtered and the filtrate was evaporated in vacuo. The residue was purified by chromatography on silica gel (eluent: $CHCl_3$/MeOH 98:2 v/v or AcOEt/MeOH 9:1 v/v) or crystalized with AcOEt to give the desired product.

Tetraethyl [(6-(benzamido)benzo[d]thiazol-2-amino)methyl]-1,1-bisphosphonate: White solid, 61% yield (chromatography, eluent: $CHCl_3$/MeOH 98: 2 v/v). 1H NMR (500 MHz, [D6] DMSO): δ = 1.13–1.97 (m, 12H, CH_3), 4.06–4.12 (m, 8H, CH_2), 5.09–5.21 (td, J_{HP} = 22.5, J_{HH} = 9.79, 1H, PCHP), 7.40 (d, J = 8.81, 1H, aromatic), 7.50–7.53 (m, 4H, aromatics), 7.92–7.94 (m, 2H, aromatics), 8.16 (d, J = 1.96, 1H, aromatic), 8.73 (d, J = 9.78, 1H, NH), 10.25 (s, 1H, NH). MS (ESI): m/z: 578 [M+Na]$^+$, 556 [M+H]$^+$; MS_2(556): m/z (%): 418 (100), 390 (92), 372 (33), 381 (33), 362 (57), 344 (73), 280 (29). 554 [M-H]$^-$.

Tetraethyl [(6-(4-bromobenzamido)benzo[d]thiazol-2-amino)methyl]-1,1-bisphosphonate: White solid, 82% yield, (AcOEt). 1H NMR(500 MHz, [D6] DMSO): δ = 1.13–1.21 (m, 12H, CH_3), 4.07–4.10 (m, 8H, CH_2), 5.15 (td, J_{HP} = 22.5, J_{HH} = 9.30, 1H, PCHP), 7.40 (d, J = 8.81, 1H, aromatic), 7.51 (d, J = 8.81, 1H, aromatic), 7.73 (d, J = 8.32, 2H, aromatics), 7.91(d, J = 8.32, 2H, aromatics), 8.16 (s, 1H, aromatic), 8.75 (d, J = 9.30, NH), 10.37 (s, 1H, NH). MS(ESI): m/z: 658[M+2+Na]$^+$, 656[M+Na]$^+$, 636[M+2+H]$^+$, 634 [M+H]$^+$; MS2(634): m/z (%): 498(100), 496 (70), 470 (55), 468 (62), 452 (40), 450 (32), 424 (55), 422 (47).

Tetraethyl [(6-(4-nitrobenzamido)benzo[d]thiazol-2-amino)methyl]-1,1-bisphosphonate: Yellow solid, 64% yield, (AcOEt). 1H NMR (500 MHz, [D6] DMSO): δ = 1.15–1.20 (m, 12H, CH_3), 4.07–4.12 (m, 8H, CH_2), 5.15 (t, J_{HP} = 22.5, 1H, PCHP), 7.43 (d, J = 8.81, 1H, aromatic), 7.51–7.53 (dd, J_1 = 8.32, J_2 = 1.96, 1H, aromatic), 8.16–8.18 (m, 3H, aromatics), 8.68 (m, 2H, aromatics), 8.8 (bs, 1H, NH), 10.59 (s, 1H, NH). MS(ESI): m/z: 623[M+Na]$^+$, 601[M+H]$^+$; MS2: m/z (%): 555(24), 463 (100), 436(18.25), 389 (69), 325 (25).

Tetraethyl [(6-(2-phenylacetamido)benzo[d]thiazol-2-amino)methyl]-1,1-bisphosphonate: White solid, 56% yield, (AcOEt). 1H NMR (500 MHz, [D6] DMSO): δ = 1.12–1.18 (m, 12H, CH_3), 3.60 (s, 2H, CH_2), 3.97–4.17 (m, 8H, CH_2), 5.13 (td, J_{HP} = 22.25, J_{HH} = 9.78, 1H, PCHP), 7.22–7.36 (m, 7H, aromatics), 8.03 (s, 1H, aromatic), 8.67 (d, J = 9.78, 1H, NH), 10.15 (s, 1H, NH). MS (ESI): m/z: 592[M+Na]$^+$, 570[M+H]+; MS2 (570): m/z (%): 432(100), 404 (81), 386 (30), 358(61), 294(24). 568[M-H]$^-$; MS2: m/z (%): 430(89), 339 (90), 310 (100), 287 (49), 259 (27), 190 (42), 185 (16), 137 (97), 107 (33).

Tetraethyl [(6-(3-phenylureido)benzo[d]thiazol-2-amino)methyl]-1,1-bisphosphonate: A solution of phenyl isothiocyanate (1.2 mmol) in anhydrous toluene (2 mL) was added to the suspension of 5b (1 mmol) in anhydrous toluene (2 mL) and the mixture was heated to reflux for 2 h. After the given time, the eluent was evaporated in vacuo and the residue was triturated with AcOEt and filtered to afford the desired product: white solid, 57% yield. 1H NMR(300 MHz, [D6]DMSO): δ = 1.12–1.20 (m, 12H, CH_3), 4.06–4.13 (m, 8H, CH_2), 5.05–5.13 (td, J_{HP} = 22.26, J_{HH} = 9.96, 1H, PCHP), 6.91–9.95 (t, J = 7.02, 1H, aromatic), 7.12–7.28 (m, 3H, aromatics), 7.34 (d, J = 8.787, 1H, aromatic), 7.43 (d, J = 8.20, 2H, aromatics), 7.88 (d, J = 2.34, 1H, aromatic), 8.62 (d, J = 9.96, 1H, NH), 8.72 (s, 1H, NH), 8.75 (s, 1H, NH). MS(ESI): m/z: 593[M+Na]$^+$, 571[M+H]+; MS_2: m/z (%): 525 (17), 433 (100), 405 (77), 378 (32), 359 (58).

Tetraethyl [(6-(1,3-dioxoisoindolin-2-yl)benzothiazol-2-amino)methyl]-1,1-bisphosphonate: A mixture of 5b (1 mmol) and phthalic anhydride (1.07 mmol) in 6 mL of glacial acetic acid (AcOH) was refluxed for 4 h. Then, the eluent was evaporated in vacuo; the residue was diluted with EtOAc and a solution of 6 M NaOH was added until pH = 6, and the layers were separated. The organic phase was washed with brine, dried over anhydrous Na_2SO_4, filtered, and the filtrate was evaporated in vacuo. The residue was purified by chromatography on silica gel (eluent: $CHCl_3$/MeOH 9: 1 v/v) to give the desired product: green solid, 54% yield. 1H NMR(500 MHz, [D6]DMSO): δ = 1.15–1.21 (m, 12H, CH_3), 4.07- 4.13 (m, 8H, CH_2), 5.14–5.25 (td, 1H, J_{HP} = 22.25, J_{HH} = 8.07, 1H, PCHP), 7.28–7.30 (dd, J_1 = 8.33, J_2 = 1.96, 1H, aromatic), 7.54(d, J = 8.33, 1H, aromatic), 7.76 (d, J = 1.96, 1H, aromatic), 7.88–7.97 (m, 4H, aromatics), 8.95 (d, J = 8.07, 1H, NH). MS (ESI): m/z: 604[M+Na]$^+$, 582[M+H]+; MS2(582): m/z (%): 536 (23), 444 (100), 417 (23), 416 (80), 398 (30), 388 (57), 370 (72), 306 (24), 297 (11). 580[M-H]$^-$; MS2: m/z (%): 137 (100), 108 (32).

Tetraethyl [(6-((L) N-Boc 2 aminopropanamido)benzothiazol-2-amino)methyl]-1,1-bisph osphonate: Boc-L-Ala (1 mmol), EDC (1 mmol) and DMAP (2 mmol) were added to the solution of 5b (1 mmol) in $CHCl_3$ (4 mL) and the mixture was stirred at room temperature on nitrogen for 12 h. After the given time, the eluent was evaporated in vacuo, the residue was diluted with EtOAc, the organic phase was washed with brine, dried over anhydrous Na_2SO_4, and filtered, and the filtrate was evaporated in vacuo. The residue was triturated with AcOEt and filtered to afford the final compound: white solid, 47% yield. 1H NMR(500 MHz, [D6]DMSO): δ = 1.12–1.20 (m, 12H, CH_3), 1.23(d, J = 6.85, 3H, CH_3), 1.36 (s, 9H, $-(CH_3)_3$), 4.07–4.13 (m, 9H, CH_2, CH), 5.07–5.19 (td, 1H, J_{HP} = 22.25, J_{HH} = 9.78, 1H, PCHP), 7.02 (d, J = 6.85, 1H, CH), 7.32 (dd, J_1 = 8.32, J_2 = 1.47, 1H, aromatic), 7.35 (d, J = 8.32, 1H, aromatic), 8.03 (d, J = 1.47, 1H, aromatic), 8.67 (d, J = 9.78, 1H, NH), 9.88 (s, 1H, NH). MS(ESI): m/z: 645[M+Na]$^+$, 623[M+H]+; MS$_2$ (623): m/z (%): 521(62), 430(18), 401 (19), 385 (21), 383 (25), 355 (18), 286 (12).

3.2.3. General Procedure for the Preparation of 1,1-Bisphosphonic Acids (1–12)

Method A: A solution of the appropriate tetraethyl bisphosphonate (1 mmol) in 4 mL 2N HCl solution was kept at reflux for 12–24 h. After removal of the aqueous phase under reduced pressure, the crude bisphosphonic acids were triturated with the opportune solvent and filtered to afford the final compounds as white solids in 20–96% yield.

Method B: Anhydrous trimethylsilylbromide (17–32 mmol) was carefully added to a solution of the corresponding tetraethyl bisphosphonate (1 mmol) in anhydrous acetonitrile (6 mL) at 0 °C under argon and the resulting mixture was stirred at room temperature for 24–48 h. After the given time, 2 mL of MeOH was added and the mixture was stirred for 5 min. The solvent was distilled off and the crude bisphosphonic acids were triturated with the opportune solvent and filtered to afford the desired compounds as white solids in 40–100% yield.

(Benzo[d]thiazol-2-amino)methyl-1,1-bisphosphonic acid (1): Method A; white solid, 72% yield; mp: >250 °C (Et_2O). 1H NMR (500 MHz, NaOD): δ = 2.8–4.00 (br, 6H, NH, OH, PCHP), 6.95 (t, 1H, J = 7.58, aromatic), 7.16 (t, J = 7.58, 1H, aromatic), 7.23 (d, J = 7.83, 1H, aromatic), 7.51 (d, J = 7.83, 1H, aromatic). MS(ESI): m/z: 323[M-H]$^-$; MS2: m/z (%): 287(26), 241 (100), 177 (33), 149 (10).

(6-fluorobenzo[d]thiazol-2-amino)methyl-1,1-bisphosphonic acid (2): Method A; white solid, 26% yield; mp: >250 °C (EtOAc and Acetone). 1H NMR (500 MHz, [D6] DMSO): δ = 3.67–5.45 (br, 4H, OH), 4.77 (t, J_{HP} = 21.28, 1H, PCHP), 7.06 (m, 1H, aromatic), 7.33–7.36 (m, 1H, aromatic), 7.59 (m, 1H, aromatic), 8.26–9.17 (b, 1H, NH). 31P NMR(500 MHz, [D6] DMSO): δ = 14.04 (d, JPH= 16.79, 2P, PCHP). MS (ESI): m/z: 341 [M-H]$^-$; MS2: m/z (%): 305(32), 259 (100), 195 (40), 167 (13).

(6-chlorobenzo[d]thiazol-2-amino)methyl-1,1-bisphosphonic acid (3): Method A; white solid, 68% yield; mp: > 250°C (Acetone/AcOEt). 1H NMR (500 MHz, [D6] DMSO): δ = 4.00–6.00 (br, 4H, OH), 4.80 (t, J_{HP} = 20.79, 1H, PCHP), 7.23 (dd, J_1 = 8.81, J_2 = 1.46, 1H, aromatic), 7.34 (d, J = 8.81, 1H, aromatic), 7.77 (d, J = 1.46, 1H, aromatic), 7.86–8.26 (bs, 1H, NH). ^{31}P NMR (500 MHz, [D6] DMSO): δ = 14.074 (d, 2P, PCHP). MS(ESI): m/z: 359[M+2-H]$^-$, 357 [M-H]$^-$; MS2: m/z (%): 323(24), 321 (41), 277 (43), 275 (100), 213 (21), 211 (44).

(6-chlorobenzo[d]thiazol-2-amino)ethyl-1,1-bisphosphonic acid (12): Method A; white solid, 96% yield; mp: > 250 °C (AcOEt). 1H NMR (500 MHz, [D6] DMSO): δ = 3.6–4.00 (br, 8H, OH, PCHP, NH, CH$_2$), 7.22 (d, J = 8.56, 1H, aromatic), 7.34 (d, J = 8.56, 1H, aromatic), 7.79 (s, 1H, aromatic). ^{31}P NMR (500 MHz, [D6] DMSO): δ = 17.40 (d, JPH= 22.86, 2P, PCHP). MS (ESI): m/z: 373[M+2-H]$^-$, 371[M-H]$^-$; MS2 m/z (%): 337(13), 335(26), 247(40), 245(100).

(6-nitrobenzo[d]thiazol-2-amino)methyl-1,1-bisphosphonic acid (4): Method B; white solid, 76% yield; mp: >250 °C (MeOH). 1H NMR (500 MHz, [D6] DMSO): δ = 4.1–5.5 (br, 5H, OH, PCHP), 7.44 (d, J = 8.81, 1H, aromatic), 8.08 (d, J = 8.81, 1H, aromatic), 8.65 (s, 1H, aromatic), 8.65 (bs, 1H, NH). ^{31}P NMR (500 MHz, [D6] DMSO): δ = 13.64 (d, J = 21.45, PCHP). MS (ESI): m/z: 368[M-H]$^-$; MS2: m/z (%): 332 (45), 305 (25), 286 (58), 222 (100).

(6-benzamidobenzo[d]thiazol-2-amino)methyl-1,1-bisphosphonic acid (5): Method B; white solid, 40% yield; mp: >250 °C(Acetone/ MeOH 3:1 v/v). 1H NMR (500 MHz, [D6] DMSO): δ = 4.80 (t, J$_{HP}$ = 20.5, 1H, PCHP), 5.00–6.4 (br, 4H, OH), 7.38 (d, J = 8.32, 1H, aromatic), 7.49–7.58 (m, 4H, aromatics), 7.93 (dd, J$_1$ = 7.34, J$_2$ = 1.46, 1H, aromatics), 8.16 (d, J = 1.46, 1H, aromatic), 8.95 (bs, 1H, NH), 10.24 (s, 1H, NH). 31P NMR (500 MHz, [D6] DMSO): δ = 13.69 (d, J = 16.78, PCHP). MS (ESI): m/z: 442 [M-H]$^-$; MS2: m/z (%): 407 (15), 406 (62), 361 (22), 360 (100), 343 (18), 342 (69), 296 (25).

(6-(4-bromobenzamido)benzo[d]thiazol-2-amino)methyl-1,1-bisphosphonic acid (6): Method B; white solid, 78% yield; mp: >250 °C ISO). 1H NMR(500 MHz, [D6] DMSO): 2.00–3.5 (br, 6H, OH, NH, PCHP), 7.22–7.88 (m, 6H, aromatics), 8.11 (s, 1H, aromatic), 10.24 (s, 1H, NH). ^{31}P NMR (500 MHz, [D6] DMSO): δ = 14.40 (d, 2P, PCHP). MS (ESI): m/z: 522[M+2-H]$^-$, 520 [M-H]$^-$; MS2: m/z (%): 486 (91), 484 (16), 440 (10), 438 (66), 422 (95.5), 420 (72).

(6-(4-nitrobenzamido)benzo[d]thiazol-2-amino)methyl-1,1-bisphosphonic acid (7): Method B; white solid, 96% yield; mp: > 250 °C (MeOH). 1H NMR (300 MHz, [D6] DMSO): δ = 4.73 (t, J$_{HP}$ = 20.33, 1H, PCHP), 4.86- 6.09 (br, 5H, OH, NH), 7.45 (d, J = 7.91, 1H, aromatic), 7.63 (d, J = 7.91, 1H, aromatic), 8.17 (d, J = 8.79, 2H, aromatic), 8.29 (s, 1H, aromatic), 8.36 (d, J = 8.79, 2H, aromatics), 10.89 (s, 1H, NH). 31P NMR (500 MHz, [D6] DMSO): δ = 12.76 (d, J$_{PH}$ = 21.24, 2P, PCHP). MS (ESI): m/z: 487[M-H]$^-$; MS2: m/z (%): 452 (13), 451 (53), 406 (17), 405 (78), 388 (24), 387 (100), 341 (25), 339 (20), 233 (14).

(6-(2-phenylacetamido)benzo[d]thiazol-2-amino)methyl-1,1-bisphosphonic acid (8): Method B; white solid, 69% yield; mp: >250 °C (ISO). 1H NMR (500 MHz, [D6] DMSO): δ = 4.43 (s, 2H, CH2), 3.90–5.10 (br, 5H, OH, PCHP), 8.01–8.14 (m, 8H, aromatics), 8.81 (s, 1H, NH), 10.17 (s, 1H, NH). MS (ESI): m/z: 456[M-H]$^-$; MS2: m/z (%): 420 (55), 375 (24), 374 (100), 357 (12), 356 (49), 310 (19.5).

(6-(3-phenylureido)benzo[d]thiazol-2-amino)methyl-1,1-bisphosphonic acid (9): Method B; white solid, 69% yield; mp:> 250 °C (ISO). 1H NMR (300 MHz, [D6] DMSO): δ = 4.76 (t, J$_{HP}$ = 19.62, 1H, PCHP), 5.00–6.00 (br, 5H, OH, NH), 6.94 (t, J = 7.03, 1H, aromatic), 7.22–7.44 (m, 6H, aromatics), 7.92 (s, 1H, aromatic), 8.66 (s, 1H, NH), 8.74 (s, 1H, NH). 31P NMR (500 MHz, [D6] DMSO): 13.02 (d, 2P, PCHP). MS (ESI): m/z: 457[M-H]$^-$; MS2: m/z (%): 421 (58), 376 (24), 375 (100), 357 (20), 311 (24), 238 (94).

(6-(1,3-dioxoisoindolin-2-yl)benzo[d]thiazol-2-amino)methyl-1,1-bisphosphonic acid (10): Method B; white solid, 58% yield; mp: >250 °C (ISO). 1H NMR(500 MHz, [D6] DMSO): δ = 4.20–6.00 (br, 5H, OH, PCHP), 6.97–7.25 (m, 1H, aromatic), 7.33–7.47 (m, 3H, aromatics), 7.71 (s, 1H, aromatic), 7.83–7.95 (m, 2H, aromatics), 8.7 (br, 1H, NH). 31P NMR (500 MHz, [D6] DMSO): δ = 13.02 (d, 2P, PCHP). MS (ESI): m/z: 468[M-H]$^-$; MS2: m/z (%): 432 (69), 386 (100), 322 (20), 197 (18).

6-((L) 2-aminopropanamido)benzo[d]thiazol-2-amino)methyl-1,1-bisphosphonic acid (11): Method B; white solid, 94% yield; mp: >250 °C (ISO). 1H NMR (500 MHz, NaOD): δ = 1.19 (d, J = 6.85, 3H, CH3), 3.43–3.47 (q, J = 6.85, 1H, CH), 4.49–4.88 (br, 9H, OH, PCHP, NH, NH, NH2), 7.08 (dd, J$_1$ = 8.80, J$_2$ = 1.96, 1H, aromatic), 7.20 (d, J = 8.8, 1H, aromatic), 7.54 (d,

J = 1.96, 1H, aromatic). 31P (500 MHz, NaOD): δ = 14.84 (d, J = 18.31, 2P, PCHP). MS (ESI): m/z: 409 [M-H]$^-$; MS2: m/z (%): 373 (41), 327 (100), 309 (51), 263 (27). Procedure for the

preparation of Tetraethyl (2-methoxyethane-1,1-diyl)bisphosphonate (12c): Diethylamine (1 mmol) was added to the suspension of paraformaldehyde (1 mmol) in 13 mL of methanol. The system was left under reflux until complete solubilization of the suspension. Subsequently, the solution obtained was stirred at room temperature and a solution of tetraethyl methylenebisphosphonate (1.6 mmol) in 2 mL of methanol was added. The system was kept at reflux for 24 h. After the given time, the solvent was evaporated in vacuo and 3 mL of toluene was added and subsequently evaporated away three times, obtaining a yellow oil which was used in the next reaction without further purification. 1H NMR(500 MHz CDCl$_3$): δ = 1.2 (t, J = 7.1, 12H, OCH$_2$CH$_3$), 2.52 (t, J$_{HP}$ = 24.00, J$_{HH}$ = 6.00, 1H, PCHP), 3.2 (s, 3H, OCH$_3$), 3.63 (m, 2H, CH$_3$OCH$_2$), 4.02 (m, J = 7.30, 8H, OCH$_2$CH$_3$). GC-MS: m/z (%): 332 (0.3), 195 (100).

Procedure for the preparation of tetraethyl ethenylidenebisphosphonate (12b): p-TsA monohydrate (0.26 mmol) was added to the solution of tetraethyl (2-methoxyethyl)bisphosphor 12c (5 mmol) in 7 mL of anhydrous toluene and the mixture was refluxed for 6 h. After the given time, the solvent was evaporated in vacuo. The residue was partitioned between CHCl$_3$ and NaHCO$_3$ and the layers were separated. The organic layer was washed with brine, dried over anhydrous Na$_2$SO$_4$, filtered, and the filtrate was evaporated in vacuo, obtaining a brown oil which was used in the next reaction without further purification: 81% yield. 1H NMR (500 MHz, CDCl3): δ = 1.34 (t, J = 7.1, 12H, CH$_3$), 4.02 (m, 8H, CH$_2$), 6.98 (d, 2H, C=CH$_2$). GC-MS: m/z (%): 301 (3.5), 171 (100), 163 (96).

Tetraethyl (((2-(6-chlorobenzo[d]thiazol-2-yl)amino)ethane-1,1-diyl)bisphosphonate) (12a): Tetraethyl ethenylidenebisphosphonate 12b (1.2 mmol) was added to the solution of 2-amino-6-chlorobenzothiazole (1.62 mmol) in 8 mL of anhydrous CHCl$_3$ and the system was stirred, under nitrogen atmosphere, at 40 °C for 27 h. After the given time, the solvent was evaporated in vacuo and the residue was purified by chromatography on silica gel (eluent: EtOAc) to give the desired product: white solid, 65% yield. 1H NMR (500 MHz, CDCl$_3$): δ = 1.24–1.76 (m, 12H, CH$_3$), 2.84 (tt, J$_{HP}$ = 23, J$_{HH}$ = 6.36, 1H, CH), 3.99–4.07 (td, J$_1$ = 16.00, J$_2$ = 6.36, 2H, CH$_2$), 4.08–4.26 (m, 8H, CH$_2$), 6.38 (bs, 1H, NH), 7.22–7.24 (dd, J$_1$ = 8.81, J$_2$ = 1.96, 1H, aromatic), 7.41 (d, J = 8.81, 1H, aromatic), 7.55 (d, J = 1.96, 1H, aromatic).

3.3. Docking Studies

All calculations were carried out using the Schrodinger Suite 2019-3 (Schrodinger, New York, NY, USA) [30]. Ligand structures were built using Maestro in a fully deprotonated form, minimized with MacroModel using the OPLS2005 force field, the PRCG algorithm, at a convergence gradient of 0.05. A Monte Carlo Multiple Minimum/Low Mode Conformational Search (MCMM/LMCS) protocol was applied for the conformational search using the automatic setup, performing 200 steps per rotatable bond. The global minimum geometry was used to follow the docking studies carried out on the four studied MMPs following the previously described procedure [13].

4. Conclusions

Over the years, inhibition of MMPs has been a much sought-after target for the treatment of bone metastases, showing promise in stopping the vicious cycle of metastatic cell growth and bone matrix degradation. Although this therapeutic strategy is plagued by numerous and severe side effects, such as musculoskeletal syndromes, recent research showed that targeting to a specific tissue, such as bone, and targeting specific MMPs which are overexpressed in the metastatic microenvironment, such as MMP-13, are viable strategies to minimize them.

Following both of these approaches, we demonstrated that a benzothiazole scaffold coupled with a bisphosphonic moiety is a versatile starting point for the development of bone-targeted MMP-13 inhibitors, with the possibility of tailoring their activity profile in order to also target other validated MMPs, mainly MMP-2, allowing for a mostly favorable therapeutic profile for the treatment of bone malignancies.

Supplementary Materials: The following are available online at https://www.mdpi.com/1424-8 247/14/2/85/s1: Figure S1. Docked pose of compound **7** (stick) in the MMP-13 binding site (grey cartoon) obtained with Glide (green C atoms) and Autodock (cyan C atoms).

Author Contributions: Conceptualization, P.T.; investigation, A.L., A.C., L.B., L.P., and M.A.; writing—original draft preparation, L.B. and L.P.; writing—review and editing, M.A., P.T., and F.L.; supervision, P.T. and F.L.; project administration, P.T. All authors have read and agreed to the published version of the manuscript.

Funding: This research received no external funding.

Institutional Review Board Statement: Not applicable.

Informed Consent Statement: Not applicable.

Data Availability Statement: Data is contained in the article.

Conflicts of Interest: The authors declare no conflict of interest.

References

1. Winer, A.; Adams, S.; Mignatti, P. Matrix Metalloproteinase Inhibitors in Cancer Therapy: Turning Past Failures into Future Successes. *Mol. Cancer Ther.* **2018**, *17*, 1147–1155. [CrossRef] [PubMed]
2. Conlon, G.A.; Murray, G.I. Recent advances in understanding the roles of matrix metalloproteinases in tumour invasion and metastasis. *J. Pathol.* **2019**, *247*, 629–640. [CrossRef] [PubMed]
3. Li, K.; Tay, F.R.; Yiu, C.K.Y. The past, present and future perspectives of matrix metalloproteinase inhibitors. *Pharmacol. Ther.* **2020**, *207*, 107465. [CrossRef]
4. Lynch, C.C. Matrix metalloproteinases as master regulators of the vicious cycle of bone metastasis. *Bone* **2011**, *48*, 44–53. [CrossRef] [PubMed]
5. Guise, T.A.; Chirgwin, J. Transforming Growth Factor-Beta in Osteolytic Breast Cancer Bone Metastases. *Clin. Orthop. Relat. Res.* **2003**, *415*, S32–S38. [CrossRef]
6. Coussens, L.M.; Fingleton, B.; Matrisian, L.M. Matrix Metalloproteinase Inhibitors and Cancer–Trials and Tribulations. *Science* **2002**, *295*, 2387–2392. [CrossRef]
7. Von Moos, R.; Costa, L.; Gonzalez-Suarez, E.; Terpos, E.; Niepel, D.; Body, J. Management of bone health in solid tumours: From bisphosphonates to a monoclonal antibody. *Cancer Treat. Rev.* **2019**, *76*, 57–67. [CrossRef]
8. Tauro, M.; Shay, G.; Sansil, S.S.; Laghezza, A.; Tortorella, P.; Neuger, A.M.; Soliman, H.; Lynch, C.C. Bone-Seeking Matrix Metalloproteinase-2 Inhibitors Prevent Bone Metastatic Breast Cancer Growth. *Mol. Cancer Ther.* **2017**, *16*, 494–505. [CrossRef]
9. Jakob, T.; Tesfamariam, Y.M.; Macherey, S.; Kuhr, K.; Adams, A.; Monsef, I.; Heidenreich, A.; Skoetz, N. Bisphosphonates or RANK-ligand-inhibitors for men with prostate cancer and bone metastases: A network meta-analysis. *Cochrane Database Syst. Rev.* **2020**, *12*. [CrossRef]
10. Wang, L.; Fang, D.; Xu, J.; Luo, R. Various pathways of zoledronic acid against osteoclasts and bone cancer metastasis: A brief review. *BMC Cancer* **2020**, *20*, 1059. [CrossRef]
11. Leuci, R.; Brunetti, L.; Laghezza, A.; Loiodice, F.; Tortorella, P.; Piemontese, L. Importance of Biometals as Targets in Medicinal Chemistry: An Overview about the Role of Zinc (II) Chelating Agents. *Appl. Sci.* **2020**, *10*, 4118. [CrossRef]
12. Tauro, M.; Laghezza, A.; Campestre, C.; Tortorella, P.; Loiodice, F.; Piemontese, L.; CaraDonna, A.; Capelli, D.; Montanari, R.; Pochetti, G.; et al. Catechol-based matrix metalloproteinase inhibitors with additional antioxidative activity. *J. Enzyme Inhib. Med. Chem.* **2016**, *31*, 25–37. [CrossRef] [PubMed]
13. Tauro, M.; Loiodice, F.; Ceruso, M.; Supuran, C.T.; Tortorella, P. Arylamino bisphosphonates: Potent and selective inhibitors of the tumor-associated carbonic anhydrase XII. *Bioorg. Med. Chem. Lett.* **2014**, *24*, 1941–1943. [CrossRef] [PubMed]
14. Tauro, M.; Loiodice, F.; Ceruso, M.; Supuran, C.T.; Tortorella, P. Dual carbonic anhydrase/matrix metalloproteinase inhibitors incorporating bisphosphonic acid moieties targeting bone tumors. *Bioorg. Med. Chem. Lett.* **2014**, *24*, 2617–2620. [CrossRef]
15. Tauro, M.; Laghezza, A.; Loiodice, F.; Agamennone, M.; Campestre, C.; Tortorella, P. Arylamino methylene bisphosphonate derivatives as bone seeking matrix metalloproteinase inhibitors. *Bioorg. Med. Chem.* **2013**, *21*, 6456–6465. [CrossRef]
16. Savino, S.; Toscano, A.; Purgatorio, R.; Profilo, E.; Laghezza, A.; Tortorella, P.; Angelelli, M.; Cellamare, S.; Scala, R.; Tricarico, D.; et al. Novel bisphosphonates with antiresorptive effect in bone mineralization and osteoclastogenesis. *Eur. J. Med. Chem.* **2018**, *158*, 184–200. [CrossRef]
17. Shay, G.; Tauro, M.; Loiodice, F.; Tortorella, P.; Sullivan, D.M.; Hazlehurst, L.A.; Lynch, C.C. Selective inhibition of matrix metalloproteinase-2 in the multiple myeloma-bone microenvironment. *Oncotarget* **2017**, *8*, 41827–41840. [CrossRef]

18. Laghezza, A.; Piemontese, L.; Loiodice, F.; Tortorella, P.; Brunetti, L.; CaraDonna, A.; Agamennone, M.; Di Pizio, A.; Pochetti, G.; Montanari, R.; et al. Bone-Seeking Matrix Metalloproteinase Inhibitors for the Treatment of Skeletal Malignancy. *Pharmaceuticals* **2020**, *13*, 113. [CrossRef]

19. Rubino, M.T.; Agamennone, M.; Campestre, C.; Campiglia, P.; Cremasco, V.; Faccio, R.; Laghezza, A.; Loiodice, F.; Maggi, D.; Panza, E.; et al. Biphenyl Sulfonylamino Methyl Bisphosphonic Acids as Inhibitors of Matrix Metalloproteinases and Bone Resorption. *ChemMedChem* **2011**, *6*, 1258–1268. [CrossRef]

20. Aureli, L.; Gioia, M.; Cerbara, I.; Monaco, S.; Fasciglione, G.F.; Marini, S.; Ascenzi, P.; Topai, A.; Coletta, M. Structural Bases for Substrate and Inhibitor Recognition by Matrix Metalloproteinases. *Curr. Med. Chem.* **2008**, *15*, 2192–2222. [CrossRef]

21. Cawston, T.E.; Wilson, A.J. Understanding the role of tissue degrading enzymes and their inhibitors in development and disease. *Best Pr. Res. Clin. Rheumatol.* **2006**, *20*, 983–1002. [CrossRef] [PubMed]

22. Shah, M.H.; Huang, D.; Blick, T.; Connor, A.; Reiter, L.A.; Hardink, J.R.; Lynch, C.C.; Waltham, M.; Thompson, E.W. An MMP13-Selective Inhibitor Delays Primary Tumor Growth and the Onset of Tumor-Associated Osteolytic Lesions in Experimental Models of Breast Cancer. *PLoS ONE* **2012**, *7*, e29615. [CrossRef] [PubMed]

23. Nannuru, K.C.; Futakuchi, M.; Varney, M.L.; Vincent, T.M.; Marcusson, E.G.; Singh, R.K. Matrix Metalloproteinase (MMP)-13 Regulates Mammary Tumor–Induced Osteolysis by Activating MMP9 and Transforming Growth Factor-β Signaling at the Tumor-Bone Interface. *Cancer Res.* **2010**, *70*, 3494–3504. [CrossRef]

24. Ohshiba, T.; Miyaura, C.; Inada, M.; Ito, A. Role of RANKL-induced osteoclast formation and MMP-dependent matrix degradation in bone destruction by breast cancer metastasis. *Br. J. Cancer* **2003**, *88*, 1318–1326. [CrossRef] [PubMed]

25. Xie, X.-W.; Wan, R.-Z.; Liu, Z.-P. Recent Research Advances in Selective Matrix Metalloproteinase-13 Inhibitors as Anti-Osteoarthritis Agents. *ChemMedChem* **2017**, *12*, 1157–1168. [CrossRef]

26. Fischer, T.; Riedl, R. Molecular Recognition of the Catalytic Zinc (II) Ion in MMP-13: Structure-Based Evolution of an Allosteric Inhibitor to Dual Binding Mode Inhibitors with Improved Lipophilic Ligand Efficiencies. *Int. J. Mol. Sci.* **2016**, *17*, 314. [CrossRef]

27. Cheng, X.-C.; Wang, Q.; Fang, H.; Xu, W.-F. Role of sulfonamide group in matrix metalloproteinase inhibitors. *Curr. Med. Chem.* **2008**, *15*, 368–373. [CrossRef]

28. Gimeno, A.; Beltrán-Debón, R.; Mulero, M.; Pujadas, G.; Garcia-Vallvé, S. Understanding the variability of the S1′ pocket to improve matrix metalloproteinase inhibitor selectivity profiles. *Drug Discov. Today* **2020**, *25*, 38–57. [CrossRef]

29. Morris, G.M.; Huey, R.; Lindstrom, W.; Sanner, M.F.; Belew, R.K.; Goodsell, D.S.; Olson, A.J. AutoDock4 and AutoDockTools4: Automated docking with selective receptor flexibility. *J. Comput. Chem.* **2009**, *30*, 2785–2791. [CrossRef]

30. Schrödinger LLC. *Schrödinger Suite 2019-3: MacroModel, Glide, SiteMap, Maestro*; Schrödinger: New York, NY, USA, 2019.

*pharmaceuticals*

Article

Coumarin Derivatives Act as Novel Inhibitors of Human Dipeptidyl Peptidase III: Combined In Vitro and In Silico Study

Dejan Agić [1,*], Maja Karnaš [1], Domagoj Šubarić [1], Melita Lončarić [2], Sanja Tomić [3], Zrinka Karačić [3], Drago Bešlo [1], Vesna Rastija [1], Maja Molnar [2], Boris M. Popović [4] and Miroslav Lisjak [1]

[1] Faculty of Agrobiotechnical Sciences Osijek, Josip Juraj Strossmayer University of Osijek, 31000 Osijek, Croatia; maja.karnas@fazos.hr (M.K.); domagoj.subaric@fazos.hr (D.Š.); drago.beslo@fazos.hr (D.B.); vesna.rastija@fazos.hr (V.R.); miroslav.lisjak@fazos.hr (M.L.)
[2] Faculty of Food Technology Osijek, Josip Juraj Strossmayer University of Osijek, 31000 Osijek, Croatia; melita.loncaric@ptfos.hr (M.L.); maja.molnar@ptfos.hr (M.M.)
[3] Division of Organic Chemistry and Biochemistry, Ruđer Bošković Institute, 10000 Zagreb, Croatia; sanja.tomic@irb.hr (S.T.); zrinka.karacic@irb.hr (Z.K.)
[4] Department of Field and Vegetable Crops, Faculty of Agriculture, University of Novi Sad, 21000 Novi Sad, Serbia; boris.popovic@polj.uns.ac.rs
* Correspondence: dejan.agic@fazos.hr

check for **updates**

Citation: Agić, D.; Karnaš, M.; Šubarić, D.; Lončarić, M.; Tomić, S.; Karačić, Z.; Bešlo, D.; Rastija, V.; Molnar, M.; Popović, B.M.; et al. Coumarin Derivatives Act as Novel Inhibitors of Human Dipeptidyl Peptidase III: Combined In Vitro and In Silico Study. *Pharmaceuticals* **2021**, *14*, 540. https://doi.org/10.3390/ph14060540

Academic Editor: Salvatore Santamaria

Received: 30 April 2021
Accepted: 2 June 2021
Published: 5 June 2021

Publisher's Note: MDPI stays neutral with regard to jurisdictional claims in published maps and institutional affiliations.

Abstract: Dipeptidyl peptidase III (DPP III), a zinc-dependent exopeptidase, is a member of the metalloproteinase family M49 with distribution detected in almost all forms of life. Although the physiological role of human DPP III (hDPP III) is not yet fully elucidated, its involvement in pathophysiological processes such as mammalian pain modulation, blood pressure regulation, and cancer processes, underscores the need to find new hDPP III inhibitors. In this research, five series of structurally different coumarin derivatives were studied to provide a relationship between their inhibitory profile toward hDPP III combining an in vitro assay with an in silico molecular modeling study. The experimental results showed that 26 of the 40 tested compounds exhibited hDPP III inhibitory activity at a concentration of 10 µM. Compound **12** (3-benzoyl-7-hydroxy-2H-chromen-2-one) proved to be the most potent inhibitor with IC_{50} value of 1.10 µM. QSAR modeling indicates that the presence of larger substituents with double and triple bonds and aromatic hydroxyl groups on coumarin derivatives increases their inhibitory activity. Docking predicts that **12** binds to the region of inter-domain cleft of hDPP III while binding mode analysis obtained by MD simulations revealed the importance of 7-OH group on the coumarin core as well as enzyme residues Ile315, Ser317, Glu329, Phe381, Pro387, and Ile390 for the mechanism of the binding pattern and compound **12** stabilization. The present investigation, for the first time, provides an insight into the inhibitory effect of coumarin derivatives on this human metalloproteinase.

Keywords: dipeptidyl peptidase III; coumarin derivatives; inhibitor; molecular modeling; metalloproteinase

1. Introduction

Dipeptidyl peptidase III (DPP III) is a zinc-hydrolase that cleaves dipeptides sequentially from the N-terminal of different bioactive peptides [1]. As a member of the metalloproteinase family M49, DPP III distribution is detected in almost all forms of life [2]. Human DPP III is a very well-characterized member of this family in terms of biochemistry, structural biology and computational chemistry [3–10]. Due to the relative non-specificity of the peptide substrates as well as the lack of selective inhibitors of the metallopeptidases of the M49 family, the physiological substrates of DPP III have not been accurately identified, and its fundamental physiological role has not been precisely determined. However, it is assumed that it is involved in post-proteasomal intracellular protein catabolism [3], defense against oxidative stress [11,12] mammalian pain modulatory system [4,13], malignant processes [14–17] and blood pressure regulation [18–20]. Because of its involvement in biological processes, hDPP III has become interesting to investigate as a potential drug target. To obtain information on the mechanism of action of hDPP III, the influence of selected mutations on the enzyme activity was tested, and research was conducted to find potential inhibitors of this mammalian metalloproteinase [21–23]. It has recently been shown that newly synthesized guanidiniocarbonylpyrrole–fluorophore conjugates could be used for enzyme sensing and bio-activity inhibiting (theragnostic) studies of DPP III [24]. In search of new inhibitors from natural sources, we previously reported that luteolin has the best inhibitory effect against hDPP III (IC50 = 22 μM) of the 17 flavonoids tested, and that the number and exact distribution of –OH groups on the flavonoid core is important for their inhibitory properties [25]. The latest study of the biological activity of cornelian cherry fruit extracts showed inhibitory activity against hDPP III with bioactive constituent pelargonidin 3-robinobioside with the best binding energy [26]. Coumarin or 2H-chromen-2-one and its derivatives represent an important group of oxygen-containing heterocycles with benzopyrone skeleton [27]. They can be isolated from plant material [28] or synthesized [29]. Coumarin derivatives possess various beneficial biological activities, for example, anticoagulant, anticancer, analgesic, anti-inflammatory, bactericidal, antifungal, anticonvulsant, anti-hypertensive, muscle relaxant, antioxidant, etc. [28]. It is known that coumarins exhibit an inhibitory effect on the enzymes such as acetylcholinesterase, β-secretase, and monoamine oxidase [30]. Studies have also shown that simple coumarin derivatives influence the activity of some zinc-dependent metalloproteinases [31–33].

Because of all mentioned above as well as our efforts to find new hDPP III inhibitors, we report the investigation of 40 structurally different coumarin compounds and provide a relationship between their inhibitory profile toward hDPP III combining in vitro assay with Quantitative Structure–Activity Relationship (QSAR) analysis. Additionally, docking and MD simulations were conducted to explore the mechanism of the most potent inhibitor binding into the active site of hDPP III.

2. Results and Discussion

2.1. DPP III Inhibitory Activity

In the current study, we evaluated forty various coumarin compounds for their inhibitory potential towards hDPP III. Results in Table 1 showed that substituted 3-acetyl-2H-chromen-2-ones with a bromo group at the C6 position (compound **1**) was the most active with the inhibition rate of 28.5%, while the presence of a hydroxyl group at the same position on compound **2** reduced (12.8%) the inhibitory potential. Shifting a hydroxyl group to C7 in **4** resulted in a slightly increased inhibitory potential (16.2%) as compared to **2** while the presence of diethylamino group at this position in compound **3** was found to be inactive. Additionally, compounds **5** and **6** which possess a hydroxyl and ethoxy group at C8 were completely inactive at the concentration of 10 μM. Compound **7** bearing unsubstituted 3-acetyl-2H-chromen-2-one showed a weak (7.8%) inhibitory activity.

The most potent inhibitory potential of substituted 3-benzoyl-2H-chromen-2-ones was obtained with compound **12** (100.0%) where the hydroxyl group is present at C7. However, the substitution of C7 with the benzoyl (**11**) and methoxy (**13**) group caused a decrease in hDPP III inhibitory (22.8% and 16.5%, respectively) activity. Moderate (67.5%) to weak (4.4%) enzyme inhibition was observed with compounds **10** and **8** which possess a hydroxyl and chloro group at the C6 position, respectively. Compound **9** with a bromo group at the C6 and C8 as well as compound **14** with an ethoxy group on C8 did not exhibit inhibition effects on enzyme activities. Unsubstituted 3-benzoyl-2H-chromen-2-one (**15**) showed only a weak inhibitory activity (9.6%) as compared to **12**.

Of the seven substituted 2-oxo-2H-chromene-3-carbonitriles tested, only compounds **21** and **18** which only differ with hydroxyl group positions (C8 and C6, respectively) moderately inhibited enzyme with an inhibition rate of 62.6% and 44.6%, respectively. In the case where the ethoxy group is at the C8 position (**22**), no inhibitory activity was observed. Furthermore, the methoxy substituent at the C6 position (**17**) gave an inhibition rate of 19.8% while its presence at the C7 position (**20**) was found to be inactive. Substitution of bromo group at C6 in **16** and benzoyl group at C7 in **19** showed almost similar inhibitory potential (7.9% and 7.1%, respectively). Unsubstituted 2-oxo-2H-chromene-3-carbonitrile (**23**) was not effective in inhibiting hDPP III.

Table 1. Structures of analysed compounds, values of experimentally determined inhibition of hDPP III (at 10 µM concentration of compounds) and calculated logarithmic values of the % inhibition of hDPP III.

Compound No.	Substituents	DPP III Inh. (%)	Log (% hDPP III Inh.) exp.	Log (% DPP III Inh.) Calc. *
1	3-acetyl; 6-bromo	28.5	1.45	1.29
2	3-acetyl; 6-hydroxy	12.8	1.11	Excl.
3	3-acetyl; 7-diethylamino	NA	0.00	-
4	3-acetyl; 7-hydroxy	16.2	1.21	1.70
5	3-acetyl; 8-ethoxy	NA	0.00	0.35
6	3-acetyl; 8-hydroxy	NA	0.00	-
7	3-acetyl	7.8	0.89	1.01
8	3-benzoyl; 6-chloro	4.4	0.64	1.01
9	3-benzoyl; 6,8-dibromo	NA	0.00	-
10	3-benzoyl; 6-hydroxy	67.5	1.83	1.95
11	3-benzoyl; 7-benzoyl	22.8	1.36	1.18
12	3-benzoyl; 7-hydroxy	100.0 (1.10 ± 0.05 µM)	2.00	1.89
13	3-benzoyl; 7-methoxy	16.5	1.22	0.78
14	3-benzoyl; 8-ethoxy	NA	0.00	-
15	3-benzoyl	9.6	0.98	0.65
16	3-cyano; 6-bromo	7.9	0.90	0.98
17	3-cyano; 6-methoxy	19.8	1.30	0.83
18	3-cyano; 6-hydroxy	44.6	1.65	1.81
19	3- cyano; 7-benzoyl	7.1	0.85	1.04
20	3-cyano; 7-methoxy	NA	0.00	-
21	3- cyano; 8-hydroxy	62.6	1.80	1.67
22	3-cyano; 8-ethoxy	NA	0.00	0.17
23	3-cyano	NA	0.00	-
24	3-ethoxycarbonyl; 6-bromo	NA	0.00	-
25	3-ethoxycarbonyl; 6-chloro	20.1	1.30	0.93
26	3-ethoxycarbonyl; 6-dihydroxyamino	59.7	1.78	1.49
27	3-ethoxycarbonyl; 6-hydroxy	66.0	1.82	1.76
28	3- ethoxycarbonyl; 6,8-dibromo	29.4	1.47	1.82
29	3-ethoxycarbonyl; 7-methoxy	NA	0.00	0.26
30	3-ethoxycarbonyl; 8-ethoxy	NA	0.00	0.37
31	3- ethoxycarbonyl	NA	0.00	-
32	3- methoxycarbonyl; 6-bromo	6.5	0.81	0.78
33	3-methoxycarbonyl; 6-dihydroxyamino	21.2	1.33	1.46
34	3-methoxycarbonyl; 6-hydroxy	23.5	1.37	1.35
35	3-methoxycarbonyl; 6-methoxy	9.9	1.00	0.64
36	3-methoxycarbonyl; 7-hydroxy	100.0 (2.14 ± 0.06 µM)	2.00	1.50
37	3-methoxycarbonyl; 7-methoxy	NA	0.00	0.56
38	3-methoxycarbonyl	2.3	0.35	0.59
39	coumarin	NA	0.00	−0.37
40	7-hydroxycoumarin	2.1	0.33	0.49

NA, no activity; Excl.; excluded as outlier; -, excluded from initial dataset; * Calculated by quantitative structure-activity relationship (QSAR) equation: log % hDPP III inh. = −4.07 + 1.85 (0.59) *EEig05x* + 1.60 (0.52) *Mor10u* + 0.56 (0.39) *nArOH*; numbers in brackets represent IC_{50} values.

Four of seven substituted 3-acetyl-2H-chromen-2-ones: **24–27** possess different groups but in the same C6 position. Compound **27** with hydroxyl group and **26** with dihydroxyamino group were found to be more active (66.0% and 59.7% respectively) than the compounds **25** (20.1%) and **24** (not active) with the chloro and bromo group, respectively. Interestingly, dibromo substituents at C6 and C8 (**28**) increased inhibitory potential (29.4%) as compared to mono substituted analog (**24**). Compounds **29** and **30** which possess methoxy group at C7 and ethoxy group at C8, respectively as well as unsubstituted 3-acetyl-2H-chromen-2-one (**31**) did not exhibit inhibitory potential.

Among the substituted methyl 2-oxo-2H-chromene-3-carboxylates, only **36** containing a hydroxyl group at C7 completely inhibited enzymatic activity. Changing the methoxy group at the same position (**37**) completely reduced inhibitory potential. The compound **34** with dihydroxyamino and compound **33** with hydroxyl group at C6 position were found to be more efficient in the inhibitory potential (23.5% and 21.2%, respectively) as compared to the methoxy (9.9%) and bromo (6.5%) substituted analogs **35** and **32**, respectively. A very weak inhibitory potential (2.3%) was found for unsubstituted methyl 2-oxo-2H-chromene-3-carboxylate (**38**). Similarly, 7-hydroxycoumarin (**39**) and coumarin (**40**) exhibited a strong decrease (2.1% and not active, respectively) in the inhibitory potential towards hDPP III.

From the above analysis, it can be concluded that the best inhibitory potential had substituted 3-benzoyl-2H-chromen-2-one (**12**) and methyl 7-hydroxy-2-oxo-2*H*-chromene-3-carboxylate (**36**) containing a hydroxyl group at position C7, where they completely inhibited the enzyme at the concentration of 10 µM with IC_{50} values of 1.10 ± 0.05 µM and 2.14 ± 0.06 µM, respectively (Table 1 and Figure 1). Additionally, comparing the structures of derivatives **12** and **36** with compounds **39** and **40** suggests that in addition to the presence of the hydroxyl group at the C7 position, the exact presence of particular substituents at the C3 position is important for increasing the inhibitory activity of tested coumarin derivatives. Furthermore, when the hydroxyl group is at the C6 and C8 positions, the compounds mostly show moderate inhibitory activity. Coumarin derivatives with a substituted bromo, chloro, and benzoyl group showed lower inhibitory potential compared to C6 hydroxy analogs. Finally, compounds with a dihydroxyamino group at the C7 position had moderate enzyme inhibition while most coumarin derivatives with a substituted methoxy and diethylamino group showed no inhibitory activity against hDPP III.

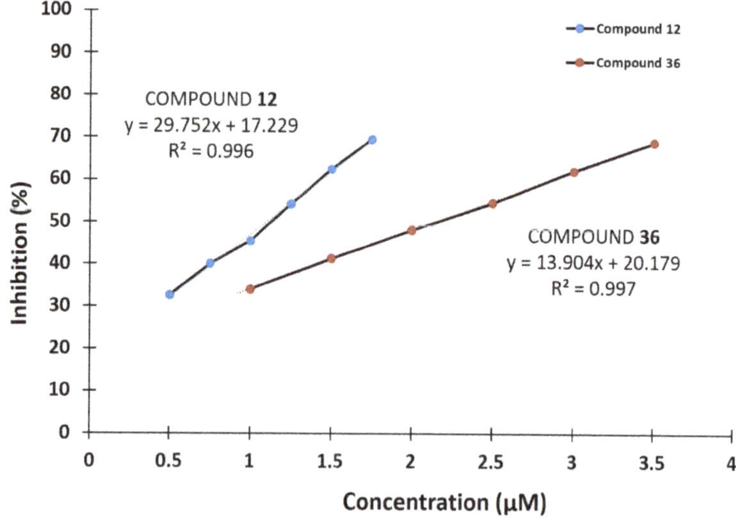

Figure 1. IC_{50} value determination of compounds **12** and **36** against hDPP III. Data points represent the average values of three determinations.

2.2. Results of the QSAR Analysis

The best QSAR model obtained for hDPP III inhibition is:

$$\log \% \text{ hDPP III inh.} = -4.03 + 1.82 \ (0.58) \ EEig05x + 1.46 \ (0.49) \ Mor10u + 0.49 \ (0.36) \ nArOH \tag{1}$$

where *EEig05x* is an edge-adjacency index descriptor weighted by edge degrees, *Mor10u* is a 3D-MoRSE descriptor (unweighted) and *nArOH* is the number of aromatic hydroxyl groups.

The model satisfied the threshold for the fitting and internal validation criteria [34], but Williams plot revealed one outlier, compound **2** (MolID in QSARINS: **24**) as shown in Figure 2. After the exclusion of this compound from the dataset, the subsequent analysis produced the improved QSAR model:

$$\log \% \text{ hDPP III inh.} = -4.07 + 1.85 \ (0.59) \ EEig05x + 1.60 \ (0.52) \ Mor10u + 0.56 \ (0.39) \ nArOH \tag{2}$$

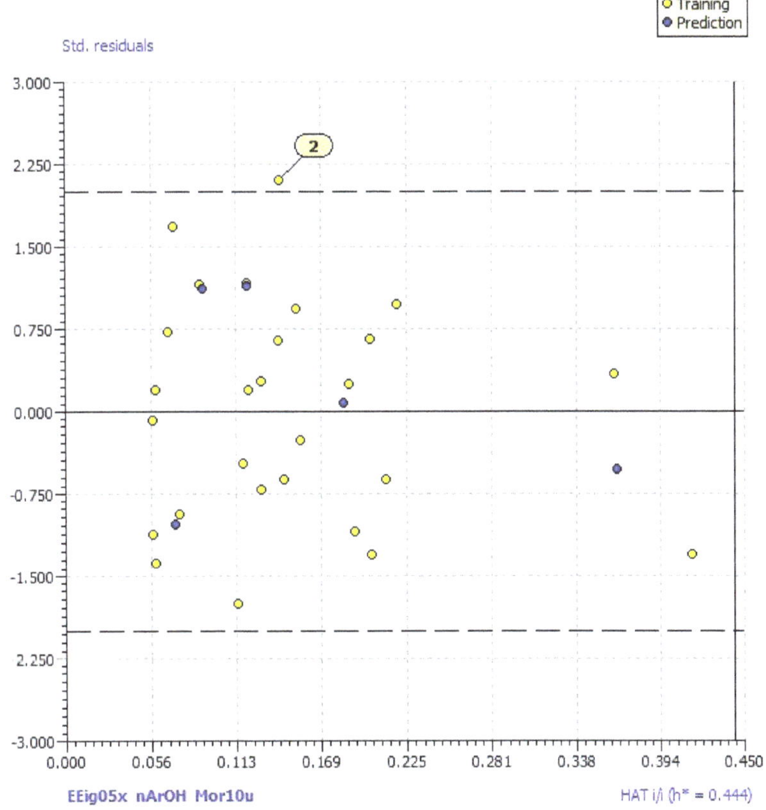

Figure 2. Williams plot (plot of standardized residuals vs. leverages (*h*) for each compound) of applicability domain of the QSAR model for hDPP III inhibition calculated by model 1. The warning leverage (*h** = 0.444) is defined as $3p'/n$ (*n* is the number of training compounds, and *p'* the number of model adjustable parameters).

The variables in Equations (1) and (2) are listed in order of relative importance by their standardized regression coefficient (β, written in brackets). The statistical parameters for both models are given in Table 2. The values of the descriptors included in the models are given in the Supplementary Materials (Table S1). The values of log % hDPP III inh.;

both experimentally obtained and calculated by Equation (2) are presented in Table 1 and Table S1.

Table 2. The statistical parameters for QSAR models.

Statistical Parameters	Model 1	Model 2
N_{tr}	27	26
N_{ex}	5	5
R^2	0.746	0.796
R^2_{adj}	0.713	0.768
s	0.352	0.323
F	22.565	28.572
K_{xx}	0.210	0.190
ΔK	0.201	0.215
$RMSE_{tr}$	0.325	0.297
MAE_{tr}	0.276	0.253
CCC_{tr}	0.855	0.886
Q^2_{LOO}	0.650	0.710
$RMSE_{cv}$	0.381	0.354
MAE_{cv}	0.326	0.302
$PRESS_{cv}$	3.923	3.258
CCC_{cv}	0.807	0.844
R^2_{Yscr}	0.115	0.121
Q^2_{Yscr}	−0.236	−0.244
$RMSE_{ext}$	0.292	0.295
MAE_{ext}	0.255	0.275
R^2_{ext}	0.795	0.785
CCC_{ext}	0.868	0.873
Q^2_{F1}	0.783	0.778
Q^2_{F2}	0.780	0.776
Q^2_{F3}	0.794	0.798
r^2_m average	0.614	0.653
r^2_m difference	0.186	0.174
Applicability domain		
N outliers	1 (2)	-
N out of app. domain	-	-

N_{tr} (number of compounds in training set); N_{ex} (number of compounds in test set); LOO (leave-one-out); R^2 (coefficient of determination); R^2_{adj} (adjusted coefficient of determination); s (standard deviation of regression); F (Fisher ratio); K_{xx} (global correlation among descriptors); ΔK (global correlation among descriptors); $RMSE_{tr}$ (root-mean-square error of the training set); MAE_{tr} (mean absolute error of the training set); CCC_{tr} (concordance correlation coefficient of the training set); Q^2_{LOO} (cross-validated explained variance); $RMSE_{cv}$ (root-mean-square error of the training set determined through the cross validated method); MAE_{cv} (mean absolute error of the internal validation set); $PRESS_{cv}$ (cross-validated predictive residual sum of squares); CCC_{cv} (concordance correlation coefficient test set using cross validation); R^2_{Yscr} (Y-scramble correlation coefficients); Q^2_{Yscr} (Y-scramble cross-validation coefficients); $RMSE_{ext}$ (root-mean-square error of the external validation set); MAE_{ext} (mean absolute error of the external validation set); R^2_{ext} (coefficient of determination of external validation set); CCC_{ext} (concordance correlation coefficient of the test set); Q^2_{F1}, Q^2_{F2}, Q^2_{F3} (predictive squared correlation coefficients); r^2_m average (average value of squared correlation coefficients between the observed and leave-one-out predicted values of the compounds with and without intercept); r^2_m difference (absolute difference between the observed and leave-one-out predicted values of the compounds with and without intercept).

The collinearity of the descriptors in the model was evaluated with a correlation matrix (Table 3) to exclude the possibility that the improved model is overfitted (correlation coefficient $R \leq 0.7$). Furthermore, low collinearity was verified with the low value of K_{xx} and ΔK being ≥ 0.05. The model satisfied fitting and internal validation criteria: R^2 and $R^2_{adj} \geq 0.60$; $CCC_{tr} \geq 0.85$; $RMSE$ and MAE close to zero; $RMSE_{tr} < RMSE_{cv}$; $Q^2_{LOO} \geq 0.50$ (with $R^2 - Q^2$ being low); high value of F (Table 2). The value of the cross-validated correlation coefficient ($Q^2_{LOO} = 0.710$) shows that model 2 has a good internal prediction power. The robustness of the improved model was confirmed with both R^2_{Yscr} and Q^2_{Yscr} values < 0.2, and $R^2_{Yscr} > Q^2_{Yscr}$ [34]. Model 2 also satisfied the following external validation criteria: $R^2_{ext} \geq 0.60$; low differences between $RMSE_{tr}$ and $RMSE_{ext}$ as well as between MAE_{tr} and MAE_{ext} and between CCC_{tr} and CCC_{ext}; Q^2_{F1}, Q^2_{F2}, and $Q^2_{F3} \geq 0.60$; r^2_m average ≥ 0.60 and r^2_m difference ≤ 0.20 indicating that this model could be used for external prediction

Table 3. Correlation matrix (with correlation coefficient values R) for descriptors used in Equation (2).

Descriptor	EEig05x	Mor10u	nArOH
EEig05x	1.000		
Mor10u	−0.264	1.000	
nArOH	−0.129	0.489	1.000

The Williams plot for model (2) showed no compounds outside the applicability domain of the model.

The descriptors from the best model were more closely observed to gain insight into the factors that contribute to the inhibitory activity of tested compounds. The first variable in Equation (2) with a high positive contribution is descriptor *EEig05x*, 5th eigenvalue from edge adjacency matrix weighted by edge degrees (the bond order of the various edges). It belongs to the edge-adjacency topological indices derived from the edge adjacency matrix, which encodes the connectivity between graph edges, and is derived from an H-depleted molecular graph of molecules. These descriptors are sensitive to the size, shape, branching, and cyclicity of molecules [35,36]. It is shown that compounds with relatively higher values of this descriptor tend to exhibit higher inhibition of hDPP III. This indicates that compounds with larger, aromatic, and substituents with a higher number of double or triple bonds may exhibit enhanced inhibition. Similar conclusions were also drawn in previous work [37].

The second variable, *Mor10u*, belongs to the 3D-MoRSE (Molecule Representation of Structures based on Electron diffraction) group of descriptors. It has a scattering parameter $s = 9 \text{ Å}^{-1}$ and since it is unweighted, treats all atoms equally [38]. The positive coefficient of *Mor10u* in Equation (2) indicates the importance of the three-dimensional arrangement of all atoms in a molecule and their pairwise distances. Compounds with higher inhibitory activity tend to have more positive values of this descriptor. Since larger molecules, with larger interatomic distances, have higher MoRSE descriptor values, this confirms the above conclusion about the *EEig05x* descriptor that larger molecules are more active.

The third variable in the equation is *nArOH*, a descriptor from the functional group counts that represents the number of aromatic hydroxyls [38]. The positive coefficient in Equation (2) indicates that the presence of aromatic hydroxyl groups contributes to the inhibition of hDPP III. This is in accordance with an earlier study, where the presence of hydrophilic regions (i.e., hydroxyl groups) in flavonoids increased their inhibitory activity against hDPP III [25]. Williams plot revealed compound **2** as an outlier since it had a high predicted residual (predicted value in model (1) was significantly higher than the experimentally obtained). The presence of the hydroxyl group at position 6 might be the reason for the increase in the estimated value according to Equation (1), as well as the high value of *Mor10u* (Table S1). However, this model equation does not consider the presence of substituents at position 3, such as the -COCH$_3$ group in this case, that may have a negative effect on inhibition since compounds with this substituent exhibited relatively low inhibition values (Table 1).

Based on the conclusions given in the QSAR analysis, structures of two modified compounds (**41**, **42**) with possible improved activity are proposed, log (% inh. hDPP III) 3.08 and 3.01, respectively (Figure 3). Values of their calculated descriptors, as well as predicted inhibitory activities of the proposed compounds calculated using Equation (2) are given in the Supplementary Materials (Table S1). Since their calculated values exceed 100% of inhibition, these compounds could be potent inhibitors at concentrations lower than 10 μM. Both compounds possess a benzoyl group at the position C-3, and two hydroxyl groups at the position C-5 and C-7 (**41**), and at the position C-6 and C-8 (**42**). Improved calculated inhibition can be attributed to the introduction of an aromatic substituent and additional hydroxyl groups, as indicated by QSAR analysis.

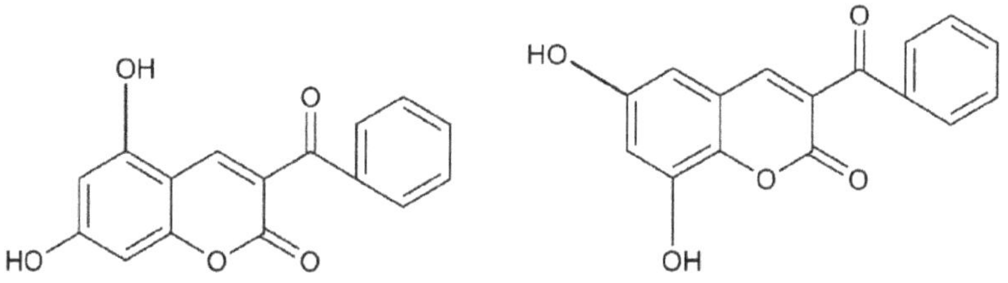

Figure 3. Structures of proposed compounds with possible enhanced inhibition of hDPP III.

2.3. Docking

In order to obtain further information on the possible interactions of the most active compound (**12**) with a semi-closed form of hDPP III, we combined docking with MD simulations. The best results regarding AutoDock Vina binding energy ($-8.6 \text{ kcal mol}^{-1}$) of enzyme–ligand complex predict that compound **12** binds to the inter-domain cleft, near the lower β sheet (residues 389–393) (Figure 4A) and enzyme active site, with the minimum distance between the catalytic Zn cation and **12** (oxygen atom at C2 of coumarin core) being ~7 Å. In this complex, the position of compound **12** closely resembles the substrate position in the hDPP III active site [9] which is accommodated similarly to the opioid peptides in the enzyme binding pocket [5]. Namely, binding of **12** into the inter-domain cleft is accompanied by its interactions mostly with amino acid residues of the hDPP III S1, S1′, S2, S2′ and S3′ substrate binding subsite (Figures S1 and 4B).

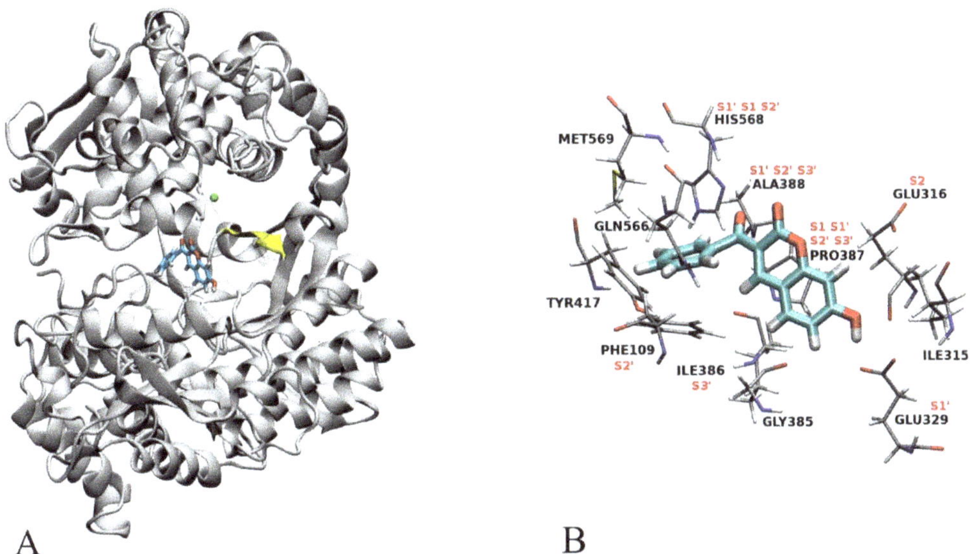

Figure 4. (**A**) Best docking pose for compound **12** in the inter-domain cleft of hDPP III. Compound **12** is shown in stick representation, the lower β sheet is colored yellow, and zinc cation is represented as a green sphere. (**B**) Potential interactions of compound **12** with amino acid residues of hDPP III as presented in the 2D scheme (Figure S1). Substrate binding subsites S1, S1′, S2, S2′ and S3′ are indicated.

2.4. MD Simulations

To prove the reliability of the best docking result, the binding mode of compound **12** in complex with hDPP III was investigated by productive MD simulations using the AMBER16 software package. Simulations of complex were performed in three replicates, each 300 ns long, and used for comparison. Dynamic behavior, protein, and ligand stability during simulations were analyzed by root mean square deviation (RMSD), while the analysis of the intermolecular interactions during MD simulations included hydrogen bonding (H-bond), native contacts, and Gibbs free energy. Representative structures of the complex were used to describe intermolecular interactions in more detail.

2.4.1. RMSD Profile

The RMSD profiles (Figure 5) calculated during the simulations for the protein backbone atoms show similar protein stability in all three runs, with only slightly higher protein stability in run 1 (average RMSD ± SD of 1.48 ± 0.18 Å, 1.76 ± 0.23 Å, and 1.96 ± 0.34 Å for run 1, 2, and 3, respectively). According to the RMSD values for the heavy atoms of compound **12** between replicates, it can be seen (Figure 6) that the stability of **12** is better in run 1 compared to the other two replicates. The average RMSD ± SD were 0.37 ± 0.15 Å, 0.46 ± 0.12 Å and 0.54 ± 0.15 Å, for run 1, run 2 and run 3, respectively.

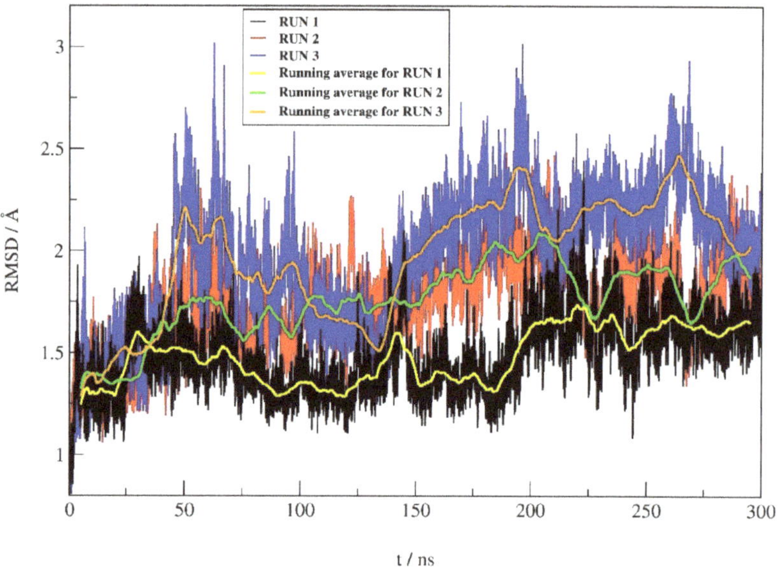

Figure 5. RMSD profile of the protein backbone atoms obtained during 300 ns of MD simulations.

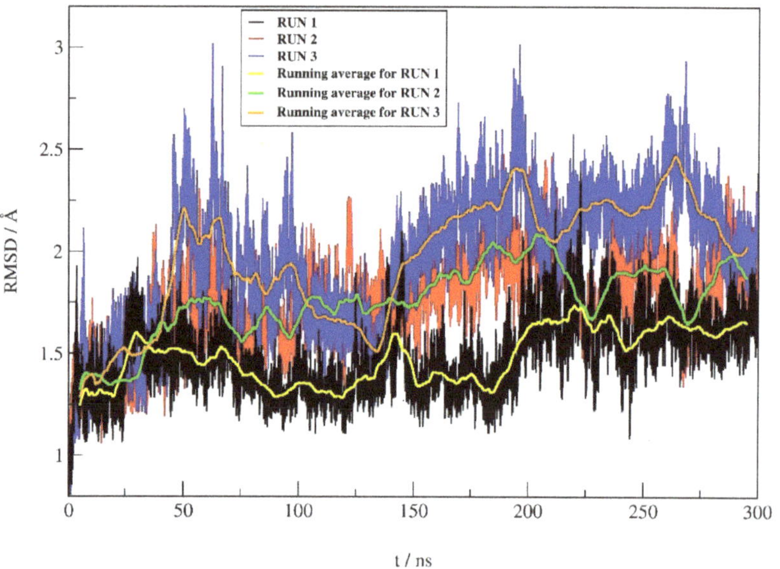

Figure 6. RMSD profile for the heavy atoms (hydrogen atoms were not considered) of compound **12** obtained from MD simulations of complex.

2.4.2. Hydrogen Bond Analysis

Hydrogen bond analysis was undertaken to investigate the stability and occupancy of hydrogen bonds between compound **12** and the key residues of the binding site of hDPP III. The results of trajectory H-bonds analysis for all three replicates are listed in Table S2.

In run 1, there were two H-bonds formed during the MD process with the occupation time >10% (Figure 7). The first H-bond is formed by the OE2 atom of Glu329 and the H-O4 of **12** with an occupation time of 99%, and the second one is formed by the H-NE2 of Gln566 and the O3 of **12** with an occupation time of 23%. Additionally, atom O2 of compound **12** forms H-bond with H-NE2 of Gln566 and H-OH of Tyr318 with the occupation time of 5% and 4%, respectively.

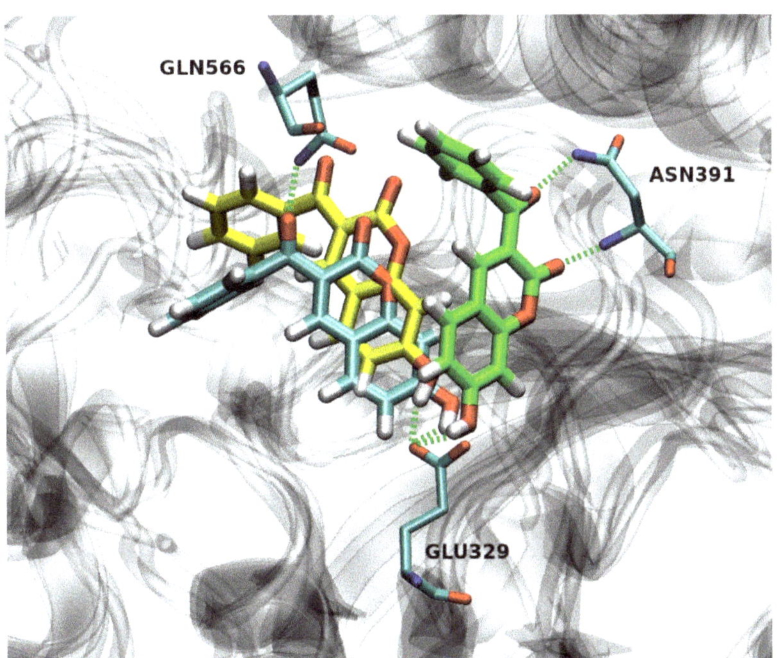

Figure 7. Overlay of the hDPP III with compound **12** in its preferable binding mode after 300 ns of MD simulation for run 1 (cyan), run 2 (green) and run 3 (yellow). Hydrogen bonds are depicted as green dashed lines.

Three hydrogen bonds were formed during the simulations in run 2 with the occupation time >10% (Figure 7). The first one is formed by the OE1 atom of Glu329 and the H-O4 of **12** with an occupation time of 85%. The second and third H-bond was formed between the H-N and H-NH2 of Asn391 and the O2 and O3 of **12** with the occupation time of 17% and 13%, respectively. One H-bond with an occupation time of 6% is formed by the H-ND2 of Asn391 and the O2 of compound **12**.

Only one hydrogen bond (with the occupation time > 10%) was formed during the simulations in run 3 by the OE2 atom of Glu329 and the H–O4 of **12** with an occupation time of 99% (Figure 7). Moreover, atom O3 of **12** forms H-bond with H–OH of Tyr417 and H-NE2 of His568, both with the occupation time of 4%, while atom O2 of **12** forms H-bond with H-NE2 of His568 with the occupation time of 3%. It is worthwhile to note that Glu329 and Gln568 are found to be constituents of hDPP III S1, S1′ and S2′ substrate-binding subsites [5]. Besides this, His568 and Asn491 are highly conserved residues among known DPP IIIs [39].

2.4.3. Native Contacts

To further investigate the interactions of hDPP III and compound **12**, we calculated the relative occupancy of native contacts during the MD simulations. Relative occupancy of native contacts is defined as the sum of fractions of native contacts during the trajectory for each residue pair relative to the total number of native contacts involved with that pair. Native contacts were defined as a distance between the atoms of enzyme residues and atoms of compound **12** within a distance cutoff of 5 Å. The fractions of native contacts were calculated using the *nativecontacts* command in the CPPTRAJ module.

Figures 8 and 9 depict the protein residues that were involved in forming native contacts with a relative occupancy of more than 30% in the run 1, 2, and 3 of the complex throughout the simulation time. In all three replicates, the protein forms native contacts with **12** through residues Ile315, Glu316, Glu329, Phe381, and Pro387, which indicates that these residues could be quite important for the stabilization of the complex. In runs 1 and 3, protein additionally forms native contacts with compound **12** through Phe 109, Ser317, and Ile386. The remaining contacts are formed in runs 2 and 3 through residues Gly389 and Ile390, and Gly385 and Tyr417, respectively. Some of the listed amino acid residues such as Glu316, Ser317, Glu329, Ile386, Pro387 and Tyr417 were found to contribute to the binding of synthetic inhibitors into the active site of hDPP III [22,24,39].

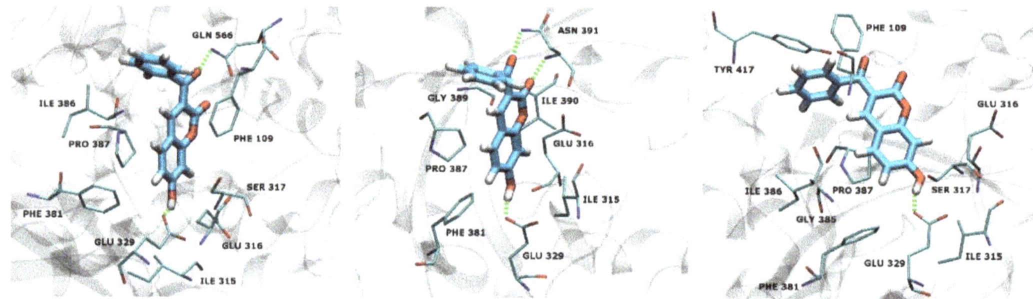

Figure 8. hDPP III residues involved in native contacts and H-bond formation with compound **12** for run 1 (**left**), run 2 (**middle**) and run 3 (**right**). H-bonds are depicted as green dashed lines, and compound **12** as light blue sticks.

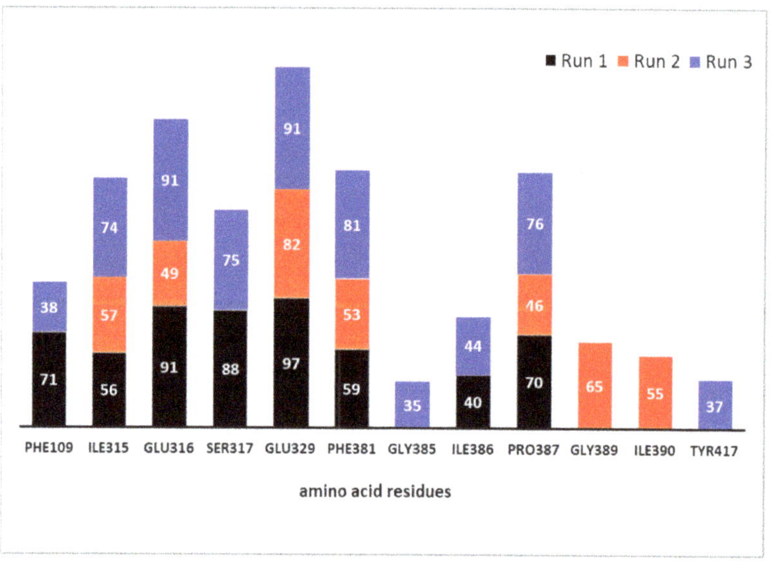

Figure 9. Native contacts between hDPP III residues and compound **12** with relative occupancy of more than 30% during 300 ns MD simulations.

2.4.4. Types of Intermolecular Interactions

A detailed analysis of different types of intermolecular interactions between replicates was performed using the Discovery Studio Visualizer. For this purpose, the extracted complex structures obtained from trajectory after 300 ns of MD simulations were optimized and used as representative. According to Figures S2–S4 representing 2D schemes of the intermolecular interaction types, compound **12** forms almost the same number of interactions with amino acid residues in all three replicates. However, in run 1, the largest number of interactions is formed between the coumarin core of **12** and the hDPP III, which is not the case in the other two replicates, where the benzoyl group of compound **12** also interacts with the amino acid residues. The above differences in the distributions of intermolecular interactions of **12** and amino acid residues relate mainly to the van der Waals interactions as shown in Figures S2–S4.

Comparing the other types of intermolecular interactions between replicates, the 2D schemes show that Glu329 and Pro387 form H-bonds and π-alkyl interactions with coumarin core in all three replicates, while additional H-bond and π-alkyl are formed with Asn391 and Ile390 in run 2, respectively. Additionally, in run 1, coumarin core forms π–π stacked interaction with Phe109 and amide π stacked interaction with Ile386. The benzoyl group of compound **12** forms one π-donor H-bond with Ile386, one π-alkyl with Ala567, and one π–π shaped interaction in run 1, run 2 and run 3, respectively.

2.4.5. MM-GBSA Free Energy Calculations

MM-GBSA calculations were used to obtain quantitative estimates of the free binding energies of compound **12** in the complex with the hDPP III for all three replicates. According to the results given in Table 4, the electrostatic contribution (ΔE_{ele}) is the most important to the ΔG_{bind} for complex in run 1 and run 2. This is in accordance with the results of MD simulations because in run 1 and run 2, compound **12** formed two and three H-bonds (with the occupation time > 10%), respectively, relative to run 3 where only one H-bond is formed. Another important contribution to the ΔG_{bind} of the complexes is the van der Waals interactions (E_{vdw}) with values similar in all three replicates. These results are in line with the observed similarities of the native contacts formed during the simulations, especially with residues Ile315, Phe381, and Pro387 in all three replicates. The unfavorable polar solvation contribution (E_{GB}) was slightly higher for complex in run 2 compared to runs 1 and 3, while the favorable nonpolar contribution (E_{SA}) had similar values for all three replicates.

Table 4. Binding free energy (kcal mol^{-1}) of the complexes obtained during the last 5 ns of MD simulations for all three replicates.

Energy Component	Run 1	Run 2	Run 3
E_{vdw}	−28.33	−27.44	−26.16
E_{ele}	−30.55	−30.76	−23.34
E_{GB}	35.94	41.26	34.69
E_{SA}	−3.44	−4.06	−3.41
ΔG_{gas}	−58.88	−58.20	−49.50
ΔG_{solv}	32.50	37.20	31.28
ΔG_{bind}	−26.36	−20.98	−18.22

E_{vdw}—van der Waals potential energy; E_{ele}—electrostatic energy; E_{GB}—polar solvation energy; E_{SA}—nonpolar solvation energy; ΔG_{gas}—gas phase free energy; ΔG_{solv}—solvation free energy; ΔG_{bind}—binding free energy.

From the estimated values of ΔG_{bind} between replicates, it can be concluded that the reason for the highest free binding energy in run 1 (−26.36 kcal mol^{-1}) is in the more favored E_{GB} and E_{ele} contribution compared to those in run 2 and run 3, respectively.

3. Material and Methods

3.1. Synthesis of Coumarin Derivatives

Synthesis and characterization of the coumarin derivatives were performed as described previously [40]. Briefly, series of coumarin derivatives were synthesized via Knoevenagel condensation starting from various substituted salicylaldehydes and ethyl acetoacetate (series 1; compounds **1–7**), ethyl benzoylacetate (series 2; compounds **8–15**), ethyl cyanoacetate (series 3; compounds **16–23**), diethyl malonate (series 4; compounds **24–31**), and dimethyl malonate (series 5, compounds **32–38**). 7-hydroxycoumarin and coumarin (compounds **39** and **40** respectively) were purchased from Sigma Chemical Co. (St. Louis, MO, USA). The structures of the tested compounds are presented in Table 1.

3.2. Heterologous Expression and Purification of Human DPP III

C-terminally truncated human DPP III was expressed and purified as described by Kumar et al. [7]. Briefly, C-terminally truncated hDPP III gene on a pET28-MHL plasmid, with an N-terminal His-tag and a TEV protease cleavage site fusion, was expressed in BL21-CodonPlus (DE3) RIL E. coli strain using 0.25 mM IPTG for induction of expression at 18 °C and 130 rpm shaking. After 20 h, cells were centrifuged and frozen at -20 °C until purification. Bacterial cells were lysed by a combination of lysozyme lysis and sonication, and the lysate was centrifuged to precipitate the cell debris. A brief DNase I treatment was performed before centrifugation to reduce the viscosity caused by DNA released by lysis. The lysate was purified on a Ni-NTA column (5 mL prepacked His-trap FF, GE Healthcare) using a buffer system with 50 mM Tris HCl, pH = 8.0, 300 mM NaCl, and increasing imidazole concentrations: 10 mM for lysis, 20 mM for wash and 300 mM for elution buffer. Fractions with hDPP III were pooled and incubated with TEV protease to remove the His-tag. hDPP III was recovered using flow-through affinity chromatography (TEV protease is His-tagged), and additionally purified on a 16/60 Superdex-200 gel-filtration column (GE Healthcare). Main fractions were pooled and desalted. Aliquots of protein in 20 mM Tris HCl buffer pH = 7.4 were stored at -80 °C until use. SDS PAGE of the purified enzyme was presented in Figure S5.

3.3. Assay of Human DPP III Activity

Purified hDPP III (1.5 nM) was preincubated with coumarin derivatives (10 μM) first for 5 min at 25 °C and then for 10 min at 37 °C in 50 mM Tris-HCl buffer, pH 7.4. The enzymatic reaction was started with Arg_2-2NA (40 μM) as a substrate, and after the 15 min incubation at 37 °C the reaction was stopped and the absorbance was measured using the spectrophotometric method described before [41]. Percentage enzyme inhibition (% inh.) was calculated by comparing the enzymatic activity without (control activity), and with inhibitor (inhibited activity) using the following formula:

$$\text{\% inh.} = [(\text{control activity} - \text{inhibited activity})/(\text{control activity})] \times 100\%$$

The IC_{50} values of selected compounds (**12** and **36**) were determined by the linear regression of the percentage of enzyme inhibition against the increasing concentrations (0.5–3.5 μM) of coumarin derivatives. The IC_{50} value is defined as the concentration of an inhibitor that caused a 50% reduction in the enzyme activity under assay conditions. The stock solutions (8 mM) of coumarin derivatives were freshly prepared in dimethyl sulfoxide and diluted with 50 mM Tris-HCl buffer, pH 7.4 buffer before assay of enzymatic activity.

3.4. Molecular Modeling
3.4.1. QSAR Analysis

Randomly ordered structures of 38 coumarin derivatives, coumarin, and 7-hydroxycoumarin (40 compounds in total) were drawn and optimized using the MM+ molecular mechanics force field [42]. Afterward, the structures were also subjected to geometry optimization using the PM3 semi-empirical method [43], using the Polak–Ribiere algorithm, until the root-mean-square gradient (RMS) was 0.1 kcal/(Åmol). Drawing and optimization of structures were performed in Avogadro 1.2.0. (University of Pittsburgh, Pittsburgh, PA, USA) [44].

Descriptor calculation for the resulted minimum energy conformations of compounds was performed with Parameter Client (Virtual Computational Chemistry Laboratory, an electronic remote version of the Dragon program) [45]. Logarithmic values of experimentally obtained hDPP III inhibition percentages were taken as response values. The generation and validation of QSAR models were performed using QSARINS 2.2.4 (University of Insubria, Varese, Italy) [46].

In order to reduce a large number of calculated descriptors, constant and semi-constant descriptors, i.e., those with a constant value for more than 85% of compounds, and descriptors that were too intercorrelated (>95%) were rejected by QSARINS. The final number of remaining descriptors was 514. Due to the high number of inactive compounds (14), 8 of them were randomly chosen and excluded from the dataset. A genetic algorithm (GA) was used to generate the best model. The number of descriptors in the multiple linear regression equation was limited to three. The splitting of compounds into the training set (n = 27 molecules) and test set (n = 5 molecules) was performed by activity sampling [47]. Compounds were ranked by their activities (from the most active to the least active compound) and then divided into five groups of the approximately same size. One compound was selected randomly from each group and assigned to the test set. The models were validated by the internal cross-validation performed using the "leave-one-out" (LOO) and Yscrambling method [46]. The following evaluation criteria were included: coefficient of determination (R^2), adjusted coefficient of determination (R^2_{adj}), cross-validated correlation coefficient (Q^2_{LOO}), inter-correlation among descriptors (K_{xx}), the difference of the correlation among the descriptors and the descriptors plus the responses (ΔK), the standard deviation of regression (s), Fisher ratio (F), root-mean-square error ($RMSE$); LOO cross-validated root-mean-square error ($RMSE_{cv}$), concordance correlation coefficient (CCC), LOO cross-validation concordance correlation coefficient (CCC_{cv}), mean absolute error of the training set (MAE), mean absolute error of the internal validation set (MAE_{cv}), and LOO cross-validated predictive residual sum of squares ($PRESS_{cv}$). QSAR model robustness was tested using the Y-randomization test, giving R^2_{Yscr} and Q^2_{Yscr} values [34]. External validation parameters included the coefficient of determination of the test set (R^2_{ext}), external validation set root-mean-square error ($RMSE_{ext}$), external validation set concordance correlation coefficient (CCC_{ext}), external validation set mean absolute error (MAE_{ext}), predictive squared correlation coefficients (Q^2_{F1}, Q^2_{F2}, Q^2_{F3}) and the average value of squared correlation coefficients between the observed and LOO predicted values of the compounds with and without intercept (r^2_m) [48].

To identify the possible outliers and compounds out of the warning leverage (h^*) in a model, a leverage plot (plot of standardized residuals vs. leverages (h); the Williams plot) was used. The warning leverage is generally defined as $3p'/n$ (n being the number of training compounds, and p' the number of model adjustable parameters [49]. Outliers in the Williams plot are compounds that have values of standardized residuals higher than two standard deviation units.

3.4.2. Preparation of the Complex Structure

The complex between the enzyme and compound **12** was built using the semi-closed conformation of hDPP III obtained earlier [8] by MD simulations of the structure available in the Protein Data Bank (PDB code: 3FVY), since it has been proved that this is the most preferable enzyme form in water solution [50]. Before the docking procedure, the protonation of histidines was checked according to their ability to form hydrogen bonds with neighboring amino acid residues. All Glu and Asp residues are negatively charged (-1) and all Arg and Lys residues are positively charged ($+1$), as expected at physiological conditions. AutoDock Vina 1.1.2 [51] was used to search for the best pose of the ligand to the enzyme active site. The docking site was defined as a cubical grid box with dimensions $75 \times 75 \times 75$ Å^3 and the center placed on the Zn^{2+}. Docking simulation was done with the standard 0.375 Å resolution and 20 conformations were generated. The complex with the best AutoDock Vina docking score was chosen for the productive MD simulations. Parameterization of the complex structure was performed by the AMBER-Tools16 modules *antechamber* and *tleap* using General Amber Force Field (GAFF) [52] and ff14SB [53] force fields to parameterize the ligand and the protein, respectively. For the zinc cation, Zn^{2+}, new hybrid bonded-nonbonded parameters were used from our previous work [54]. The complex was dipped into the truncated octahedral box filled with TIP3P water molecules with a margin distance of 11 Å. Besides water molecules, 24 sodium ions were added to neutralize the system and placed in the vicinity of charged amino acids at the protein surface.

3.4.3. Molecular Dynamics Simulations

Before the productive MD simulations, the complex was energy-minimized in three cycles to eliminate or reduce the energy constraints. Firstly, 1500 steps of minimization were performed, where the first 450 steps were of the steepest descent method, and the rest was the conjugate gradient. Both the protein atoms and the metal were constrained using a harmonic potential of force constant 32 kcal/(mol Å^2), to equilibrate water molecules. Secondly, 2500 steps were performed and only the first 470 steps of steepest descent were used. The metal and protein backbone were constrained with 32 kcal/(mol Å^2). Finally, in the third cycle, the same number of minimization steps was as in the first cycle, and both protein backbone and metal were constrained with 10 and 32 kcal/(mol Å^2), respectively. Next, the minimized system was heated from 0 to 300 K during 30 ps using a canonical ensemble (NVT), and then equilibrated 80 ps during which the initial constraints on the protein and the metal ion were used. This was followed by another equilibration stage of 100 ps, during which the initial constraints on the protein and the metal ion were removed and the water density was adjusted. The time step during the periods of heating and the water density adjustment was 1 fs. The equilibrated system was then subjected to 300 ns of the productive MD simulations (in three replicates) at constant temperature and pressure (300 K and 1 atm) using the NPT ensemble, without any constraints. The temperature was held constant using Langevin dynamics with a collision frequency of 1 ps^{-1}. Bonds involving hydrogen atoms were constrained using the SHAKE algorithm [55]. Simulations of the complex were performed within the AMBER16 software package [56]. The time step used for the productive MD simulations was set to 2 fs and the trajectory files were collected every 10 ps for the subsequent analysis. Trajectory analysis and the binding free energies (ΔG_{bind}) evaluation was performed by the CPPTRAJ module and MMPBSA.py script, respectively, from the AmberTools16 program package and examined visually using VMD 1.9.3 [57] and Discovery Studio Visualizer, version 20.1.0.19295 (BIOVIA, San Diego, CA, USA) software [58].

4. Conclusions

In summary, the potential hDPP III inhibitory activity of a series of coumarin derivatives was investigated for the first time by combining in vitro and in silico approaches. Compound **12** (3-benzoyl-7-hydroxy-2H-chromen-2-one) was found to be the most potent inhibitory molecule with IC$_{50}$ value of 1.10 μM. The productive MD simulations indicate that H-bonds between the 7-OH group of compound **12** and the carboxyl group of Glu329 as well as van der Waals interactions with Ile315, Ser317, Phe381, Pro387, and Ile390 are important for the mechanism of binding. According to the results of QSAR and binding mode analyses, two new compounds with possible improved activity were proposed. The discovery of coumarin derivatives as hDPP III inhibitors may provide new clues to the relationship between the chemical structure and biological activity of these naturally occurring compounds and their derivatives, and provide guidelines for the development of novel coumarin scaffolds as potent inhibitors of this mammalian metalloproteinase.

Supplementary Materials: The following are available online at https://www.mdpi.com/article/10.3390/ph14060540/s1. Table S1: The values of the descriptors included in QSAR model equation (2): log % hDPP III inh. = −4.07 + 1.85 (0.59) *EEig05x* + 1.60 (0.52) *Mor10u* + 0.56 (0.39) *nArOH*; Table S2: Detailed analysis of hydrogen bonds between compound **12** (LIG) and hDPP III residues during MD simulations for run 1, 2 and 3 obtained by *hbond* command in CPPTRAJ module; Figure S1: 2D diagram of compound **12** interactions with the hDPP III residues for the best docking pose; Figure S2: 2D diagram of compound **12** interactions with the hDPP III residues for run 1; Figure S3: 2D diagram of compound **12** interactions with the hDPP III residues for run 2; Figure S4: 2D diagram of compound **12** interactions with the hDPP III residues for run 3; Figure S5: SDS-PAGE demostrating purity of hDPP III sample on a 10% gel: M. PageRuler Prestained protein marker. with 72 kDa band in red; lane 1. hDPP III sample after affinity chromatography; lanes 2–5. fractions of the main hDPP III peak after gel-filtration.

Author Contributions: Conceptualization, D.A.; Methodology, D.A., V.R., M.M.; Software, S.T., D.A., and M.K.; Validation, D.A. and V.R.; Formal Analysis, D.A. and M.K.; Investigation, D.A., M.K., D.Š.; Resources, M.L. (Melita Lončarić), Z.K., D.B., B.M.P. and M.L. (Miroslav Lisjak); Data Curation, D.A. and M.K.; Writing–Original Draft Preparation, D.A., M.K., and S.T.; Visualization, D.A. and M.K.; Supervision, D.A.; Project Administration, S.T. and M.M.; Funding Acquisition, S.T., M.M., D.A.,V.R. and D.B.; Writing-review and editing, D.A., M.K., S.T., V.R., Z.K. All authors have read and agreed to the published version of the manuscript.

Funding: This work has been supported by the Croatian Science Foundation under the projects IP-2018-01-2936 (S.T.) and UIP-2017-05-6593 (M.M.).

Institutional Review Board Statement: Not applicable.

Informed Consent Statement: Not applicable.

Data Availability Statement: The data presented in this study are available on request from the corresponding author.

Acknowledgments: The authors acknowledge the Isabella cluster (http://www.srce.unizg.hr/en/isabella/, accessed on 9 September 2020) for computer time.

Conflicts of Interest: The authors declare no conflict of interest.

References

1. Chen, J.M.; Barrett, A.J. Dipeptidyl-peptidase III. In *Handbook of Proteolytic Enzymes*, 2nd ed.; Barrett, A.J., Rawlings, N.D., Woessner, J.F., Eds.; Elsevier Academic Press: Amsterdam, The Netherlands, 2004; Volume 1, pp. 809–812.
2. Rawlings, N.D.; Barrett, A.J.; Thomas, P.D.; Huang, X.; Bateman, A.; Finn, R.D. The MEROPS database of proteolytic enzymes, their substrates and inhibitors in 2017 and a comparison with peptidases in the PANTHER database. *Nucleic Acids Res.* **2018**, *46*, D624–D632. [CrossRef]
3. Abramić, M.; Zubanović, M.; Vitale, L. Dipeptidyl Peptidase III from Human Erythrocytes. *Biol. Chem. Hoppe Seyler* **1988**, *369*, 29–38. [CrossRef]
4. Baršun, M.; Jajčanin, N.; Vukelić, B.; Špoljarić, J.; Abramić, M. Human dipeptidyl peptidase III acts as a post-proline-cleaving enzyme on endomorphins. *Biol. Chem.* **2007**, *388*, 343–348. [CrossRef]
5. Bezerra, G.A.; Dobrovetsky, E.; Viertlmayr, R.; Dong, A.; Binter, A.; Abramic, M.; Macheroux, P.; Dhe-Paganon, S.; Gruber, K. Entropy-driven binding of opioid peptides induces a large domain motion in human dipeptidyl peptidase III. *Proc. Natl. Acad. Sci. USA* **2012**, *109*, 6525–6530. [CrossRef] [PubMed]
6. Karačić, Z.; Špoljarić, J.; Rozman, M.; Abramić, M. Molecular determinants of human dipeptidyl peptidase III sensitivity to thiol modifying reagents. *Biol. Chem.* **2012**, *393*, 1523–1532. [CrossRef]
7. Kumar, P.; Reithofer, V.; Reisinger, M.; Wallner, S.; Pavkov-Keller, T.; Macheroux, P.; Gruber, K. Substrate complexes of human dipeptidyl peptidase III reveal the mechanism of enzyme inhibition. *Sci. Rep.* **2016**, *6*, 23787. [CrossRef] [PubMed]
8. Tomić, A.; González, M.; Tomić, S. The Large Scale Conformational Change of the Human DPP III–Substrate Prefers the "Closed" Form. *J. Chem. Inf. Model.* **2012**, *52*, 1583–1594. [CrossRef] [PubMed]
9. Tomić, A.; Tomić, S. Hunting the human DPP III active conformation: Combined thermodynamic and QM/MM calculations. *Dalton Trans.* **2014**, *43*, 15503–15514. [CrossRef]
10. Tomić, A.; Kovačević, B.; Tomić, S. Concerted nitrogen inversion and hydrogen bonding to Glu451 are responsible for protein-controlled suppression of the reverse reaction in human DPP III. *Phys. Chem. Chem. Phys.* **2016**, *18*, 27245–27256. [CrossRef]
11. Liu, Y.; Kern, J.T.; Walker, J.R.; Johnson, J.A.; Schultz, P.G.; Luesch, H. A genomic screen for activators of the antioxidant response element. *Proc. Natl. Acad. Sci. USA* **2007**, *104*, 5205–5210. [CrossRef]
12. Matić, S.; Kekez, I.; Tomin, M.; Bogár, F.; Šupljika, F.; Kazazić, S.; Hanić, M.; Jha, S.; Brkić, H.; Bourgeois, B.; et al. Binding of dipeptidyl peptidase III to the oxidative stress cell sensor Kelch-like ECH-associated protein 1 is a two-step process. *J. Biomol. Struct. Dyn.* **2020**, 1–12. [CrossRef]
13. Sato, H.; Kimura, K.; Yamamoto, Y.; Hazato, T. Activity of DPP III in human cerebrospinal fluid derived from patients with pain. *Masui. Jpn. J. Anesthesiol.* **2003**, *52*, 257–263.
14. Šimaga, Š.; Babić, D.; Osmak, M.; Ilić-Forko, J.; Vitale, L.; Miličić, D.; Abramić, M. Dipeptidyl peptidase III in malignant and non-malignant gynaecological tissue. *Eur. J. Cancer* **1998**, *34*, 399–405. [CrossRef]
15. Šimaga, Š.; Babić, D.; Osmak, M.; Šprem, M.; Abramić, M. Tumor cytosol dipeptidyl peptidase III activity is increased with histological aggressiveness of ovarian primary carcinomas. *Gynecol. Oncol.* **2003**, *91*, 194–200. [CrossRef]

16. Šafranko, Ž.M.; Sobočanec, S.; Šarić, A.; Jajčanin-Jozić, N.; Krsnik, Ž.; Aralica, G.; Balog, T.; Abramić, M. The effect of 17β-estradiol on the expression of dipeptidyl peptidase III and heme oxygenase 1 in liver of CBA/H mice. *J. Endocrinol. Investig.* **2015**, *38*, 471–479. [CrossRef]

17. Prajapati, S.C.; Chauhan, S.S. Human dipeptidyl peptidase III mRNA variant I and II are expressed concurrently in multiple tumor derived cell lines and translated at comparable efficiency in vitro. *Mol. Biol. Rep.* **2016**, *43*, 457–462. [CrossRef]

18. Allard, M.; Simonnet, G.; Dupouy, B.; Vincent, J.D. Angiotensin II Inactivation Process in Cultured Mouse Spinal Cord Cells. *J. Neurochem.* **1987**, *48*, 1553–1559. [CrossRef] [PubMed]

19. Pang, X.; Shimizu, A.; Kurita, S.; Zankov, D.P.; Takeuchi, K.; Yasuda-Yamahara, M.; Kume, S.; Ishida, T.; Ogita, H. Novel Therapeutic Role for Dipeptidyl Peptidase III in the Treatment of Hypertension. *Hypertension* **2016**, *68*, 630–641. [CrossRef] [PubMed]

20. Jha, S.; Taschler, U.; Domenig, O.; Poglitsch, M.; Bourgeois, B.; Pollheimer, M.; Pusch, L.M.; Malovan, G.; Frank, S.; Madl, T.; et al. Dipeptidyl peptidase 3 modulates the renin–angiotensin system in mice. *J. Biol. Chem.* **2020**, *295*, 13711–13723. [CrossRef] [PubMed]

21. Matovina, M.; Agić, D.; Abramić, M.; Matić, S.; Karačić, Z.; Tomić, S. New findings about human dipeptidyl peptidase III based on mutations found in cancer. *RSC Adv.* **2017**, *7*, 36326–36334. [CrossRef]

22. Matić, J.; Šupljika, F.; Tir, N.; Piotrowski, P.; Schmuck, C.; Abramić, M.; Piantanida, I.; Tomić, S. Guanidiniocarbonyl-pyrrole-aryl conjugates as inhibitors of human dipeptidyl peptidase III: Combined experimental and computational study. *RSC Adv.* **2016**, *6*, 83044–83052. [CrossRef]

23. Šmidlehner, T.; Karačić, Z.; Tomić, S.; Schmuck, C.; Piantanida, I. Fluorescent cyanine-guanidiniocarbonyl-pyrrole conjugate with pH-dependent DNA/RNA recognition and DPP III fluorescent labelling and inhibition properties. *Mon. Chem. Chem. Mon.* **2018**, *149*, 1307–1313. [CrossRef]

24. Ćehić, M.; Sajko, J.S.; Karačić, Z.; Piotrowski, P.; Šmidlehner, T.; Jerić, I.; Schmuck, C.; Piantanida, I.; Tomić, S. The guanidiniocarbonylpyrrole–fluorophore conjugates as theragnostic tools for dipeptidyl peptidase III monitoring and inhibition. *J. Biomol. Struct. Dyn.* **2019**, *38*, 3790–3800. [CrossRef] [PubMed]

25. Agić, D.; Brkić, H.; Tomić, S.; Karačić, Z.; Špoljarević, M.; Lisjak, M.; Bešlo, D.; Abramić, M. Validation of flavonoids as potential dipeptidyl peptidase III inhibitors: Experimental and computational approach. *Chem. Biol. Drug Des.* **2017**, *89*, 619–627. [CrossRef] [PubMed]

26. Blagojević, B.; Agić, D.; Serra, A.T.; Matić, S.; Matovina, M.; Bijelić, S.; Popović, B.M. An in vitro and in silico evaluation of bioactive potential of cornelian cherry (Cornus mas L.) extracts rich in polyphenols and iridoids. *Food Chem.* **2021**, *335*, 127619. [CrossRef]

27. Detsi, A.; Kontogiorgis, C.; Hadjipavlou-Litina, D. Coumarin derivatives: An updated patent review (2015–2016). *Expert Opin. Ther. Pat.* **2017**, *27*, 1201–1226. [CrossRef]

28. Borges, F.; Roleira, F.; Milhazes, N.; Santana, L.; Uriarte, E. Simple Coumarins and Analogues in Medicinal Chemistry: Occurrence, Synthesis and Biological Activity. *Curr. Med. Chem.* **2005**, *12*, 887–916. [CrossRef] [PubMed]

29. Lončarić, M.; Gašo-Sokač, D.; Jokić, S.; Molnar, M. Recent Advances in the Synthesis of Coumarin Derivatives from Different Starting Materials. *Biomolecules* **2020**, *10*, 151. [CrossRef]

30. Anand, P.; Singh, B.; Singh, N. A review on coumarins as acetylcholinesterase inhibitors for Alzheimer's disease. *Bioorg. Med. Chem.* **2012**, *20*, 1175–1180. [CrossRef]

31. Parellada, J.; Suárez, G.; Guinea, M. Inhibition of Zinc Metallopeptidases by Flavonoids and Related Phenolic Compounds: Structure-Activity Relationships. *J. Enzym. Inhib.* **1998**, *13*, 347–359. [CrossRef]

32. Ali, Y.; Seong, S.H.; Jung, H.A.; Choi, J.S. Angiotensin-I-Converting Enzyme Inhibitory Activity of Coumarins from Angelica decursiva. *Molecules* **2019**, *24*, 3937. [CrossRef]

33. Xu, X.-M.; Zhang, Y.; Qu, D.; Feng, X.-W.; Chen, Y.; Zhao, L. Osthole suppresses migration and invasion of A549 human lung cancer cells through inhibition of matrix metalloproteinase-2 and matrix metallopeptidase-9 in vitro. *Mol. Med. Rep.* **2012**, *6*, 1018–1022. [CrossRef]

34. Gramatica, P. Principles of QSAR models validation: Internal and external. *QSAR Comb. Sci.* **2007**, *26*, 694–701. [CrossRef]

35. Todeschini, R.; Consonni, V. *Handbook of Molecular Descriptors*, 1st ed.; Wiley VCH: Weinheim, Germany, 2000; pp. 91–92, 131–132.

36. Estrada, E.; Ramírez, A. Edge Adjacency Relationships and Molecular Topographic Descriptors. Definition and QSAR Applications. *J. Chem. Inf. Comput. Sci.* **1996**, *36*, 837–843. [CrossRef]

37. Rastija, V.; Agić, D.; Tomić, S.; Nikolic, S.; Hranjec, M.; Karminski-Zamola, G.; Abramić, M. Synthesis, QSAR, and Molecular Dynamics Simulation of Amidino-substituted Benzimidazoles as Dipeptidyl Peptidase III Inhibitors. *Acta Chim. Slov.* **2015**, *62*, 867–878. [CrossRef] [PubMed]

38. Schuur, J.H.; Selzer, P.; Gasteiger, J. The Coding of the Three-Dimensional Structure of Molecules by Molecular Transforms and Its Application to Structure-Spectra Correlations and Studies of Biological Activity. *J. Chem. Inf. Comput. Sci.* **1996**, *36*, 334–344. [CrossRef]

39. Abramić, M.; Špoljarić, J.; Šimaga, Š. Prokaryotic homologs help to define consensus sequences in peptidase family M49. *Period. Biol.* **2004**, *106*, 161–168.

40. Lončarić, M.; Sušjenka, M.; Molnar, M. An Extensive Study of Coumarin Synthesis via Knoevenagel Condensation in Choline Chloride Based Deep Eutectic Solvents. *Curr. Org. Synth.* **2020**, *17*, 98–108. [CrossRef]

41. Špoljarić, J.; Salopek-Sondi, B.; Makarević, J.; Vukelić, B.; Agić, D.; Šimaga, Š.; Jajčanin-Jozić, N.; Abramić, M. Absolutely conserved tryptophan in M49 family of peptidases contributes to catalysis and binding of competitive inhibitors. *Bioorg. Chem.* **2009**, *37*, 70–76. [CrossRef]

42. Hocquet, A.; Langgård, M. An Evaluation of the MM+ Force Field. *J. Mol. Model.* **1998**, *4*, 94–112. [CrossRef]

43. Stewart, J.J.P. Optimization of parameters for semiempirical methods I. Method. *J. Comput. Chem.* **1989**, *10*, 209–220. [CrossRef]

44. Hanwell, M.D.; E Curtis, D.; Lonie, D.C.; Vandermeersch, T.; Zurek, E.; Hutchison, G.R. Avogadro: An advanced semantic chemical editor, visualization, and analysis platform. *J. Chemin.* **2012**, *4*, 17. [CrossRef] [PubMed]

45. Tetko, I.V.; Gasteiger, J.; Todeschini, R.; Mauri, A.; Livingstone, D.; Ertl, P.; Palyulin, V.; Radchenko, E.; Zefirov, N.S.; Makarenko, A.; et al. Virtual Computational Chemistry Laboratory–Design and Description. *J. Comput. Mol. Des.* **2005**, *19*, 453–463. [CrossRef] [PubMed]

46. Gramatica, P.; Chirico, N.; Papa, E.; Cassani, S.; Kovarich, S. QSARINS: A new software for the development, analysis, and validation of QSAR MLR models. *J. Comput. Chem.* **2013**, *34*, 2121–2132. [CrossRef]

47. Golbraikh, A.; Shen, M.; Xiao, Z.; Xiao, Y.-D.; Lee, K.-H.; Tropsha, A. Rational selection of training and test sets for the development of validated QSAR models. *J. Comput. Mol. Des.* **2003**, *17*, 241–253. [CrossRef] [PubMed]

48. Chirico, N.; Gramatica, P. Real External Predictivity of QSAR Models. Part 2. New Intercomparable Thresholds for Different Validation Criteria and the Need for Scatter Plot Inspection. *J. Chem. Inf. Model.* **2012**, *52*, 2044–2058. [CrossRef]

49. Eriksson, L.; Jaworska, J.; Worth, A.P.; Cronin, M.; McDowell, R.M.; Gramatica, P. Methods for reliability and uncertainty assessment and for applicability evaluations of classification- and regression-based QSARs. *Environ. Health Perspect.* **2003**, *111*, 1361–1375. [CrossRef]

50. Tomić, A.; Berynskyy, M.; Wade, R.C.; Tomić, S. Molecular simulations reveal that the long range fluctuations of human DPP III change upon ligand binding. *Mol. BioSyst.* **2015**, *11*, 3068–3080. [CrossRef] [PubMed]

51. Trott, O.; Olson, A.J. AutoDock Vina: Improving the speed and accuracy of docking with a new scoring function, efficient optimization, and multithreading. *J. Comput. Chem.* **2010**, *31*, 455–461. [CrossRef] [PubMed]

52. Wang, J.; Wolf, R.M.; Caldwell, J.W.; Kollman, P.A.; Case, D.A. Development and testing of a general amber force field. *J. Comput. Chem.* **2004**, *25*, 1157–1174. [CrossRef]

53. Maier, J.A.; Martinez, C.; Kasavajhala, K.; Wickstrom, L.; Hauser, K.E.; Simmerling, C. ff14SB: Improving the Accuracy of Protein Side Chain and Backbone Parameters from ff99SB. *J. Chem. Theory Comput.* **2015**, *11*, 3696–3713. [CrossRef] [PubMed]

54. Tomić, A.; Horvat, G.; Ramek, M.; Agić, D.; Brkić, H.; Tomić, S. New Zinc Ion Parameters Suitable for Classical MD Simulations of Zinc Metallopeptidases. *J. Chem. Inf. Model.* **2019**, *59*, 3437–3453. [CrossRef] [PubMed]

55. Ryckaert, J.-P.; Ciccotti, G.; Berendsen, H.J.C. Numerical integration of the cartesian equations of motion of a system with constraints: Molecular dynamics of n-alkanes. *J. Comput. Phys.* **1977**, *23*, 327–341. [CrossRef]

56. Case, D.; Betz, R.; Cerutti, D.; Cheatham, T.E., III; Darden, T.; Duke, R.E.; Giese, T.; Gohlke, H.; Goetz, A.W.; Homeyer, N.; et al. *AMBER 2016*; University of California: San Francisco, CA, USA, 2016.

57. Humphrey, W.; Dalke, A.; Schulten, K. VMD: Visual molecular dynamics. *J. Mol. Graph.* **1996**, *14*, 33–38. [CrossRef]

58. Dassault Systèmes BIOVIA. *Discovery Studio Visualizer*; Release 2019; Dassault Systèmes: San Diego, CA, USA, 2019.

pharmaceuticals

Review

Targeting Cartilage Degradation in Osteoarthritis

Oliver McClurg, Ryan Tinson and Linda Troeberg *

Norwich Medical School, University of East Anglia, Bob Champion Research and Education Building, Rosalind Franklin Road, Norwich NR4 7UQ, UK; O.Mcclurg@uea.ac.uk (O.M.); rtinson@googlemail.com (R.T.)
* Correspondence: L.Troeberg@uea.ac.uk; Tel.: +44-1603-591910

Abstract: Osteoarthritis is a common, degenerative joint disease with significant socio-economic impact worldwide. There are currently no disease-modifying drugs available to treat the disease, making this an important area of pharmaceutical research. In this review, we assessed approaches being explored to directly inhibit metalloproteinase-mediated cartilage degradation and to counteract cartilage damage by promoting growth factor-driven repair. Metalloproteinase-blocking antibodies are discussed, along with recent clinical trials on FGF18 and Wnt pathway inhibitors. We also considered dendrimer-based approaches being developed to deliver and retain such therapeutics in the joint environment. These may reduce systemic side effects while improving local half-life and concentration. Development of such targeted anabolic therapies would be of great benefit in the osteoarthritis field.

Keywords: osteoarthritis; cartilage; metalloproteinase; targeting

Citation: McClurg, O.; Tinson, R.; Troeberg, L. Targeting Cartilage Degradation in Osteoarthritis. *Pharmaceuticals* **2021**, *14*, 126. https://doi.org/10.3390/ph14020126

Academic Editor: Salvatore Santamaria
Received: 23 January 2021
Accepted: 1 February 2021
Published: 5 February 2021

Publisher's Note: MDPI stays neutral with regard to jurisdictional claims in published maps and institutional affiliations.

1. Introduction

Osteoarthritis (OA) is a degenerative joint disorder that affects over 250 million people worldwide [1,2], causing pain, stiffness, and impaired movement. OA most commonly occurs in the knee, hip, and hand joints [2,3], but can also affect other articulating joints. The disease is characterized by progressive loss of cartilage, which impairs smooth articulation of opposing bones in the joint, with remodeling of subchondral bone (including sclerosis and osteophyte formation) and other joint tissues (such as the meniscus, ligaments, synovium, and intrapatellar fat pad) also contributing to joint degeneration [4–6].

There are currently no disease-modifying OA drugs (DMOADs) available, with treatment limited to analgesia for early-stage disease and surgical joint replacement for late-stage disease. The development of effective drugs to treat OA is thus of utmost importance. Much research in this area is focused on cartilage, and aims to identify approaches for either stopping cartilage breakdown or promoting cartilage repair [2,7].

Cartilage consists of a single cell type, the chondrocytes, which synthesize and are embedded in a relatively large volume of extracellular matrix (ECM) that is critical for the tissue's mechanical properties [8]. The two most abundant components of the cartilage ECM are type II collagen and aggrecan, which confer tensile strength and resistance to compression, respectively. In healthy cartilage, anabolic synthesis of ECM components is balanced with their catabolic turnover to maintain joint homeostasis. In OA, changes in the mechanical environment of the joint (e.g., by injury or ageing) shift this balance towards degradation, with breakdown of type II collagen and aggrecan by metalloproteinases leading to progressive joint damage [9,10]. Expression of several of these metalloproteinases, as well as pro-inflammatory mediators, is induced by mechanical destabilization of the joint [9]. Additional factors including chondrocyte senescence [11], oxidative stress [12–15], and/or inflammation can also increase metalloproteinase expression and tip the balance towards cartilage breakdown.

Collagenases such as matrix metalloproteinase 13 (MMP-13) are thought to be important for degradation of type II collagen in OA [16–18], while aggrecanases such as adamalysin with thrombospondin motifs 5 (ADAMTS-5) drive aggrecan loss [19,20]. Activity of these metalloproteinases is increased in OA through a variety of molecular mechanisms, including increased expression [16,21], decreased endocytic clearance [22,23], and a reduction in levels of their endogenous inhibitor, tissue inhibitor of metalloproteinases 3 (TIMP-3) [24]. These enzymes have been the target of multiple academic and industry DMOAD programs, but this approach has been challenging, due to high homology with several homologous metalloproteinases that have important homeostatic roles in processes such as wound healing and cell migration.

An alternative approach to OA therapy is to boost the intrinsic repair capacity of cartilage, through delivery of growth factors that can promote ECM synthesis and cartilage repair. For example, growth factors in the fibroblast growth factor (FGF), transforming growth factor β (TGF-β), and Wnt families are known to have important roles in cartilage development, and can also promote repair of damaged adult cartilage. However, development of growth factor-based DMOADs has also been challenging, due to complexities in their signaling pathways and unexpected effects including fibrosis and osteophyte formation [25].

Here we provided an update on recent developments and clinical trials in the search for metalloproteinase- and growth factor-focused DMOADs, as well as approaches being explored for delivery and retention of potential OA therapeutics in the joint. Such delivery strategies are increasingly seen as a promising way to limit effects of potential DMOADs to the joint and thus to reduce adverse systemic effects.

2. Novel Strategies for Inhibiting Cartilage Breakdown

2.1. Inhibiting ADAMTS Activity

The role of MMPs and ADAMTSs in OA cartilage degradation is well appreciated, making these enzymes attractive targets for DMOAD development [26–30]. ADAMTS-5 has received particular attention [26–28], because reversible degradation of aggrecan is thought to precede irreversible collagen loss [31,32], and ADAMTS-5 is thought to be the critical "aggrecanase" in both murine and human OA [19,20,33].

As with MMP inhibitors before them [34,35], design of ADAMTS-5 inhibitors has been challenging, with high homology between metalloproteinase active sites leading to off-target inhibition of metalloproteinases with homeostatic functions in processes such as angiogenesis, cell migration, and wound healing [26–28]. Strategies to improve selectivity include development of bi-specific, cross-domain antibody scaffolds. This approach has been utilized by Merck Serono, whose bi-specific nanobody against ADAMTS-5 (M6495) has been shown to protect against surgically-induced murine OA in vivo [36]. M6495 was recently out-licensed to Novartis, and phase II trials are expected. Targeting exosites outside of the metalloproteinase active site is another promising approach, recently shown to enable development of an inhibitor with high selectivity for ADAMTS-5 over ADAMTS-4 [37].

In addition to adverse off-target effects, metalloproteinase inhibitors can also potentially have unwanted on-target effects, arising from expression of target aggrecanases and collagenases outside of the joint. For example, expression of ADAMTS-5 in the vasculature and heart valves is thought to underlie cardiovascular side-effects observed with the anti-ADAMTS-5 inhibitory antibody GSK2394002 [38]. Potential roles of ADAMTS-5 in wound healing, glucose metabolism, inflammation, and neural plasticity (reviewed in [28]) warrant further investigation. Similarly, MMP-13 has physiological roles in wound healing [39], muscle regeneration [40], and fracture healing [41,42] that should be kept in mind when designing MMP-13 inhibitors for OA therapy.

Strategies other than direct inhibition of metalloproteinase activity are also being explored. For example, siRNAs against *Adamts5* [43] and *Mmp13* [43,44] have both been successfully used to reduce cartilage degradation in rodent OA models. ADAMTS-5 and MMP-13 are also post-translationally regulated by endocytosis and lysosomal degradation, mediated by the low-density lipoprotein receptor-related protein 1 (LRP1) scavenger receptor [22,23,45], and this protective endocytic process is impaired in OA cartilage [22,23,46]. Antibodies inhibiting shedding of LRP1 were found to inhibit cartilage degradation in vitro [46].

2.2. Augmenting Levels of Endogenous TIMPs

Another potential approach is to enhance levels of the endogenous inhibitors of MMPs and ADAMTSs in cartilage. Among the 4 mammalian tissue inhibitors of metalloproteinases (TIMPs), TIMP-3 is the only one that effectively inhibits both MMPs and ADAMTSs [47], with TIMP-1, TIMP-2, and TIMP-4 having more restricted inhibitory profiles and largely targeting MMPs (reviewed in [48]). Although there is no significant change in mRNA expression of TIMP-3 in OA cartilage [24,49], levels of the protein itself are reduced [24]. Subsequent studies in our group showed that, as with ADAMTS-5 and MMP-13, TIMP-3 levels are also post-translationally regulated by LRP1-mediated endocytosis [50]. It is not currently clear how endocytosis of metalloproteinases and TIMP-3 is coordinated in cartilage, although factors such as their differential affinity for extracellular matrix glycosaminoglycans and localization in tissue are likely to be key [51], seeing as their affinity for LRP1 is not markedly different [52]. Mutants of TIMP-3 with reduced affinity for LRP1 were found to have longer half-lives in cartilage and improved chondroprotective activity [53]. Sulfated glycans and glycan mimetics that block TIMP-3 binding to LRP1 were found to be similarly protective in cartilage explant assays [54,55] and murine OA models [56], raising the possibility that small molecule inhibitors of TIMP-3 binding to LRP1 may inhibit cartilage matrix degradation. TIMP-3-directed approaches would have to overcome the same issues of specificity and joint targeting as discussed for metalloproteinase-directed approaches.

3. Promoting Cartilage Repair with Anabolic Growth Factors

An alternative approach to directly targeting metalloproteinases is to stimulate cartilage repair through delivery of growth factors that inhibit metalloproteinase-mediated cartilage degradation while promoting anabolic repair pathways (reviewed by [57]). Here we discuss progress and recent or current clinical trials in this area.

3.1. Platelet-Rich Plasma Therapy

Platelet-rich plasma (PRP) therapy for OA involves intra-articular injection with a preparation of autologous plasma containing high platelet levels. Once in the joint, platelets are activated by abundant cartilage ECM proteins, leading to release of their cytoplasmic components, including several anabolic growth factors such as TGF-β1, platelet-derived growth factor (PDGF), insulin-like growth factor (IGF), and FGF2, which promote aggrecan and collagen synthesis, while reducing expression and activity of catabolic metalloproteases [58].

In vitro studies on the effects of PRP on chondrocyte gene expression have yielded mixed results. For example, PRP prepared from porcine blood was found to have anabolic effects on porcine chondrocytes cultured in alginate beads, stimulating an increase in DNA, glycosaminoglycan, and type II collagen content relative to cells cultured with platelet-poor plasma or 10% fetal bovine serum [59]. Similarly, PRP has been shown to increase expression of *COL2A1* [60,61] and *ACAN* [60,61] in human OA chondrocytes, while decreasing interleukin-1β (IL-1β)-induced *ADAMTS4* expression [60]. However, other studies have found PRP to have no beneficial effects on *COL2A1* expression in human OA chondrocytes [62], or have reported adverse effects such as dedifferentiation of chondrocytes to a fibroblast-like phenotype [63]. The reason for these discrepancies is not clear, but may result from differences in the method of PRP preparation. For example, comparison of two commercially available kits showed that white blood cell levels were higher in some preparations than others, with increased neutrophil levels correlating with higher concentrations of pro-inflammatory cytokines such as IL-1β [64], which may promote MMP activation and cartilage breakdown rather than repair.

Clinical trials of PRP therapy have also shown mixed results, although comparison between randomized control trials (RCT) trials of PRP are complicated by variation in their control arms, with some comparing PRP to saline [65], and others comparing it with hyaluronan (HA) [66–68]. Among studies that compared PRP with HA, both positive and negative results have been reported. For example, PRP has been shown to reduce Western Ontario and McMaster Universities Osteoarthritis Index (WOMAC) scores more significantly than HA for up to 6 months [66] and to improve self-reported pain more effectively than HA for up to 12 months [67], while other studies have found PRP to be no more effective than HA at improving various patient-reported outcome measures at 2, 6, and 12 months after treatment [68]. As with in vitro studies of PRP, these different RCT outcomes may reflect differences in PRP preparation, with double filtration of plasma more likely to remove leukocytes than centrifugation [69].

RCTs of PRP have largely assessed patient-reported outcome measures, and there is relatively little information on whether PRP can alter structural joint outcomes. A recently completed but as yet unpublished RCT (NCT03491761) compared the effects of PRP and HA on knee OA by measuring objective changes such as cartilage thickness in addition to subjective measures such as WOMAC scoring. Such studies are likely to shed light on whether PRP can directly promote cartilage repair and regeneration in vivo. An additional challenge in this area is the requirement for cheap and efficient methods of producing PRP fractions with low leukocyte levels.

3.2. FGF18 Promotes Cartilage Anabolism

While FGF2 has been shown to have both anabolic and catabolic effects on chondrocytes (see Section 3.5, Considerations around Receptor Expression and Downstream Signaling), the current literature supports a chondroprotective, anabolic role for FGF18. For example, in vitro studies showed that FGF18 increased proliferation and proteoglycan production by primary human and porcine chondrocytes [70], and in vivo studies have demonstrated that intra-articular injection of FGF18 significantly reduced cartilage degeneration in rat pre-clinical OA models [71,72].

Following on from these promising studies, pharmaceutical interest in FGF18 has grown. Nordic Biosciences and Merck/EMD Serano developed a truncated form of FGF18, named sprifermin, which lacks the signal peptide and 11 C-terminal amino acids [73]. This modified form of FGF18 retains biological activity, with sprifermin shown to dose-dependently stimulate proliferation of cultured human and porcine chondrocytes, and to increase glycosaminoglycan and type II collagen accumulation, while decreasing *ADAMTS5* expression [73].

A number of human clinical trials on sprifermin have now been completed, with intra-articular injection of 3–300 µg of the growth factor followed by evaluation of joint structural parameters and patient-reported outcome measures for up to 2 years [74–76].

Dahlberg et al. conducted a first-in-human double-blind RCT with dose escalation that established safety in humans [74]. Overall, treatment-emergent adverse events were not increased in the sprifermin-treated cohort compared to placebo. Twice as many acute inflammatory reactions were seen in the sprifermin-treated cohort (12.7%) as in the placebo group (5.5%), but overall sprifermin was deemed tolerable. The study found no difference in Mankin scores, joint space width or semi-quantitative MRI parameters between the cohorts, although the study was not sufficiently powered for such comparisons with only 55 participants [74].

A second dose-escalating RCT failed to meet its primary endpoint, with no significant change in the thickness of cartilage in the central medial femorotibial compartment detected by quantitative magnetic resonance imaging (MRI) at 6 and 12 months [75]. However, sprifermin did cause a statistically significant reduction in the loss of total and lateral femorotibial cartilage thickness and in joint space width narrowing in the lateral femorotibial compartment compared with the placebo group. WOMAC pain scores improved significantly in both the treatment and control arms, although the improvement was significantly less in the sprifermin-treated cohort than in the placebo group [75].

Finally, the FORWARD phase II RCT completed in 2019 reported statistically significant and dose-dependent increases in total femorotibial cartilage thickness after 2 years in cohorts treated with 100 µg of sprifermin every 6 or 12 months compared to placebo [76]. Furthermore, significant increases in cartilage thickness were observed in both the medial and lateral femorotibial compartments, along with significant dose-dependent effects on joint space width in the lateral, but not medial, compartment in cohorts treated with 100 µg of sprifermin. Lower doses of sprifermin did not cause significant improvements. As in the previous sprifermin RCT [75], there was no statistically significant change in WOMAC pain scores between treatment cohorts, with analgesia use similar across treatment groups [76]. There was also no significant improvement in function or stiffness sub-scores.

Taken together, these RCTs support sprifermin having a chondroprotective effect, with positive effects on cartilage thickness and joint space narrowing. Effects on pain and stiffness have not been demonstrated, reflecting the broader question of whether OA pain correlates with structural joint changes (see Section 5, Broader Challenges in DMOAD Development).

3.3. Wnt Pathway Inhibition

Wnt signaling is required for cartilage and bone development and homeostasis, but sustained or elevated Wnt signaling in chondrocytes promotes their proliferation and hypertrophic differentiation, with deleterious effects on cartilage homeostasis (reviewed in [77–79]). Inhibition of Wnt signaling is thus being explored as a potential therapy for OA. This approach could have the added benefit of inducing differentiation of mesenchymal stem cells (MSCs) into chondrocytes, since Wnt signaling in MSCs promotes their differentiation into osteoblasts rather than chondrocytes [80]. MSCs are present in elevated numbers in the synovial fluid of OA patients [81], suggesting that while OA joints have the potential to repair, the osteoarthritic environment does not support differentiation of resident MSCs into cartilage-forming chondrocytes. Inhibitors of Wnt signaling may thus both inhibit cartilage degradation and harness its potential for repair.

SM04690 (Lorecivivint, Samumed), a small molecule inhibitor of Wnt signaling, has progressed from in vitro assessment to human clinical trials. In vitro, SM04690 inhibited Wnt pathway activation and induced differentiation of human MSCs into chondrocytes while inhibiting expression of catabolic metalloproteinases [82,83]. In a rat anterior cruciate ligament transection (ACLT) model of OA, intra-articular injection of SM04690 a week after surgery significantly reduced cartilage degradation 12 weeks later [83]. The mechanism of action is proposed to be through inhibition of the intranuclear kinases CDC-like kinase 2 (CLK2) and dual-specificity tyrosine phosphorylation-regulated kinase 1A (DYRK1A), without affecting β-catenin [84].

A subsequent phase I RCT showed that SM04690 is well tolerated and safe in humans [85]. Participants with moderate to severe OA received a single intra-articular dose of SM04690 ($n = 48$ treated with 0.03 to 0.23 mg SM04690, compared with $n = 11$ in the placebo arm), with safety and efficacy evaluated 24 weeks later. Adverse effects potentially related to treatment were observed primarily in those receiving the highest dose of SM04690 (0.23 mg), and included arthralgia, joint swelling and stiffness (in 5 subjects), as well as gastrointestinal effects (in 4 subjects) [85].

In a phase IIa clinical trial of SM04960 completed in 2018, clinically meaningful improvements in WOMAC pain and function scores were seen in all groups, including the placebo group who received an intra-articular injection of PBS [86]. The trial thus did not meet its primary endpoint of a significant reduction in the WOMAC pain score compared with placebo 13 weeks after a single intra-articular dose of SM04960. However, WOMAC pain scores were significantly lower at the 52-week time point in the group treated with 0.07 mg of SM0496 than in the placebo group [86]. Medial joint space width was also significantly improved at 52 weeks in this treatment group compared with placebo [86].

The Wnt inhibitors SAH-Bcl9 and StAx-35R are also currently under early investigation in OA [87]. These compounds have been shown to inhibit Wnt signaling in cancers, and also inhibited Wnt3a-induced downregulation of chondrogenic markers such as *COL2A1* and *SOX9* in human OA cartilage explants in vitro [87].

Effective and sustained therapeutic targeting of this pathway is likely to be challenging, with multiple ligands and antagonists influencing both canonical and non-canonical Wnt signaling (reviewed in [78,79]), and potential reciprocal antagonism between these pathways shaping the effect on the chondrocyte phenotype [88]. Further investigation of these complexities is required to understand the consequences of Wnt modulation in vivo.

3.4. TGF-β1 Supplementation

TGF-β1 is the most extensively studied growth factor in cartilage, with in vivo evaluation of its anabolic effects stretching back 3 decades (reviewed in [25]). Genetic evidence for a chondroprotective role for TGF-β1 is strong, with a number of mutations in its signaling cascade associated with increased OA risk [25]. For example, the D-14 polymorphism in asporin, which reduces TGF-β signaling, has been found to increase OA susceptibility in some populations [89,90], and a *SMAD3* single nucleotide polymorphism (SNP) has been linked to increased hip and knee OA in a European 527-strong cohort [91]. These findings are supported by in vivo studies, which showed, for example, accelerated OA in mice that overexpress a dominant negative form of TGF-β receptor II (*Tgfbr2*) [92] or that lack *Tgfbr2* [93] or *Smad3* [94]. However, in addition to its beneficial effects on cartilage, the anabolic effects of TGF-β1 are undesirable in other joint tissues, leading to the synovial fibrosis and osteophyte formation reported in a number of studies (reviewed in [25]).

Mont and colleagues have conducted a number of clinical trials to assess treatment of knee OA by intra-articular injection of allogeneic chondrocytes transduced to express TGF-β1. A small phase I trial in 12 patients concluded the treatment was safe, and reported improved range of motion and pain scores after 6 and 12 months [95]. Subsequent phase IIa and II RCTs indicated improvements in function and pain [96,97]. Fibrosis and osteophyte formation were not observed, which the authors ascribed to the localized and controlled expression of TGF-β1 achieved by the cell-mediated delivery procedure [97]. Adverse effects such as itching, warm sensations, and knee effusion were limited to the injection site and resolved within a few days [96]. A rat model of allogenic chondrocyte implantation detected infiltrating immune cells in deep but not superficial zones of cartilage [98], suggesting that further studies are warranted to investigate immune responses to allogenic chondrocytes injected into late-stage OA joints.

A more recent double-blinded, placebo-controlled randomized phase II clinical trial in 102 patients ($n = 67$ in treatment arm, with $n = 35$ receiving placebo) found significantly less progression of cartilage damage by 3T MRI at 52 weeks, along with significant improvements in function and pain scores compared with placebo at 12, 52, and 72 weeks, using the International Knee Documentation Committee (IKDC) and visual analogue scale (VAS) scoring systems [99]. The improvement in IKDC scores was maintained at 104 weeks [99]. These results were supported by an independent phase II RCT in 86 patients, which also found reduced OA progression by MRI at 12 months [100]. Further trials in larger cohorts are required to ascertain whether these improvements in function and pain are maintained, or whether repeat administration of TGF-β1-expressing cells is required, with potential increased risk of fibrosis and osteophyte formation. A phase III trial is currently underway (NCT03203330). Strategies such as co-administration of SMAD7, which has been shown to block synovial fibrosis while enabling anabolic TGF-β1 signaling in chondrocytes [101], may be beneficial.

3.5. Considerations around Receptor Expression and Downstream Signaling

Many of the growth factors discussed here can signal through more than one cell surface receptor, leading to potentially disparate downstream biological effects. It is thus essential to understand receptor expression patterns and signaling pathways if growth factors are to be used therapeutically.

For example, there are 4 FGF receptors (FGFRs), of which FGFR1 and FGFR3 are thought to be most important in cartilage. Some studies have concluded that FGF2 has catabolic effects on chondrocytes, inhibiting proteoglycan accumulation and aggrecan expression in vitro and ex vivo and inducing MMP-13 expression [102–104]. Other studies have ascribed a protective role to FGF2, with $Fgf2^{-/-}$ mice developing accelerated spontaneous and surgically-induced OA, which could be reversed by addition of exogenous FGF2 [105]. Subsequent studies in human cartilage explants indicated that FGF2 could suppress IL-1-induced aggrecanase activity [106]. An elegant explanation for this disparity was put forward by Yan et al., who showed that the effects of FGF2 on the joint depended on which FGFR was activated, with the catabolic effects of FGF2 resulting from signaling through FGFR1 and not FGFR3 [107]. In OA, FGF2 is thought to shift towards catabolic signaling, with increased activation of FGFR1 relative to FGFR3. The molecular mechanism for this is yet to be elucidated, but may involve changes in relative FGFR expression or alteration in sulfation of heparan sulfate (HS) [108,109] that could alter formation and/or stability of trimolecular FGF2:FGFR:HS signaling complexes. FGF18 preferentially signals through FGFR3 [107], favoring anabolic rather than catabolic signaling.

Similarly, TGF-β1 signaling through its type 1 receptor activin receptor-like kinase 5 (ALK5) leads to phosphorylation of SMAD3, inducing transcription of SMAD3-target genes such as *COL2* and *ACAN* [110], while signaling through ALK1 leads to phosphorylation of SMAD1/5/8 and transcription of catabolic genes such as *COLX* and *MMP13* [111]. ALK5 expression levels decrease with age, and the ALK1/ALK5 ratio is elevated in both OA and healthy aged cartilage [111,112], potentially favoring catabolic gene expression upon ageing.

4. Targeting Therapies to Cartilage

4.1. Strategies for Delivery to the Cartilage Matrix or to Chondrocytes

In addition to inhibiting cartilage loss and stimulating cartilage repair, a successful DMOAD must be able to reach its molecular target in cartilage and be retained at effective levels in the tissue. The negatively-charged cartilage extracellular matrix acts as a barrier to entry of many molecules, especially those with lipophilic properties. Several groups have identified the potential of cartilage ECM to act as a reservoir rather than as a barrier to entry, and are developing targeting strategies that exploit the composition and high negative fixed charge density of the matrix to deliver and retain DMOADs in cartilage.

Here, we discuss some of these strategies, with particular focus on peptide-based moieties that bind to abundant and selectively expressed cartilage ECM molecules such as type II collagen and aggrecan (Table 1). These can be combined with protein therapeutics by standard protein engineering techniques and/or dendrimers to improve cartilage targeting and half-life at the desired site of action (Table 2), while reducing systemic and off-target effects and toxicity. These approaches may also be of use in rheumatoid arthritis, where the significant levels of inflammation and angiogenesis further promote delivery of potential therapies to the inflamed joint [113].

4.1.1. Targeting Type II Collagen

Type II collagen is selectively expressed in cartilage and has a low rate of turnover in adult cartilage, making it attractive for targeting potential DMOADs to cartilage.

Table 1. Peptides tested for targeting of potential osteoarthritis (OA) therapies to cartilage.

Peptide	Identified by	Binds to	Delivers	In Vivo Efficacy	Delivery
WYRGRL [114]	Screening of phage display peptide libraries against collagenase D-treated bovine cartilage pieces	Type II collagen	Nanoparticles [114,119]; dexamethasone [115]; pepstatin A via DOTAM scaffold [116]; HA-binding peptide [117,118]	72-fold higher cartilage targeting compared with scrambled peptide [114]; 14-fold more retention of DOTAM-pepstatin A in murine knee joints, with ex vivo reduction in cathepsin D activity [116]	Intra-articular [114,116,118]
HSNGLPL [120]	Screening of phage display peptide libraries against TGFβ1	TGFβ1	TGFβ1 [120]	Increases cartilage regeneration in rabbit full thickness defect model [120]	Intra-articular during surgery to create cartilage defect [120]
DWRVIIPPRPSA [121]	Screening of phage display peptide libraries against rabbit cartilage pieces	Chondrocytes	DNA vector [121]; siRNA targeting *Hif2a* [122]	Higher uptake by chondrocytes than scrambled peptide [121]. Reduced cartilage damage in murine OA model than scrambled peptide [122]	Intra-articular [122]
RLDPTSYLRTFW and HDSQLEALIKFM [123]	Screening of phage display peptide libraries against cultured chondrocytes	Chondrocytes, at least in part via binding to aggrecan			
KRKKKGKGLGKKRDPSLRKYK [124]	Sequence taken from heparin-binding domain of HB-EGF [124]	Heparin in vitro, binding to HS in vivo not shown [124]	Fusion protein consisting of IGF-1 fused with a heparin-binding domain [124]	Increased in retention of IGF-1 and proteoglycan synthesis in cartilage in vivo. Reduced cartilage damage in rat knee OA model [124]	Intra-articular [124]

Rothenfluh et al. [114] used phage display of peptide libraries to select a peptide (with the sequence WYRGRL) that selectively binds to type II collagen. While cross-reactivity with type I collagen was not directly evaluated in this study, strong cartilage targeting was observed in mice in vivo after intra-articular injection [114]. For example, the signal from WYRGRL-targeted fluorescent nanoparticles was 72-fold higher after 48 h than the signal from nanoparticles bearing a scrambled peptide [114]. The WYRGRL peptide has subsequently been used to deliver other cargo to cartilage, including dexamethasone [115], pepstatin A [116], and an HA-binding peptide [117,118] (Table 1).

A recently-reported collagen-targeting strategy utilizes avimers, which are artificial binding proteins that can be engineered to bind with high affinity to target molecules [125]. The avimer scaffold is based on protein A domains found in various cell surface receptors, and in vitro exon shuffling and phage display of these sequences generates binding moieties with high affinity and in vivo stability [125]. Hulme et al. used an avimer phage display library to generate avimers which bind to type II collagen with high affinity, enabling retention in rat knees for a month after intra-articular injection [126]. Fusion of the avimer with IL-1Ra generated a construct that was able to block IL-1 activity in rat knee joints in vivo, even when administered a week before the IL-1 challenge [126].

Phage display has also been utilized to generate single-chain variable fragments (scFv) that bind to type II collagen which has been post-translationally modified by reactive oxygen species (ROS) [127]. These antibodies selectively bound to damaged rheumatoid and osteoarthritic but not normal murine cartilage [127], indicating they can selectively target areas of joint damage. Cartilage binding was retained after coupling to payloads such as an MMP-cleavable form of viral IL-10 [128] and soluble TNF receptor II [127]. The fragments also enabled in vivo imaging of murine OA cartilage damage, with significant increases in signal 8 weeks after surgical destabilization of the medial meniscus (DMM) [129]. However, the relatively large size of scFv fragments (~27 kDa) makes avimers (~4 kDa) and peptides (850 Da for WYRGRL) more attractive for construction of targeted DMOADs.

4.1.2. Targeting Aggrecan

Aggrecan is another abundant cartilage ECM molecule, with a high fixed charge density due its many chondroitin and keratan sulfate moieties. This property has been exploited for cartilage targeting, through strategies that use electrostatic interactions to increase binding and retention of positively charged molecules in the cartilage ECM [130–133] (Table 2). Cationic carriers that have been evaluated for cartilage delivery include peptides such as RRRR(AARRR)$_3$R [131] and proteins such as avidin [130]. Avidin-conjugated dexamethasone was found to inhibit IL-1-driven aggrecan breakdown in cartilage explant cultures more effectively than soluble dexamethasone [134], illustrating the potential of this approach. Heparin-binding domains, such as those found in growth factors including FGF18, are cationic at neutral pH, and are thus also well-suited for cartilage delivery. For example, the heparin-binding domain of heparin-binding epidermal growth factor (HB-EGF) has been used to increase retention of IGF-1 in cartilage in vivo, with increased therapeutic efficacy in a rat medial meniscal tear model of OA [124]. One caveat of cationic delivery strategies is that they must be designed to support weak, reversible interactions with the cartilage matrix, as a high net positive charge can favor tight binding that limits penetrability [131].

Table 2. Examples of scaffolds used for targeting of potential OA therapies to cartilage.

Strategy	Identified by	Binds to	Delivers	In Vivo Efficacy	Delivery
Cationic carriers (avidin, peptides, etc.) [130–132,134]	Various	Negatively-charged cartilage matrix	Dexamethasone [134]	In vitro: Improved retention of cargo in cartilage explants [134]	Intra-articular [135]
scFv	Screening of scFv phage display library against ROS-modified type II collagen [127]	ROS-modified type II collagen [127]	MMP-cleavable form of viral IL-10 [128]; soluble TNF receptor II [127]; anti-inflammatory extracellular vesicles [136]	Reduced inflammation in RA models [127,128]; in vivo imaging of murine OA [129]	Intra-peritoneal [127,128]; intravenous [136]
Avimer [126]	Screening of avimer phage display library against rat and human type II collagen	Type II collagen	IL-1Ra [126]	Blocked IL-1 activity in rat knee joints when administered at same time as IL-1, and also when administered 1 week before	Intra-articular [126]
Metalloproteinase-activatable prodrugs	Use of latency-associated peptide of TGFβ1 [137]	Cleaved by activating MMPs and ADAMTSs	IFNβ [137,138]; TIMP-3 [139]	Reduced joint swelling [137], in vivo targeting, and therapeutic efficacy in CIA model [138]	Intramuscular [136], intraperitoneal [138]

4.1.3. Targeting Chondrocytes

While some potential DMOADs are designed to act on targets found in the cartilage ECM (e.g., secreted enzymes), others may have intracellular targets that require targeting to chondrocytes rather than to the cartilage matrix. A chondrocyte-binding peptide with the sequence DWRVIIPPRPSA was identified by screening phage display peptide libraries against rabbit cartilage pieces [121]. The exact molecular target of this peptide has not been reported, but it was found to bind more than a scrambled peptide to human and rabbit chondrocytes, and also bound more to chondrocytes than to synovial cells [121]. Confocal analysis indicated cellular uptake of the peptide, enabling successful delivery of a DNA vector to chondrocytes in vivo, driving expression of green fluorescent protein and luciferase [121].

4.2. Increasing Sophistication to Tailor Avidity and Enable DMOAD Latency

Polymers and dendrimers have several advantages for cartilage targeting, including potential for increased avidity (e.g., through substitution with multiple copies of a targeting moiety [116]) and in vivo imaging (e.g., through inclusion of a fluorphor [116,140] or gadolinium MRI contrast agent [141]). Here we discuss recent progress with multivalent dendrimer scaffolds such as 1,4,7,10-tetraazacyclododecane-1,4,7,10-tetraacetic acid amide (DOTAM) and polyamidoamine (PAMAM), with nanoparticles and polymers (e.g., poly(lactic-co-glycolic acid)(PLGA) recently reviewed elsewhere [57,142,143].

The DOTAM scaffold contains a multivalent tetrapodal core with flexible polyethylene linkers that can be decorated with targeting peptides and/or cationic groups to promote cartilage retention [116]. For example, the scaffold has been coupled to the type II collagen-targeting peptide WYRGRL to deliver the cathepsin D inhibitor pepstatin A to cartilage [116]. In vivo retention of this conjugate increased as the number of WYRGRL peptides attached was increased from 1 to 3 [116]. WYRGRL-derivatized DOTAM has also been combined with gadolinium and Cy5.5 to enable in vivo dual MRI/near-infrared imaging of OA cartilage in a pre-clinical rat meniscal tear model [141]. Clinical translation for human imaging would require further safety profiling.

PANAM dendrimers are similarly multivalent, enabling a combination of desirable targeting and imaging moieties. They are additionally positively charged, favoring electrostatic binding to the negatively charged cartilage ECM. Derivatization with PEG can be used to shield this positive charge in a tuneable manner, enabling tight control of matrix binding affinity. Such PEG-PANAM dendrimers have been used to deliver IGF-1 in vivo, achieving a 10-fold increase in residence times in rat knee joints and improved chondroprotection after surgical induction of OA [144]. Derivatization of PANAM dendrimers with targeting peptides such as the DWRVIIPPRPSA chondrocyte affinity peptide [121,145] may further improve their targeting efficacy.

Another approach to cartilage targeting has been to create prodrugs that should only be activated in areas of the body where metalloproteinase activity is high, such as in OA joints. For example, IFNβ has been linked to the latency-associated peptide (LAP) of TGF-β1 via the MMP-cleavage sequence PLGLWA [137]. Intramuscular gene delivery of this construct reduced joint swelling in the collagen-induced arthritis model [137]. Linking IFNβ and LAP via the ADAMTS-cleavable sequence DVQEFRGVTAVIR improved in vivo targeting and therapeutic efficacy in this model [138]. Similar constructs may also be useful for delivery of cargo in OA. For example, a LAP-TIMP-3 construct can be activated by protease activity in the synovial fluid of OA patients [139], and may be useful for blocking metalloproteinase-mediated OA cartilage degradation.

5. Broader Challenges in DMOAD Development

In addition to target identification, there are a number of additional factors to consider for the development of a successful disease-modifying OA drug.

A successful DMOAD must be able to penetrate the highly-charged cartilage matrix to reach its target site. The pore size of the collagen network is ~100 nm, with glycosaminoglycan chains on aggrecan spaced 2–4 nm apart. Molecules up to 16 nm in diameter and 500 kDa in size have been shown to penetrate the cartilage matrix (reviewed in [133,146]), but entry is highly dependent on molecular charge, as discussed in Section 4.1 (Strategies for delivery to the cartilage matrix or to chondrocytes). As OA worsens, the charge and porosity of the cartilage matrix changes progressively, with consequent effects on DMOAD delivery and retention [119,147]. Targeting can be lost when matrix molecules are degraded, or, conversely, entry and retention can increase when cartilage is damaged. For example, cationic, collagen type II-targeted PLGA nanoparticles were found to accumulate more in OA than in normal cartilage in a rat collagenase-induced OA model, at a stage when proteoglycan loss was observed in 25–50% of the cartilage matrix [119]. Cartilage damage in early and mid-stage OA may thus promote DMOAD entry and retention, although the window of therapeutic opportunity is likely to close as cartilage damage progresses.

The route of DMOAD administration is also a critical consideration. While many, but not all, molecules can pass from the blood stream into synovial fluid, the avascular nature of cartilage makes delivery and pharmacokinetics difficult to predict. DMOAD RCTs thus generally utilize intra-articular injection, although this has associated clinical risks and does not substantially increase the half-life of non-targeted DMOADs, as solutes are cleared from the joint space within 1–5 h of intra-articular injection [148]. Administration routes may have to alter if targeting strategies are introduced, depending on their molecular properties and mechanism of action.

Molecular mechanisms driving cartilage degradation and repair appear to be largely conserved in human and murine OA [149], but the substantial difference between cartilage thickness in the two species means that mice are unlikely to be a good model for evaluating targeting strategies such as those discussed here. For example, the half-life of fluorescently labeled avidin is 5–6-times shorter in rat than in rabbit cartilage [135].

One of the greatest obstacles in DMOAD development is the lack of a robust and rapid outcome measure for therapeutic efficacy, with joint space width measurements lacking sensitivity and specificity. Improved measures of OA progression, such as accurate biomarkers [150] and novel imaging modalities [151], will greatly assist in DMOAD development and are the focus of considerable attention in the field.

Another key issue is whether OA pain correlates with structural joint changes [152]. For example, in a recent longitudinal study of 600 participants, Felson and colleagues [153] demonstrated that cartilage loss over 24 months correlated significantly with worsening of WOMAC knee pain over 24 and 36 months, but found that the effect size was small. This not only presents a challenge for RCT design, but calls into question whether successful DMOADs can be developed based on chondroprotection alone. Clinical trials of sprifermin, for example, have shown a significant effect on cartilage thickness, but not on pain [75,76]. Analysis is complicated by the strong placebo effect of intra-articular injection on patient-reported pain outcome measures [154,155], which has been observed in trials for sprifermin [75] and SM04690 [86], and by variations in scoring systems used to measure pain and in their practical application [156,157]. Promisingly, preclinical studies indicate that ADAMTS-5 inhibitors may have an analgesic effect [158], as aggrecan degradation by ADAMTS-5 generates a peptide that promotes OA pain via toll-like receptor 2 activation [159].

6. Conclusions

DMOAD development remains an important and challenging area, with substantial research effort focused on inhibiting metalloproteinase-mediated cartilage degradation and on promoting cartilage repair. A number of potential therapies have recently progressed to clinical trial, indicating that investments in fundamental research in the area are bearing fruit and delivering strong targets for drug development. Strategies that target these potential DMOADs to cartilage may help overcome the challenges of OA drug delivery, by utilizing the cartilage matrix as a drug reservoir while reducing potential systemic toxicity.

This is of particular importance given the chronic nature of OA and the high prevalence of co-morbidities.

Author Contributions: All authors provided intellectual input, drafted sections of the manuscript, and approved the final manuscript. All authors have read and agreed to the published version of the manuscript.

Funding: This research was funded by Versus Arthritis grants 21294 and 21776.

Conflicts of Interest: The authors declare no conflict of interest. The funders had no role in the design of the study; in the collection, analyses, or interpretation of data; in the writing of the manuscript; or in the decision to publish the results.

References

1. James, S.L.; Abate, D.; Abate, K.H.; Abay, S.M.; Abbafati, C.; Abbasi, N.; Abbastabar, H.; Abd-Allah, F.; Abdela, J.; Abdelalim, A.; et al. Global, regional, and national incidence, prevalence, and years lived with disability for 354 Diseases and Injuries for 195 countries and territories, 1990–2017: A systematic analysis for the Global Burden of Disease Study 2017. *Lancet* **2018**, *392*, 1789–1858. [CrossRef]
2. Hunter, D.J.; Bierma-Zeinstra, S. Osteoarthritis. *Lancet* **2019**, *393*, 1745–1759. [CrossRef]
3. Prieto-Alhambra, D.; Judge, A.; Javaid, M.K.; Cooper, C.; Diez-Perez, A.; Arden, N.K. Incidence and risk factors for clinically diagnosed knee, hip and hand osteoarthritis: Influences of age, gender and osteoarthritis affecting other joints. *Ann. Rheum. Dis.* **2014**, *73*, 1659–1664. [CrossRef] [PubMed]
4. Donell, S. Subchondral bone remodelling in osteoarthritis. *EFORT Open Rev.* **2019**, *4*, 221–229. [CrossRef]
5. Belluzzi, E.; Stocco, E.; Pozzuoli, A.; Granzotto, M.; Porzionato, A.; Vettor, R.; De Caro, R.; Ruggieri, P.; Ramonda, R.; Rossato, M.; et al. Contribution of Infrapatellar Fat Pad and Synovial Membrane to Knee Osteoarthritis Pain. *Biomed. Res. Int.* **2019**, *2019*, 6390182. [CrossRef] [PubMed]
6. Englund, M.; Guermazi, A.; Lohmander, S.L. The Role of the Meniscus in Knee Osteoarthritis: A Cause or Consequence? *Radiol. Clin. N. Am.* **2009**, *47*, 703–712. [CrossRef] [PubMed]
7. Latourte, A.; Kloppenburg, M.; Richette, P. Emerging pharmaceutical therapies for osteoarthritis. *Nat. Rev. Rheumatol.* **2020**, *16*, 673–688. [CrossRef] [PubMed]
8. Heinegård, D.; Saxne, T. The role of the cartilage matrix in osteoarthritis. *Nat. Rev. Rheumatol.* **2011**, *7*, 50–56. [CrossRef] [PubMed]
9. Burleigh, A.; Chanalaris, A.; Gardiner, M.D.; Driscoll, C.; Boruc, O.; Saklatvala, J.; Vincent, T.L. Joint immobilization prevents murine osteoarthritis and reveals the highly mechanosensitive nature of protease expression in vivo. *Arthritis Rheumatol.* **2012**, *64*, 2278–2288. [CrossRef]
10. Troeberg, L.; Nagase, H. Proteases involved in cartilage matrix degradation in osteoarthritis. *Biochim. Biophys. Acta* **2012**, *1824*, 133–145. [CrossRef] [PubMed]
11. Coryell, P.R.; Diekman, B.O.; Loeser, R.F. Mechanisms and therapeutic implications of cellular senescence in osteoarthritis. *Nat. Rev. Rheumatol.* **2020**, *17*, 47–57. [CrossRef]
12. Ansari, M.Y.; Ahmad, N.; Voleti, S.; Wase, S.J.; Novak, K.; Haqqi, T.M. Mitochondrial dysfunction triggers a catabolic response in chondrocytes via ROS-mediated activation of the JNK/AP1 pathway. *J. Cell Sci.* **2020**, *133*, jcs247353. [CrossRef]
13. Zahan, O.M.; Serban, O.; Gherman, C.; Fodor, D. The evaluation of oxidative stress in osteoarthritis. *Med. Pharm. Rep.* **2020**, *93*, 12–22. [CrossRef]
14. Ansari, M.Y.; Ahmad, N.; Haqqi, T.M. Oxidative stress and inflammation in osteoarthritis pathogenesis: Role of polyphenols. *Biomed. Pharmacother.* **2020**, *129*, 110452. [CrossRef] [PubMed]
15. Bolduc, J.A.; Collins, J.A.; Loeser, R.F. Reactive oxygen species, aging and articular cartilage homeostasis. *Free Radic. Biol. Med.* **2019**, *132*, 73–82. [CrossRef]
16. Billinghurst, R.C.C.; Dahlberg, L.; Ionescu, M.; Reiner, A.; Bourne, R.; Rorabeck, C.; Mitchell, P.; Hambor, J.; Diekmann, O.; Tschesche, H.; et al. Enhanced cleavage of type II collagen by collagenases in osteoarthritic articular cartilage. *J. Clin. Investig.* **1997**, *99*, 1534–1545. [CrossRef] [PubMed]
17. Little, C.B.; Barai, A.; Burkhardt, D.; Smith, S.M.; Fosang, A.J.; Werb, Z.; Shah, M.; Thompson, E.W. Matrix metalloproteinase 13-deficient mice are resistant to osteoarthritic cartilage erosion but not chondrocyte hypertrophy or osteophyte development. *Arthritis Rheum.* **2009**, *60*, 3723–3733. [CrossRef] [PubMed]
18. Neuhold, L.A.; Killar, L.; Zhao, W.; Sung, M.L.A.; Warner, L.; Kulik, J.; Turner, J.; Wu, W.; Billinghurst, C.; Meijers, T.; et al. Postnatal expression in hyaline cartilage of constitutively active human collagenase-3 (MMP-13) induces osteoarthritis in mice. *J. Clin. Investig.* **2001**, *107*, 35–44. [CrossRef]
19. Stanton, H.; Rogerson, F.M.; East, C.J.; Golub, S.B.; Lawlor, K.E.; Meeker, C.T.; Little, C.B.; Last, K.; Farmer, P.J.; Campbell, I.K.; et al. ADAMTS5 is the major aggrecanase in mouse cartilage in vivo and in vitro. *Nature* **2005**, *434*, 648–652. [CrossRef]
20. Glasson, S.S.; Askew, R.; Sheppard, B.; Carito, B.; Blanchet, T.; Ma, H.L.; Flannery, C.R.; Peluso, D.; Kanki, K.; Yang, Z.; et al. Deletion of active ADAMTS5 prevents cartilage degradation in a murine model of osteoarthritis. *Nature* **2005**, *434*, 644–648. [CrossRef] [PubMed]

21. Bau, B.; Gebhard, P.M.; Haag, J.; Knorr, T.; Bartnik, E.; Aigner, T. Relative messenger RNA expression profiling of collagenases and aggrecanases in human articular chondrocytes in vivo and in vitro. *Arthritis Rheum.* **2002**, *46*, 2648–2657. [CrossRef] [PubMed]
22. Yamamoto, K.; Troeberg, L.; Scilabra, S.D.; Pelosi, M.; Murphy, C.L.; Strickland, D.K.; Nagase, H. LRP-1-mediated endocytosis regulates extracellular activity of ADAMTS-5 in articular cartilage. *FASEB J.* **2013**, *27*, 511–521. [CrossRef]
23. Walling, H.W.; Raggatt, L.J.; Irvine, D.W.; Barmina, O.Y.; Toledano, J.E.; Goldring, M.B.; Hruska, K.A.; Adkisson, H.D.; Burdge, R.E.; Gatt, C.J.; et al. Impairment of the collagenase-3 endocytotic receptor system in cells from patients with osteoarthritis. *Osteoarthr. Cartil.* **2003**, *11*, 854–863. [CrossRef]
24. Morris, K.J.; Cs-Szabo, G.; Cole, A.A. Characterization of TIMP-3 in human articular talar cartilage. *Connect. Tissue Res.* **2010**, *51*, 478–490. [CrossRef] [PubMed]
25. Blaney Davidson, E.N.; van der Kraan, P.M.; van den Berg, W.B. TGF-β and osteoarthritis. *Osteoarthr. Cartil.* **2007**, *15*, 597–604. [CrossRef] [PubMed]
26. Malfait, A.M.; Tortorella, M.D. The "elusive DMOAD": Aggrecanase inhibition from laboratory to clinic. *Clin. Exp. Rheumatol.* **2019**, *120*, 130–134.
27. Dancevic, C.M.; McCulloch, D.R. Current and emerging therapeutic strategies for preventing inflammation and aggrecanase-mediated cartilage destruction in arthritis. *Arthritis Res. Ther.* **2014**, *16*, 429. [CrossRef] [PubMed]
28. Santamaria, S. ADAMTS-5: A difficult teenager turning 20. *Int. J. Exp. Pathol.* **2020**, *101*, 4–20. [CrossRef] [PubMed]
29. Mehana, E.S.E.; Khafaga, A.F.; El-Blehi, S.S. The role of matrix metalloproteinases in osteoarthritis pathogenesis: An updated review. *Life Sci.* **2019**, *234*, 116786. [CrossRef]
30. Xie, X.W.; Wan, R.Z.; Liu, Z.P. Recent Research Advances in Selective Matrix Metalloproteinase-13 Inhibitors as Anti-Osteoarthritis Agents. *ChemMedChem* **2017**, *12*, 1157–1168. [CrossRef] [PubMed]
31. Pratta, M.A.; Yao, W.; Decicco, C.; Tortorella, M.D.; Liu, R.Q.; Copeland, R.A.; Magolda, R.; Newton, R.C.; Trzaskos, J.M.; Arner, E.C. Aggrecan Protects Cartilage Collagen from Proteolytic Cleavage. *J. Biol. Chem.* **2003**, *278*, 45539–45545. [CrossRef]
32. Karsdal, M.A.; Madsen, S.H.; Christiansen, C.; Henriksen, K.; Fosang, A.J.; Sondergaard, B.C. Cartilage degradation is fully reversible in the presence of aggrecanase but not matrix metalloproteinase activity. *Arthritis Res. Ther.* **2008**, *10*, R63. [CrossRef]
33. Ismail, H.M.; Yamamoto, K.; Vincent, T.L.; Nagase, H.; Troeberg, L.; Saklatvala, J. Interleukin-1 acts via the JNK-2 signaling pathway to induce aggrecan degradation by human chondrocytes. *Arthritis Rheumatol.* **2015**, *67*, 1826–1836. [CrossRef] [PubMed]
34. Coussens, L.M.; Fingleton, B.; Matrisian, L.M. Matrix metalloproteinase inhibitors and cancer: Trials and tribulations. *Science* **2002**, *295*, 2387–2392. [CrossRef] [PubMed]
35. Fields, G.B. The Rebirth of Matrix Metalloproteinase Inhibitors: Moving Beyond the Dogma. *Cells* **2019**, *8*, 984. [CrossRef]
36. Brenneis, C.; Serruys, B.; Van Belle, T.; Poelmans, S.; Kleinschmidt-Doerr, K.; Guehring, H.; Michaelis, M.; Lindemann, S. Structural and symptomatic benefit of a half-live extended, systemically applied anti-ADAMTS-5 inhibitor (M6495). *Osteoarthr. Cartil.* **2018**, *26*, S299–S300. [CrossRef]
37. Santamaria, S.; Cuffaro, D.; Nuti, E.; Ciccone, L.; Tuccinardi, T.; Liva, F.; D'Andrea, F.; de Groot, R.; Rossello, A.; Ahnström, J. Exosite inhibition of ADAMTS-5 by a glycoconjugated arylsulfonamide. *Sci. Rep.* **2021**, *11*, 949. [CrossRef]
38. Larkin, J.; Lohr, T.; Elefante, L.; Shearin, J.; Matico, R.; Su, J.-L.; Xue, Y.; Liu, F.; Rossman, E.I.; Renninger, J.; et al. The highs and lows of translational drug development: Antibody-mediated inhibition of ADAMTS-5 for osteoarthritis disease modification. *Osteoarthr. Cartil.* **2014**, *22*, S483–S484. [CrossRef]
39. Hattori, N.; Mochizuki, S.; Kishi, K.; Nakajima, T.; Takaishi, H.; D'Armiento, J.; Okada, Y. MMP-13 plays a role in keratinocyte migration, angiogenesis, and contraction in mouse skin wound healing. *Am. J. Pathol.* **2009**, *175*, 533–546. [CrossRef] [PubMed]
40. Smith, L.R.; Kok, H.J.; Zhang, B.; Chung, D.; Spradlin, R.A.; Rakoczy, K.D.; Lei, H.; Boesze-Battaglia, K.; Barton, E.R. Matrix metalloproteinase 13 from satellite cells is required for efficient muscle growth and regeneration. *Cell. Physiol. Biochem.* **2020**, *54*, 333–353. [CrossRef]
41. Kosaki, N.; Takaishi, H.; Kamekura, S.; Kimura, T.; Okada, Y.; Minqi, L.; Amizuka, N.; Chung, U.-I.; Nakamura, K.; Kawaguchi, H.; et al. Impaired bone fracture healing in matrix metalloproteinase-13 deficient mice. *Biochem. Biophys. Res. Commun.* **2007**, *354*, 846–851. [CrossRef] [PubMed]
42. Behonick, D.J.; Xing, Z.; Lieu, S.; Buckley, J.M.; Lotz, J.C.; Marcucio, R.S.; Werb, Z.; Miclau, T.; Colnot, C. Role of matrix metalloproteinase 13 in both endochondral and intramembranous ossification during skeletal regeneration. *PLoS ONE* **2007**, *2*, e1150. [CrossRef] [PubMed]
43. Chu, X.; You, H.; Yuan, X.; Zhao, W.; Li, W.; Guo, X. Protective effect of lentivirus-mediated sirna targeting adamts-5 on cartilage degradation in a rat model of osteoarthritis. *Int. J. Mol. Med.* **2013**, *31*, 1222–1228. [CrossRef] [PubMed]
44. Hoshi, H.; Akagi, R.; Yamaguchi, S.; Muramatsu, Y.; Akatsu, Y.; Yamamoto, Y.; Sasaki, T.; Takahashi, K.; Sasho, T. Effect of inhibiting MMP13 and ADAMTS5 by intra-articular injection of small interfering RNA in a surgically induced osteoarthritis model of mice. *Cell Tissue Res.* **2017**, *368*, 379–387. [CrossRef]
45. Yamamoto, K.; Okano, H.; Miyagawa, W.; Visse, R.; Shitomi, Y.; Santamaria, S.; Dudhia, J.; Troeberg, L.; Strickland, D.K.; Hirohata, S.; et al. MMP-13 is constitutively produced in human chondrocytes and co-endocytosed with ADAMTS-5 and TIMP-3 by the endocytic receptor LRP1. *Matrix Biol.* **2016**, *56*, 57–73. [CrossRef] [PubMed]
46. Yamamoto, K.; Scavenius, C.; Santamaria, S.; Botkjaer, K.A.; Dudhia, J.; Troeberg, L.; Itoh, Y.; Murphy, G.; Enghild, J.J.; Nagase, H. Inhibition of LRP1 shedding reverses cartilage degradation in osteoarthritis. *Osteoarthr. Cartil.* **2018**, *26* (Suppl. S1), S22. [CrossRef]

47. Kashiwagi, M.; Tortorella, M.; Nagase, H.; Brew, K. TIMP-3 Is a Potent Inhibitor of Aggrecanase 1 (ADAM-TS4) and Aggrecanase 2 (ADAM-TS5). *J. Biol. Chem.* **2001**, *276*, 12501–12504. [CrossRef]
48. Brew, K.; Nagase, H. The tissue inhibitors of metalloproteinases (TIMPs): An ancient family with structural and functional diversity. *Biochim. Biophys. Acta Mol. Cell Res.* **2010**, *1803*, 55–71. [CrossRef] [PubMed]
49. Davidson, R.K.; Waters, J.G.; Kevorkian, L.; Darrah, C.; Cooper, A.; Donell, S.T.; Clark, I.M. Expression profiling of metalloproteinases and their inhibitors in synovium and cartilage. *Arthritis Res. Ther.* **2006**, *8*, R124. [CrossRef]
50. Scilabra, S.D.; Troeberg, L.; Yamamoto, K.; Emonard, H.; Thøgersen, I.; Enghild, J.J.; Strickland, D.K.; Nagase, H. Differential regulation of extracellular tissue inhibitor of metalloproteinases-3 levels by cell membrane-bound and shed low density lipoprotein receptor-related protein 1. *J. Biol. Chem.* **2013**, *288*, 332–342. [CrossRef]
51. Troeberg, L.; Lazenbatt, C.; Anower-E-Khuda, M.F.; Freeman, C.; Federov, O.; Habuchi, H.; Habuchi, O.; Kimata, K.; Nagase, H. Sulfated glycosaminoglycans control the extracellular trafficking and the activity of the metalloprotease inhibitor TIMP-3. *Chem. Biol.* **2014**, *21*, 1300–1309. [CrossRef] [PubMed]
52. Carreca, A.P.; Pravatà, V.M.; Markham, M.; Bonelli, S.; Murphy, G.; Nagase, H.; Troeberg, L.; Scilabra, S.D. TIMP-3 facilitates binding of target metalloproteinases to the endocytic receptor LRP-1 and promotes scavenging of MMP-1. *Sci. Rep.* **2020**, *10*, 12067. [CrossRef] [PubMed]
53. Doherty, C.M.; Visse, R.; Dinakarpandian, D.; Strickland, D.K.; Nagase, H.; Troeberg, L. Engineered tissue inhibitor of metalloproteinases-3 variants resistant to endocytosis have prolonged chondroprotective activity. *J. Biol. Chem.* **2016**, *291*, 22160–22172. [CrossRef] [PubMed]
54. Troeberg, L.; Fushimi, K.; Khokha, R.; Emonard, H.; Ghosh, P.; Nagase, H. Calcium pentosan polysulfate is a multifaceted exosite inhibitor of aggrecanases. *FASEB J.* **2008**, *22*, 3515–3524. [CrossRef] [PubMed]
55. Chanalaris, A.; Doherty, C.M.; Marsden, B.D.; Bambridge, G.; Wren, S.P.; Nagase, H.; Troeberg, L. Suramin inhibits osteoarthritic cartilage degradation by increasing extracellular levels of chondroprotective tissue inhibitor of metalloproteinases 3. *Mol. Pharmacol.* **2017**, *92*, 459–468. [CrossRef]
56. Guns, L.-A.; Monteagudo, S.; Kvasnytsia, M.; Kerckhofs, G.; Vandooren, J.; Opdenakker, G.; Lories, R.J.; Cailotto, F. Suramin increases cartilage proteoglycan accumulation in vitro and protects against joint damage triggered by papain injection in mouse knees in vivo. *RMD Open* **2017**, *3*, e000604. [CrossRef]
57. Patel, J.M.; Saleh, K.S.; Burdick, J.A.; Mauck, R.L. Bioactive factors for cartilage repair and regeneration: Improving delivery, retention, and activity. *Acta Biomater.* **2019**, *93*, 222–238. [CrossRef]
58. Everts, P.; Onishi, K.; Jayaram, P.; Lana, J.F.; Mautner, K. Platelet-rich plasma: New performance understandings and therapeutic considerations in 2020. *Int. J. Mol. Sci.* **2020**, *21*, 7794. [CrossRef] [PubMed]
59. Akeda, K.; An, H.S.; Okuma, M.; Attawia, M.; Miyamoto, K.; Thonar, E.J.M.A.; Lenz, M.E.; Sah, R.L.; Masuda, K. Platelet-rich plasma stimulates porcine articular chondrocyte proliferation and matrix biosynthesis. *Osteoarthr. Cartil.* **2006**, *14*, 1272–1280. [CrossRef]
60. Van Buul, G.M.; Koevoet, W.L.M.; Kops, N.; Bos, P.K.; Verhaar, J.A.N.; Weinans, H.; Bernsen, M.R.; Van Osch, G.J.V.M. Platelet-rich plasma releasate inhibits inflammatory processes in osteoarthritic chondrocytes. *Am. J. Sports Med.* **2011**, *39*, 2362–2370. [CrossRef] [PubMed]
61. Chien, C.S.; Ho, H.O.; Liang, Y.C.; Ko, P.H.; Sheu, M.T.; Chen, C.H. Incorporation of exudates of human platelet-rich fibrin gel in biodegradable fibrin scaffolds for tissue engineering of cartilage. *J. Biomed. Mater. Res. Part B Appl. Biomater.* **2012**, *100B*, 948–955. [CrossRef]
62. Pettersson, S.; Wetterö, J.; Tengvall, P.; Kratz, G. Human articular chondrocytes on macroporous gelatin microcarriers form structurally stable constructs with blood-derived biological glues in vitro. *J. Tissue Eng. Regen. Med.* **2009**, *3*, 450–460. [CrossRef]
63. Gaissmaier, C.; Fritz, J.; Krackhardt, T.; Flesch, I.; Aicher, W.K.; Ashammakhi, N. Effect of human platelet supernatant on proliferation and matrix synthesis of human articular chondrocytes in monolayer and three-dimensional alginate cultures. *Biomaterials* **2005**, *26*, 1953–1960. [CrossRef]
64. Sundman, E.A.; Cole, B.J.; Fortier, L.A. Growth factor and catabolic cytokine concentrations are influenced by the cellular composition of platelet-rich plasma. *Am. J. Sports Med.* **2011**, *39*, 2135–2140. [CrossRef] [PubMed]
65. Patel, S.; Dhillon, M.S.; Aggarwal, S.; Marwaha, N.; Jain, A. Treatment with platelet-rich plasma is more effective than placebo for knee osteoarthritis: A prospective, double-blind, randomized trial. *Am. J. Sports Med.* **2013**, *41*, 356–364. [CrossRef] [PubMed]
66. Cerza, F.; Carnì, S.; Carcangiu, A.; Di Vavo, I.; Schiavilla, V.; Pecora, A.; De Biasi, G.; Ciuffreda, M. Comparison between hyaluronic acid and platelet-rich plasma, intra-articular infiltration in the treatment of gonarthrosis. *Am. J. Sports Med.* **2012**, *40*, 2822–2827. [CrossRef] [PubMed]
67. Sánchez, M.; Guadilla, J.; Fiz, N.; Andia, I. Ultrasound-guided platelet-rich plasma injections for the treatment of osteoarthritis of the hip. *Rheumatology* **2012**, *51*, 144–150. [CrossRef] [PubMed]
68. Filardo, G.; Kon, E.; Di Martino, A.; Di Matteo, B.; Merli, M.L.; Cenacchi, A.; Fornasari, P.M.; Marcacci, M. Platelet-rich plasma vs hyaluronic acid to treat knee degenerative pathology: Study design and preliminary results of a randomized controlled trial. *BMC Musculoskelet. Disord.* **2012**, *13*, 229. [CrossRef] [PubMed]
69. Dhurat, R.; Sukesh, M. Principles and methods of preparation of platelet-rich plasma: A review and author's perspective. *J. Cutan. Aesthet. Surg.* **2014**, *7*, 189–197. [CrossRef] [PubMed]

70. Ellsworth, J.L.; Berry, J.; Bukowski, T.; Claus, J.; Feldhaus, A.; Holderman, S.; Holdren, M.S.; Lum, K.D.; Moore, E.E.; Raymond, F.; et al. Fibroblast growth factor-18 is a trophic factor for mature chondrocytes and their progenitors. *Osteoarthr. Cartil.* **2002**, *10*, 308–320. [CrossRef]
71. Mori, Y.; Saito, T.; Chang, S.H.; Kobayashi, H.; Ladel, C.H.; Guehring, H.; Chung, U.-I.; Kawaguchi, H. Identification of fibroblast growth factor-18 as a molecule to protect adult articular cartilage by gene expression profiling. *J. Biol. Chem.* **2014**, *289*, 10192–101200. [CrossRef] [PubMed]
72. Moore, E.E.; Bendele, A.M.; Thompson, D.L.; Littau, A.; Waggie, K.S.; Reardon, B.; Ellsworth, J.L. Fibroblast growth factor-18 stimulates chondrogenesis and cartilage repair in a rat model of injury-induced osteoarthritis. *Osteoarthr. Cartil.* **2005**, *13*, 623–631. [CrossRef] [PubMed]
73. Gigout, A.; Guehring, H.; Froemel, D.; Meurer, A.; Ladel, C.; Reker, D.; Bay-Jensen, A.C.; Karsdal, M.A.; Lindemann, S. Sprifermin (rhFGF18) enables proliferation of chondrocytes producing a hyaline cartilage matrix. *Osteoarthr. Cartil.* **2017**, *25*, 1858–1867. [CrossRef] [PubMed]
74. Dahlberg, L.E.; Aydemir, A.; Muuraheinen, N.; Gühring, H.; Edebo, H.F.; Krarup-Jensen, N.; Ladel, C.H.; Jurvelin, J.S. A first-in-human, double-blind, randomised, placebo-controlled, dose ascending study of intra-articular rhFGF18 (sprifermin) in patients with advanced knee osteoarthritis. *Clin. Exp. Rheumatol.* **2016**, *34*, 445–450. [PubMed]
75. Lohmander, L.S.; Hellot, S.; Dreher, D.; Krantz, E.F.W.; Kruger, D.S.; Guermazi, A.; Eckstein, F. Intraarticular sprifermin (recombinant human fibroblast growth factor 18) in knee osteoarthritis: A randomized, double-blind, placebo-controlled trial. *Arthritis Rheumatol.* **2014**, *66*, 1820–1831. [CrossRef]
76. Hochberg, M.C.; Guermazi, A.; Guehring, H.; Aydemir, A.; Wax, S.; Fleuranceau-Morel, P.; Reinstrup Bihlet, A.; Byrjalsen, I.; Ragnar Andersen, J.; Eckstein, F. Effect of Intra-Articular Sprifermin vs Placebo on Femorotibial Joint Cartilage Thickness in Patients with Osteoarthritis: The FORWARD Randomized Clinical Trial. *JAMA J. Am. Med. Assoc.* **2019**, *322*, 1360–1370. [CrossRef] [PubMed]
77. Usami, Y.; Gunawardena, A.T.; Iwamoto, M.; Enomoto-Iwamoto, M. Wnt signaling in cartilage development and diseases: Lessons from animal studies. *Lab. Investig.* **2016**, *96*, 186–196. [CrossRef] [PubMed]
78. Lories, R.J.; Corr, M.; Lane, N.E. To Wnt or not to Wnt: The bone and joint health dilemma. *Nat. Rev. Rheumatol.* **2013**, *9*, 328–339. [CrossRef]
79. Dell'Accio, F.; Cailotto, F. Pharmacological blockade of the WNT-beta-catenin signaling: A possible first-in-kind DMOAD. *Osteoarthr. Cartil.* **2018**, *26*, 4–6. [CrossRef] [PubMed]
80. Hill, T.P.; Später, D.; Taketo, M.M.; Birchmeier, W.; Hartmann, C. Canonical Wnt/β-catenin signaling prevents osteoblasts from differentiating into chondrocytes. *Dev. Cell* **2005**, *8*, 727–738. [CrossRef]
81. Kim, Y.S.; Lee, H.J.; Yeo, J.E.; Kim, Y.I.; Choi, Y.J.; Koh, Y.G. Isolation and Characterization of Human Mesenchymal Stem Cells Derived from Synovial Fluid in Patients with Osteochondral Lesion of the Talus. *Am. J. Sports Med.* **2015**, *43*, 399–406. [CrossRef] [PubMed]
82. Barroga, C.; Hu, Y.; Deshmukh, V.; Hood, J. Discovery of an intra-articular injection small molecule inhibitor of the Wnt pathway (SM04690) as a potential disease modifying treatment for knee osteoarthritis. *Arthritis Rheumatol.* **2015**, *67* (Suppl. S10). Available online: https://acrabstracts.org/abstract/discovery-of-an-intra-articular-injection-small-molecule-inhibitor-of-the-wnt-pathway-sm04690-as-a-potential-disease-modifying-treatment-for-knee-osteoarthritis/ (accessed on 23 January 2021).
83. Deshmukh, V.; Hu, H.; Barroga, C.; Bossard, C.; KC, S.; Dellamary, L.; Stewart, J.; Chiu, K.; Ibanez, M.; Pedraza, M.; et al. A small-molecule inhibitor of the Wnt pathway (SM04690) as a potential disease modifying agent for the treatment of osteoarthritis of the knee. *Osteoarthr. Cartil.* **2018**, *26*, 18–27. [CrossRef] [PubMed]
84. Deshmukh, V.; O'Green, A.L.; Bossard, C.; Seo, T.; Lamangan, L.; Ibanez, M.; Ghias, A.; Lai, C.; Do, L.; Cho, S.; et al. Modulation of the Wnt pathway through inhibition of CLK2 and DYRK1A by lorecivivint as a novel, potentially disease-modifying approach for knee osteoarthritis treatment. *Osteoarthr. Cartil.* **2019**, *27*, 1347–1360. [CrossRef] [PubMed]
85. Yazici, Y.; McAlindon, T.E.; Fleischmann, R.; Gibofsky, A.; Lane, N.E.; Kivitz, A.J.; Skrepnik, N.; Armas, E.; Swearingen, C.J.; DiFrancesco, A.; et al. A novel Wnt pathway inhibitor, SM04690, for the treatment of moderate to severe osteoarthritis of the knee: Results of a 24-week, randomized, controlled, phase 1 study. *Osteoarthr. Cartil.* **2017**, *25*, 1598–1606. [CrossRef]
86. Yazici, Y.; McAlindon, T.E.; Gibofsky, A.; Lane, N.E.; Clauw, D.; Jones, M.; Bergfeld, J.; Swearingen, C.J.; DiFrancesco, A.; Simsek, I.; et al. Lorecivivint, a Novel Intraarticular CDC-like Kinase 2 and Dual-Specificity Tyrosine Phosphorylation-Regulated Kinase 1A Inhibitor and Wnt Pathway Modulator for the Treatment of Knee Osteoarthritis: A Phase II Randomized Trial. *Arthritis Rheumatol.* **2020**, *72*, 1694–1706. [CrossRef] [PubMed]
87. Held, A.; Glas, A.; Dietrich, L.; Bollmann, M.; Brandstädter, K.; Grossmann, T.N.; Lohmann, C.H.; Pap, T.; Bertrand, J. Targeting β-catenin dependent Wnt signaling via peptidomimetic inhibitors in murine chondrocytes and OA cartilage. *Osteoarthr. Cartil.* **2018**, *26*, 818–823. [CrossRef] [PubMed]
88. Nalesso, G.; Sherwood, J.; Bertrand, J.; Pap, T.; Ramachandran, M.; de Bari, C.; Pitzalis, C.; Dell'Accio, F. WNT-3A modulates articular chondrocyte phenotype by activating both canonical and noncanonical pathways. *J. Cell Biol.* **2011**, *193*, 511–564. [CrossRef] [PubMed]
89. Kizawa, H.; Kou, I.; Iida, A.; Sudo, A.; Miyamoto, Y.; Fukuda, A.; Mabuchi, A.; Kotani, A.; Kawakami, A.; Yamamoto, S.; et al. An aspartic acid repeat polymorphism in asporin inhibits chondrogenesis and increases susceptibility to osteoarthritis. *Nat. Genet.* **2005**, *37*, 138–144. [CrossRef] [PubMed]

90. Jiang, Q.; Shi, D.; Yi, L.; Ikegawa, S.; Wang, Y.; Nakamura, T.; Qiao, D.; Liu, C.; Dai, J. Replication of the association of the aspartic acid repeat polymorphism in the asporin gene with knee-osteoarthritis susceptibility in Han Chinese. *J. Hum. Genet.* **2006**, *51*, 1068–1072. [CrossRef]

91. Valdes, A.M.; Spector, T.D.; Tamm, A.A.; Kisand, K.; Doherty, S.A.; Dennison, E.M.; Mangino, M.; Tamm, A.A.; Kerna, I.; Hart, D.J.; et al. Genetic variation in the SMAD3 gene is associated with hip and knee osteoarthritis. *Arthritis Rheum.* **2010**, *62*, 2347–2352. [CrossRef]

92. Serra, R.; Johnson, M.; Filvaroff, E.H.; LaBorde, J.; Sheehan, D.M.; Derynck, R.; Moses, H.L. Expression of a truncated, kinase-defective TGF-β type II receptor in mouse skeletal tissue promotes terminal chondrocyte differentiation and osteoarthritis. *J. Cell Biol.* **1997**, *139*, 541–545. [CrossRef] [PubMed]

93. Shen, J.; Li, J.; Wang, B.; Jin, H.; Wang, M.; Zhang, Y.; Yang, Y.; Im, H.J.; O'Keefe, R.; Chen, D. Deletion of the transforming growth factor β Receptor type II gene in articular chondrocytes leads to a progressive osteoarthritis-like phenotype in mice. *Arthritis Rheum.* **2013**, *65*, 3107–3119. [CrossRef]

94. Yang, X.; Chen, L.; Xu, X.; Li, C.; Huang, C.; Deng, C.X. TGF-β/Smad3 signals repress chondrocyte hypertrophic differentiation and are required for maintaining articular cartilage. *J. Cell Biol.* **2001**, *153*, 35–46. [CrossRef]

95. Ha, C.W.; Noh, M.J.; Choi, K.B.; Lee, K.H. Initial phase i safety of retrovirally transduced human chondrocytes expressing transforming growth factor-beta-1 in degenerative arthritis patients. *Cytotherapy* **2012**, *14*, 247–256. [CrossRef]

96. Ha, C.W.; Cho, J.J.; Elmallah, R.K.; Cherian, J.J.; Kim, T.W.; Lee, M.C.; Mont, M.A. A Multicenter, Single-Blind, Phase IIa Clinical Trial to Evaluate the Efficacy and Safety of a Cell-Mediated Gene Therapy in Degenerative Knee Arthritis Patients. *Hum. Gene Ther. Clin. Dev.* **2015**, *26*, 125–130. [CrossRef] [PubMed]

97. Lee, M.C.; Ha, C.-W.; Elmallah, R.K.; Cherian, J.J.; Cho, J.J.; Kim, T.W.; Bin, S.-I.; Mont, M.A. A placebo-controlled randomised trial to assess the effect of TGF-β1-expressing chondrocytes in patients with arthritis of the knee. *Bone Jt. J.* **2015**, *97-B*, 924–932. [CrossRef] [PubMed]

98. Osiecka-Iwan, A.; Hyc, A.; Moskalewski, S. Immunosuppression and rejection of cartilage formed by allogeneic chondrocytes in rats. *Cell Transplant.* **1999**, *8*, 627–636. [CrossRef] [PubMed]

99. Lee, B.; Parvizi, J.; Bramlet, D.; Romness, D.W.; Guermazi, A.; Noh, M.; Sodhi, N.; Khlopas, A.; Mont, M.A. Results of a Phase II Study to Determine the Efficacy and Safety of Genetically Engineered Allogeneic Human Chondrocytes Expressing TGF-β1. *J. Knee Surg.* **2020**, *33*, 167–172. [CrossRef] [PubMed]

100. Guermazi, A.; Kalsi, G.; Niu, J.; Crema, M.D.; Copeland, R.O.; Orlando, A.; Noh, M.J.; Roemer, F.W. Structural effects of intra-articular TGF-β1 in moderate to advanced knee osteoarthritis: MRI-based assessment in a randomized controlled trial. *BMC Musculoskelet. Disord.* **2017**, *18*, 461. [CrossRef] [PubMed]

101. Blaney Davidson, E.N.; Vitters, E.L.; van den Berg, W.B.; van der Kraan, P.M. TGF β-induced cartilage repair is maintained but fibrosis is blocked in the presence of Smad7. *Arthritis Res. Ther.* **2006**, *8*, R65. [CrossRef] [PubMed]

102. Im, H.J.; Li, X.; Muddasani, P.; Kim, G.H.; Davis, F.; Rangan, J.; Forsyth, C.B.; Ellman, M.; Thonar, E.J.M.A. Basic fibroblast growth factor accelerates matrix degradation via a neuro-endocrine pathway in human adult articular chondrocytes. *J. Cell. Physiol.* **2008**, *215*, 452–463. [CrossRef] [PubMed]

103. Yan, D.; Chen, D.; Im, H.J. Fibroblast growth factor-2 promotes catabolism via FGFR1-Ras-Raf-MEK1/2-ERK1/2 axis that coordinates with the PKCδ pathway in human articular chondrocytes. *J. Cell. Biochem.* **2012**, *113*, 2856–2865. [CrossRef]

104. Wang, X.; Manner, P.A.; Horner, A.; Shum, L.; Tuan, R.S.; Nuckolls, G.H. Regulation of MMP-13 expression by RUNX2 and FGF2 in osteoarthritic cartilage. *Osteoarthr. Cartil.* **2004**, *12*, 963–973. [CrossRef]

105. Chia, S.L.; Sawaji, Y.; Burleigh, A.; McLean, C.; Inglis, J.; Saklatvala, J.; Vincent, T. Fibroblast growth factor 2 is an intrinsic chondroprotective agent that suppresses ADAMTS-5 and delays cartilage degradation in murine osteoarthritis. *Arthritis Rheum.* **2009**, *60*, 2019–2027. [CrossRef]

106. Sawaji, Y.; Hynes, J.; Vincent, T.; Saklatvala, J. Fibroblast growth factor 2 inhibits induction of aggrecanase activity in human articular cartilage. *Arthritis Rheum.* **2008**, *58*, 3498–3509. [CrossRef]

107. Yan, D.; Chen, D.; Cool, S.M.; van Wijnen, A.J.; Mikecz, K.; Murphy, G.; Im, H.J. Fibroblast growth factor receptor 1 is principally responsible for fibroblast growth factor 2-induced catabolic activities in human articular chondrocytes. *Arthritis Res. Ther.* **2011**, *13*, R130. [CrossRef] [PubMed]

108. Chuang, C.Y.; Lord, M.S.; Melrose, J.; Rees, M.D.; Knox, S.M.; Freeman, C.; Iozzo, R.V.; Whitelock, J.M. Heparan sulfate-dependent signaling of fibroblast growth factor 18 by chondrocyte-derived perlecan. *Biochemistry* **2010**, *49*, 5524–5532. [CrossRef]

109. Chanalaris, A.; Clarke, H.; Guimond, S.E.; Vincent, T.L.; Turnbull, J.E.; Troeberg, L. Heparan Sulfate Proteoglycan Synthesis Is Dysregulated in Human Osteoarthritic Cartilage. *Am. J. Pathol.* **2019**, *189*, 632–647. [CrossRef] [PubMed]

110. Zhu, Y.; Tao, H.; Jin, C.; Liu, Y.; Lu, X.; Hu, X.; Wang, X. Transforming growth factor-β1 induces type II collagen and aggrecan expression via activation of extracellular signal-regulated kinase 1/2 and Smad2/3 signaling pathways. *Mol. Med. Rep.* **2015**, *12*, 5573–5579. [CrossRef] [PubMed]

111. Blaney Davidson, E.N.; Remst, D.F.G.; Vitters, E.L.; van Beuningen, H.M.; Blom, A.B.; Goumans, M.-J.; van den Berg, W.B.; van der Kraan, P.M. Increase in ALK1/ALK5 Ratio as a Cause for Elevated MMP-13 Expression in Osteoarthritis in Humans and Mice. *J. Immunol.* **2009**, *182*, 7937–7945. [CrossRef] [PubMed]

112. Van Der Kraan, P.M.; Goumans, M.J.; Blaney Davidson, E.; Ten Dijke, P. Age-dependent alteration of TGF-β signalling in osteoarthritis. *Cell Tissue Res.* **2012**, *347*, 257–265. [CrossRef] [PubMed]

113. Schultz, C. Targeting the extracellular matrix for delivery of bioactive molecules to sites of arthritis. *Br. J. Pharmacol.* **2019**, *176*, 26–37. [CrossRef]

114. Rothenfluh, D.A.; Bermudez, H.; O'Neil, C.P.; Hubbell, J.A. Biofunctional polymer nanoparticles for intra-articular targeting and retention in cartilage. *Nat. Mater.* **2008**, *7*, 248–254. [CrossRef]

115. Formica, F.A.; Barreto, G.; Zenobi-Wong, M. Cartilage-targeting dexamethasone prodrugs increase the efficacy of dexamethasone. *J. Control. Release* **2019**, *295*, 118–129. [CrossRef] [PubMed]

116. Hu, H.Y.; Lim, N.H.; Ding-Pfennigdorff, D.; Saas, J.; Wendt, K.U.; Ritzeler, O.; Nagase, H.; Plettenburg, O.; Schultz, C.; Nazare, M. DOTAM Derivatives as Active Cartilage-Targeting Drug Carriers for the Treatment of Osteoarthritis. *Bioconj. Chem.* **2015**, *26*, 383–388. [CrossRef]

117. Singh, A.; Corvelli, M.; Unterman, S.A.; Wepasnick, K.A.; McDonnell, P.; Elisseeff, J.H. Enhanced lubrication on tissue and biomaterial surfaces through peptide-mediated binding of hyaluronic acid. *Nat. Mater.* **2014**, *13*, 988–995. [CrossRef] [PubMed]

118. Faust, H.J.; Sommerfeld, S.D.; Rathod, S.; Rittenbach, A.; Ray Banerjee, S.; Tsui, B.M.W.; Pomper, M.; Amzel, M.L.; Singh, A.; Elisseeff, J.H. A hyaluronic acid binding peptide-polymer system for treating osteoarthritis. *Biomaterials* **2018**, *183*, 93–101. [CrossRef]

119. Brown, S.B.; Wang, L.; Jungels, R.R.; Sharma, B. Effects of cartilage-targeting moieties on nanoparticle biodistribution in healthy and osteoarthritic joints. *Acta Biomater.* **2020**, *101*, 469–483. [CrossRef]

120. Shah, R.N.; Shah, N.A.; Lim, M.M.D.R.; Hsieh, C.; Nuber, G.; Stupp, S.I. Supramolecular design of self-assembling nanofibers for cartilage regeneration. *Proc. Natl. Acad. Sci. USA* **2010**, *107*, 3293–3298. [CrossRef] [PubMed]

121. Pi, Y.; Zhang, X.; Shi, J.; Zhu, J.; Chen, W.; Zhang, C.; Gao, W.; Zhou, C.; Ao, Y. Targeted delivery of non-viral vectors to cartilage in vivo using a chondrocyte-homing peptide identified by phage display. *Biomaterials* **2011**, *32*, 6324–6332. [CrossRef] [PubMed]

122. Pi, Y.; Zhang, X.; Shao, Z.; Zhao, F.; Hu, X.; Ao, Y. Intra-articular delivery of anti-Hif-2α siRNA by chondrocyte-homing nanoparticles to prevent cartilage degeneration in arthritic mice. *Gene Ther.* **2015**, *22*, 439–448. [CrossRef] [PubMed]

123. Cheung, C.S.F.; Lui, J.C.; Baron, J. Identification of chondrocyte-binding peptides by phage display. *J. Orthop. Res.* **2013**, *31*, 1053–1058. [CrossRef] [PubMed]

124. Loffredo, F.S.; Pancoast, J.R.; Cai, L.; Vannelli, T.; Dong, J.Z.; Lee, R.T.; Patwari, P. Targeted delivery to cartilage is critical for in vivo efficacy of insulin-like growth factor 1 in a rat model of osteoarthritis. *Arthritis Rheumatol.* **2014**, *66*, 1247–1255. [CrossRef] [PubMed]

125. Silverman, J.; Lu, Q.; Bakker, A.; To, W.; Duguay, A.; Alba, B.M.; Smith, R.; Rivas, A.; Li, P.; Le, H.; et al. Multivalent avimer proteins evolved by exon shuffling of a family of human receptor domains. *Nat. Biotechnol.* **2005**, *23*, 1556–1561. [CrossRef] [PubMed]

126. Hulme, J.T.; D'Souza, W.N.; McBride, H.J.; Yoon, B.R.P.; Willee, A.M.; Duguay, A.; Thomas, M.; Fan, B.; Dayao, M.R.; Rottman, J.B.; et al. Novel protein therapeutic joint retention strategy based on collagen-binding Avimers. *J. Orthop. Res.* **2018**, *36*, 1238–1247. [CrossRef]

127. Hughes, C.; Faurholm, B.; Dell'Accio, F.; Manzo, A.; Seed, M.; Eltawil, N.; Marrelli, A.; Gould, D.; Subang, C.; Al-Kashi, A.; et al. Human single-chain variable fragment that specifically targets arthritic cartilage. *Arthritis Rheum.* **2010**, *62*, 1007–1016. [CrossRef]

128. Hughes, C.; Sette, A.; Seed, M.; D'Acquisto, F.; Manzo, A.; Vincent, T.L.; Lim, N.H.; Nissim, A. Targeting of viral interleukin-10 with an antibody fragment specific to damaged arthritic cartilage improves its therapeutic potency. *Arthritis Res. Ther.* **2014**, *16*, R151. [CrossRef]

129. Lim, N.H.; Vincent, T.L.; Nissim, A. In vivo optical imaging of early osteoarthritis using an antibody specific to damaged arthritic cartilage. *Arthritis Res. Ther.* **2015**, *17*, 376. [CrossRef]

130. Bajpayee, A.G.; Wong, C.R.; Bawendi, M.G.; Frank, E.H.; Grodzinsky, A.J. Avidin as a model for charge driven transport into cartilage and drug delivery for treating early stage post-traumatic osteoarthritis. *Biomaterials* **2014**, *35*, 538–549. [CrossRef]

131. Vedadghavami, A.; Wagner, E.K.; Mehta, S.; He, T.; Zhang, C.; Bajpayee, A.G. Cartilage penetrating cationic peptide carriers for applications in drug delivery to avascular negatively charged tissues. *Acta Biomater.* **2019**, *93*, 258–269. [CrossRef]

132. Sterner, B.; Harms, M.; Wöll, S.; Weigandt, M.; Windbergs, M.; Lehr, C.M. The effect of polymer size and charge of molecules on permeation through synovial membrane and accumulation in hyaline articular cartilage. *Eur. J. Pharm. Biopharm.* **2016**, *101*. [CrossRef] [PubMed]

133. Bajpayee, A.G.; Grodzinsky, A.J. Cartilage-targeting drug delivery: Can electrostatic interactions help? *Nat. Rev. Rheumatol.* **2017**, *13*, 183–193. [CrossRef]

134. Bajpayee, A.G.; Quadir, M.A.; Hammond, P.T.; Grodzinsky, A.J. Charge based intra-cartilage delivery of single dose dexamethasone using Avidin nano-carriers suppresses cytokine-induced catabolism long term. *Osteoarthr. Cartil.* **2016**, *24*, 71–81. [CrossRef]

135. Bajpayee, A.G.; Scheu, M.; Grodzinsky, A.J.; Porter, R.M. A rabbit model demonstrates the influence of cartilage thickness on intra-articular drug delivery and retention within cartilage. *J. Orthop. Res.* **2015**, *33*, 660–667. [CrossRef] [PubMed]

136. Topping, L.M.; Thomas, B.L.; Rhys, H.I.; Tremoleda, J.L.; Foster, M.; Seed, M.; Voisin, M.B.; Vinci, C.; Law, H.L.; Perretti, M.; et al. Targeting Extracellular Vesicles to the Arthritic Joint Using a Damaged Cartilage-Specific Antibody. *Front. Immunol.* **2020**, *11*, 10. [CrossRef] [PubMed]

137. Adams, G.; Vessillier, S.; Dreja, H.; Chernajovsky, Y. Targeting cytokines to inflammation sites. *Nat. Biotechnol.* **2003**, *21*, 1314–1320. [CrossRef] [PubMed]

138. Mullen, L.; Adams, G.; Foster, J.; Vessillier, S.; Köster, M.; Hauser, H.; Layward, L.; Gould, D.; Chernajovsky, Y. A comparative study of matrix metalloproteinase and aggrecanase mediated release of latent cytokines at arthritic joints. *Ann. Rheum. Dis.* **2014**, *79*, 1728–1736. [CrossRef]

139. Alberts, B.M.; Sacre, S.M.; Bush, P.G.; Mullen, L.M. Engineering of TIMP-3 as a LAP-fusion protein for targeting to sites of inflammation. *J. Cell. Mol. Med.* **2019**, *23*, 1617–1621. [CrossRef] [PubMed]

140. Hu, H.Y.; Vats, D.; Vizovisek, M.; Kramer, L.; Germanier, C.; Wendt, K.U.; Rudin, M.; Turk, B.; Plettenburg, O.; Schultz, C. In vivo imaging of mouse tumors by a lipidated cathepsin S substrate. *Angew. Chem. Int. Ed.* **2014**, *53*, 7669–7673. [CrossRef]

141. Hu, H.Y.; Lim, N.H.; Juretschke, H.P.; Ding-Pfennigdorff, D.; Florian, P.; Kohlmann, M.; Kandira, A.; Peter Von Kries, J.; Saas, J.; Rudolphi, K.A.; et al. In vivo visualization of osteoarthritic hypertrophic lesions. *Chem. Sci.* **2015**, *6*, 6256–6261. [CrossRef] [PubMed]

142. Kavanaugh, T.E.; Werfel, T.A.; Cho, H.; Hasty, K.A.; Duvall, C.L. Particle-based technologies for osteoarthritis detection and therapy. *Drug Deliv. Transl. Res.* **2016**, *6*, 132–147. [CrossRef]

143. Brown, S.; Kumar, S.; Sharma, B. Intra-articular targeting of nanomaterials for the treatment of osteoarthritis. *Acta Biomater.* **2019**, *93*, 239–257. [CrossRef]

144. Geiger, B.C.; Wang, S.; Padera, R.F.; Grodzinsky, A.J.; Hammond, P.T. Cartilage-penetrating nanocarriers improve delivery and efficacy of growth factor treatment of osteoarthritis. *Sci. Transl. Med.* **2018**, *10*, eaat8800. [CrossRef]

145. Hu, Q.; Chen, Q.; Yan, X.; Ding, B.; Chen, D.; Cheng, L. Chondrocyte affinity peptide modified PAMAM conjugate as a nanoplatform for targeting and retention in cartilage. *Nanomedicine* **2018**, *13*, 749–767. [CrossRef] [PubMed]

146. Didomenico, C.D.; Lintz, M.; Bonassar, L.J. Molecular transport in articular cartilage—What have we learned from the past 50 years? *Nat. Rev. Rheumatol.* **2018**, *14*, 393–403. [CrossRef]

147. Byun, S.; Sinskey, Y.L.; Lu, Y.C.S.; Ort, T.; Kavalkovich, K.; Sivakumar, P.; Hunziker, E.B.; Frank, E.H.; Grodzinsky, A.J. Transport of anti-il-6 antigen binding fragments into cartilage and the effects of injury. *Arch. Biochem. Biophys.* **2013**, *532*, 15–22. [CrossRef] [PubMed]

148. Owen, S.; Francis, H.; Roberts, M. Disappearance kinetics of solutes from synovial fluid after intra-articular injection. *Br. J. Clin. Pharmacol.* **1994**, *38*, 349–355. [CrossRef]

149. Vincent, T.L. Of mice and men: Converging on a common molecular understanding of osteoarthritis. *Lancet Rheumatol.* **2020**, *2*, e633–e645. [CrossRef]

150. Mobasheri, A.; Bay-Jensen, A.C.; van Spil, W.E.; Larkin, J.; Levesque, M.C. Osteoarthritis Year in Review 2016: Biomarkers (biochemical markers). *Osteoarthr. Cartil.* **2017**, *25*, 199–208. [CrossRef] [PubMed]

151. Lim, N.H.; Wen, C.; Vincent, T.L. Molecular and structural imaging in surgically induced murine osteoarthritis. *Osteoarthr. Cartil.* **2020**, *28*, 874–884. [CrossRef]

152. Oo, W.M.; Hunter, D.J. Disease modification in osteoarthritis: Are we there yet? *Clin. Exp. Rheumatol.* **2019**, *37*, S135–S140.

153. Bacon, K.; Lavalley, M.P.; Jafarzadeh, S.R.; Felson, D. Does cartilage loss cause pain in osteoarthritis and if so, how much? *Ann. Rheum. Dis.* **2020**, *79*, 1105–1110. [CrossRef] [PubMed]

154. Rosseland, L.A.; Helgesen, K.G.; Breivik, H.; Stubhaug, A. Moderate-to-severe pain after knee arthroscopy is relieved by intraarticular saline: A randomized controlled trial. *Anesth. Analg.* **2004**, *98*, 1546–1551. [CrossRef] [PubMed]

155. Abhishek, A.; Doherty, M. Mechanisms of the placebo response in pain in osteoarthritis. *Osteoarthr. Cartil.* **2013**, *21*, 1229–1235. [CrossRef] [PubMed]

156. Copsey, B.; Thompson, J.Y.; Vadher, K.; Ali, U.; Dutton, S.J.; Fitzpatrick, R.; Lamb, S.E.; Cook, J.A. Problems persist in reporting of methods and results for the WOMAC measure in hip and knee osteoarthritis trials. *Qual. Life Res.* **2019**, *28*, 335–343. [CrossRef] [PubMed]

157. Woolacott, N.F.; Corbett, M.S.; Rice, S.J.C. The use and reporting of WOMAC in the assessment of the benefit of physical therapies for the pain of osteoarthritis of the knee: Findings from a systematic review of clinical trials. *Rheumatology* **2012**, *51*, 1440–1446. [CrossRef]

158. Miller, R.E.; Tran, P.B.; Ishihara, S.; Larkin, J.; Malfait, A.M. Therapeutic effects of an anti-ADAMTS-5 antibody on joint damage and mechanical allodynia in a murine model of osteoarthritis. *Osteoarthr. Cartil.* **2016**, *24*, 299–306. [CrossRef]

159. Miller, R.E.; Ishihara, S.; Tran, P.B.; Golub, S.B.; Last, K.; Miller, R.J.; Fosang, A.J.; Malfait, A.M. An aggrecan fragment drives osteoarthritis pain through Toll-like receptor 2. *JCI Insight* **2018**, *3*, e95704. [CrossRef]

MDPI
St. Alban-Anlage 66
4052 Basel
Switzerland
www.mdpi.com

Pharmaceuticals Editorial Office
E-mail: pharmaceuticals@mdpi.com
www.mdpi.com/journal/pharmaceuticals